Jacques-Eric Bergez • Elise Audouin
Olivier Therond
Editors

Agroecological Transitions: From Theory to Practice in Local Participatory Design

Editors
Jacques-Eric Bergez
AGIR, Université de Toulouse, INRA
Castanet-Tolosan, France

Elise Audouin
AGIR, Université de Toulouse, INRA
Castanet-Tolosan, France

Olivier Therond
LAE, Université de Lorraine, INRA
Colmar, France

https://doi.org/10.1007/978-3-030-01953-2

Foreword

The agroecological transition is a popular topic these days – perhaps even a bit too much so for us not to be concerned about it being a somewhat fuzzy concept. Moreover, it seems to be too broad an umbrella to cover the wide variety of research actions directed at decreasing the environmental damage related to agricultural activity or to increasing the ecologisation of systems. Despite these research actions, one acute question often remains: is the change to negotiate in order to move in this direction feasible, and if so, how? It implies the need to challenge 40 years of growth in labour productivity as the driver of agricultural development and of the competitiveness of supply chains; that means to challenge 40 years of a dominant sociotechnical regime that has shaped our ways of thinking about and steering agricultural activity (the expression of genetic potential by controlling the environment; the standardisation of products, with a reduction in the range of products), along with the means to do so (substituting labour with capital *and* simplifying and rationalising forms of managing herds and land).

The book, *Agroecological Transitions: From Theory to Practice in Local Participatory Design*, addresses this topic of change in an original way, not on the level of the farm or that of supply chains, but instead by considering the embeddedness of agricultural activity within rural territories. In this case, the agroecological transition is not an all-encompassing topic; it is a research object. Moreover, it is interdisciplinary, in other words, an object shared by researchers from various disciplines (from agronomy and livestock farming sciences to economics and ergonomics). And it is a complex research object, for several reasons: the diversity of the entities composing it (which are at the basis of the different chapters from types of farming to the governance of collective action), the processes at play in moving towards more self-sufficiency and the permanence of uncertainty. The transition is a multiform change process that does not simply clarify what would be good or possible to do over the medium term, but that also involves supporting humans, techniques, systemic coherence, products and services in a process that can be renegotiated as difficulties are encountered and new knowledge emerges. It is a change process in which the control of processes is not a given, the path is not linear, and adaptive behaviour is essential. In this case, the territory is not only a place of research or farming, an environment that is favourable to agricultural activity and its diversity, to varying degrees. It is also a set of heterogeneous actors involved or concerned by the agricultural activity and that which it produces (animal and plant products, services). Some of them will drive collective dynamics and actions, while others will be private or professional actors advising these farmers. Finally, the territory is a diversified set of farming models embedded within local and global food systems, whose relationship to nature, the agri-food transformation and places of consumption set them apart. However, it is also a place where everybody must be taken into account in the change process.

This book is not only an original way of looking at the transition constructed by a group of researchers; it also proposes a methodology to support change involving researchers and local actors, set into motion in two small regions in Southwestern France and subject to reflexive analysis and the evaluation of effects by researchers outside the group implementing the transition. *TATA-BOX*[1]: a toolbox? No doubt it is much more than that, with reference points for the use and evolution of these tools and frameworks to analyse the path to take over time and adaptation to local situations.

This book is an essential contribution to the development of research on *territorial agroecology* and the transition "in action". Readers take heed: before us is a document that marks a milestone in the way of organising thinking on the agricultural transition, which remains an open field of research that has been little explored, both at INRA and elsewhere. Following an agronomic engineering tradition, the book may also contribute to establishing a sound and balanced dialogue with landscape ecologists, who are another major scientific actor in territorial approaches to agroecology.

Lastly, I would like to give credit to this synthesis put together by the UMR AGIR (Agroecology, Innovations and Territories), with the support of the ANR, by a group of researchers from different teams from this unit. From their discipline-specific social science, humanities and biotechnical science perspectives, they all share an interest in actors and their actions, as well as complex systems with human and technical components. Not only must this action to unify a research unit be lauded, but its broader scope marking the national and European landscape must be highlighted and congratulated.

Enjoy reading it!

Science for Action and Development Division
INRA
Saint Genès Champanelle, France

Benoît Dedieu

[1] Cf Chapter 2: TATA-BOX at a glance

Acknowledgements

This book is based on research funded by the *Agence National de la Recherche* [French National Research Agency], which was carried out under the AGROBIOSPHERE programme (n°ANR-13-AGRO-0006), as part of the project "*Transition agroécologique des territoires: une boite à outil pour concevoir et mettre en œuvre une transition agroécologique des territoires agricoles avec les acteurs locaux (TATA-BOX)*" (Agroecological Transition of Territories, a toolbox for designing and implementing an agroecological transition of agricultural territories with local actors).

We would like to thank:

- All the scientists involved in the conception, testing and evaluation of the participatory design methodology and of associated boundary objects.
- The scientists who agreed to provide their insight in this book through critical analysis; their inputs are a valuable contribution for authors and readers alike.
- Local project partners, particularly the French Government Agency for regional and rural development (*Pôle d'Equilibre Territorial et Rural, PETR*) of the Centre Ouest Aveyron and Pays Midi-Quercy areas, for their commitment and regular feedback on the project.
- The 57 local stakeholders, for the time they devoted to participatory workshops and then, a posteriori, methodology evaluation.

The diversity of participants in the project, both scientists and stakeholders, resulted in a rich transdisciplinary process and outputs.

Contents

Contributors

Valérie Angeon URZ, INRA, Petit-Bourg, Guadeloupe, France

Ecodéveloppement, INRA, Avignon, France

Elise Audouin AGIR, Université de Toulouse, INRA, Castanet-Tolosan, France

Jean-Marc Barbier Innovation, INRA, CIRAD, Montpellier SupAgro, Univ Montpellier, Montpellier, France

Flore Barcellini CRTD, CNAM, Paris, France

Cécile Barnaud DYNAFOR, Université de Toulouse, INRA, Castanet-Tolosan, France

Aurélien Bénel Institut Charles Delaunay -TechCICO, Université de Technologie de Troyes, Troyes, France

Jacques-Eric Bergez AGIR, Université de Toulouse, INRA, Castanet-Tolosan, France

Jean-Pierre Cahier Institut Charles Delaunay -TechCICO, Université de Technologie de Troyes, Troyes, France

Ariane Chabert AGIR, Université de Toulouse, INRA, Castanet-Tolosan, France

Marie Chizallet LISIS, INRA, EPNC, ESIEE, Université Paris-Est Marne La Vallée, Marne La Vallée, France

CRTD, CNAM, Paris, France

Jean-Philippe Choisis DYNAFOR, Université de Toulouse, INRA, Castanet-Tolosan, France

Célia Cholez AGIR, Université de Toulouse, INRA, Castanet-Tolosan, France

Sarah Clément Territoires, Université Clermont Auvergne, AgroParisTech, INRA, Irstea, VetAgro Sup, Aubière, France

Nathalie Couix AGIR, Université de Toulouse, INRA, Castanet-Tolosan, France

Thomas Debril UMR AGIR, Université de Toulouse, INRA, Castanet-Tolosan, France

Benoît Dedieu Science for Action and Development Division, INRA, Saint Genès Champanelle, France

Jean-Pierre Del Corso LEREPS, Université de Toulouse, ENSFEA, Toulouse, France

Michel Duru AGIR, Université de Toulouse, INRA, Castanet-Tolosan, France

Charles A. Francis Department of Agronomy and Horticulture, University of Nebraska-Lincoln, Lincoln, NE, USA

Department of Plant Science (IPV), Norwegian University of Life Sciences, Aas, Norway

Danielle Galliano AGIR, Université de Toulouse, INRA, Castanet-Tolosan, France

LEREPS, Université de Toulouse, ENSFEA Toulouse, France

Nathalie Girard AGIR, Université de Toulouse, INRA, Castanet-Tolosan, France

Amélie Gonçalves AGIR, Université de Toulouse, INRA, Castanet-Tolosan, France

LEREPS, Université de Toulouse, ENSFEA Toulouse, France

Laurent Hazard AGIR, Université de Toulouse, INRA, Castanet-Tolosan, France

Laure Hossard Innovation, INRA, CIRAD, Montpellier SupAgro, Univ Montpellier, Montpellier, France

Sylvie Lardon Territoires, Université Clermont Auvergne, AgroParisTech, INRA, Irstea, VetAgro Sup, Aubière, France

Lola Leveau TSCF, Irstea, Centre de Clermont-Ferrand, Aubière, France

Geir Lieblein Department of Plant Science (IPV), Norwegian University of Life Sciences, Aas, Norway

Daniele Magda AGIR, Université de Toulouse, INRA, Castanet-Tolosan, France

Marie-Angélina Magne AGIR, Université de Toulouse, INRA, ENSFEA, Castanet-Tolosan, France

Marie-Benoît Magrini AGIR, Université de Toulouse, INRA, Castanet-Tolosan, France

Guillaume Martin AGIR, Université de Toulouse, INRA, Castanet-Tolosan, France

Catherine Milou LEREPS, Université de Toulouse, ENSFEA, Toulouse, France

Coopérative Qualisol, Castelsarrasin, France

Claude Monteil DYNAFOR, Université de Toulouse, INRA, Castanet-Tolosan, France

Marc Moraine Innovation, INRA, CIRAD, Montpellier SupAgro, Univ Montpellier, Montpellier, France

Nathalie Peyrard MIAT, Université de Toulouse, INRA, Castanet-Tolosan, France

François Pinet TSCF, Irstea, Centre de Clermont-Ferrand, Aubière, France

Gaël Plumecocq AGIR, Université de Toulouse, INRA, Castanet-Tolosan, France

LEREPS, Université de Toulouse, ENSFEA, Toulouse, France

Lorène Prost LISIS, INRA, EPNC, ESIEE, Université Paris-Est Marne La Vallée, Marne La Vallée, France

Nathalie Raulet-Croset IAE de Paris, Université Paris 1 Panthéon Sorbonne, Paris, France

Julie Ryschawy AGIR, Université de Toulouse, INRA, Castanet-Tolosan, France

Régis Sabbadin MIAT, Université de Toulouse, INRA, Castanet-Tolosan, France

Pascal Salembier Institut Charles Delaunay -TechCICO, Université de Technologie de Troyes, Troyes, France

Nicolas Salliou DYNAFOR, Université de Toulouse, INRA, Castanet-Tolosan, France

Jean-Pierre Sarthou AGIR, Université de Toulouse, INRA, Castanet-Tolosan, France

Vincent Soulignac TSCF, Irstea, Centre de Clermont-Ferrand, Aubière, France

Marie Taverne Territoires, Université Clermont Auvergne, AgroParisTech, INRA, Irstea,VetAgro Sup, Aubière, France

Vincent Thenard AGIR, Université de Toulouse, INRA, Castanet-Tolosan, France

Olivier Therond LAE, Université de Lorraine, INRA, Colmar, France

Jean-Marc Touzard Innovation, INRA, CIRAD, Montpellier SupAgro, Univ Montpellier, Montpellier, France

Pierre Triboulet AGIR, Université de Toulouse, INRA, Castanet-Tolosan, France

LEREPS, Université de Toulouse, ENSFEA, Toulouse, France

Abbreviations

AET	AgroEcological Transition
AGREST	AGRoEcology of the production System in the Territory
AIS	Advisory and Information Systems
AKS	Agricultural Knowledge System
AMAP	*Associations pour le Maintien de l'Agriculture Paysanne* (Associations for the Maintenance of Peasant Agriculture)
ASaT	Agricultural Systems in a Territory
CA	Conservation Agriculture
CIVAM	*Centre d'Initiatives pour Valoriser l'Agriculture et le Milieu rural* (Rural Environment and Farming Development Initiative Centres)
CMS	Content Management System
CNAM	*Conservatoire National des Arts et Métiers* (National Conservatory of Arts and Crafts)
DTF	Duru, Therond, Farès
EMA	Ecological Modernisation of Agriculture
ENSAT	*Ecole Nationale Supérieure Agronomique de Toulouse* (National High School for Agronomy of Toulouse)
ENSFEA	*Ecole Nationale Supérieure de Formation de l'Enseignement Agricole* (National High School of Agricultural Training and Education)
ESR	Efficiency, Substitution, Redesign
FS	Farming Systems
GAEC	*Groupement agricole d'exploitation en commun* (Joint Agricultural Group)
GIEE	*Groupe d'Intérêt Economique et Environnemental* (Economic and Environmental Interest Group)
GVC	Global Value Chains
HVE	*Haute Valeur Environnementale* (High Environmental Value)
ICT	Information and Communication Technology
IDAE	*Indicateurs de Durabilité des Exploitations agricoles* (Farm Sustainability Indicators)
IMS	Image Management System

INRA	*Institut National de la Recherche Agronomique* (National Institute for Agricultural Research)
IRSTEA	*Institut national de Recherche en Sciences et Technologies pour l'Environnement et l'Agriculture* (National Research Institute of Science and Technology for Environment and Agriculture)
KBBE	Knowledge-Based Bio-Economy
KM	Knowledge Management
LCA	Life Cycle Assessment
LCCA	Life Cycle Costing Assessment
LFS	Livestock Farming System
MDP	Markov Decision Processes
MLP	Multi-Level-Perspective
MAAF	*Ministre de l'Agriculture, de l'Agroalimentaire et de la Forêt* (Minister of Agriculture and Agri-Food and Forest)
MR	Material Resources
MR-F	Material Resources system of the Farm
MR-NT	Material Resources for management of the Natural Resources
MR-PC	Material Resources of Processing Chains
NR	Natural Resources
PBL	Problem-Based Learning
PETR	*Pôle d'Equilibre Territorial et Rural* (Territorial and Rural Balance Pole)
POMDP	Partially Observable Markov Decision Processes
RG	Reflectivity Group
RL	Reinforcement Learning
SC	Supply Chain
SCIC	*Société Coopérative d'Interêt Collectif* (Social Cooperatives of General Interest)
SCoT	*Schéma de Cohérence Territoriale* (Territorial Coherence Scheme)
SCT	Simplified Cropping Techniques
SES	Social Ecological System
SLCA	Social Life Cycle Assessment
STS	Sociotechnical System
STSoc	Science, Technology, and Society
TAES	Territorial AgroEcological System
TAET	Territorial AgroEcological Transition
tTAES	transition to Territorial AgroEcological Systems
TPD	Transition with Participatory Design
UAA	Utilised Agricultural Area
UTT	*Université de Technologie de Troyes* (University of Technology of Troyes)

Part I
Introduction

Introduction

Jacques-Eric Bergez and Olivier Therond

Abstract Impact of agriculture on environment and human health, energy crisis and climate change enjoin policy-makers and farmers to rethink the model of agricultural production. One way is to promote a strong ecologisation of agriculture reducing inputs using ecosystem services at field, farm and landscape level and new managements. Designing and implementating such agricultural model needs to deeply change the management of farming systems, natural resources and food–chain while dealing with a wide range of environmental and societal changes. To accompany this change agricultural actors and researchers require new tools. Based on the concept of ecological transition, the TATA-BOX project will propose a methodology and a set of methods and tools to help local agricultural stakeholders to develop a vision of the desirable transition of local agricultural systems and to steer it. As part of the adaptive and transition management paradigms, the project will propose an epistemological move to better match current needs of participatory research (hybridization between hard and soft sciences). The case-study will be the Tarn river watershed where water and biodiversity resources are at stake and where some collective dynamics toward agroecology have already started.

After World War, the productivist model of agriculture led to the standardisation of production methods and consequently to a decrease in the specific cognitive resources necessary to implement them. It also contributed to the specialisation of territories as a function of their comparative advantages (Lamine 2011). In the 1990s, the development of the concepts of sustainability and multifunctionality challenged the monolithic logic of the productivist model. Objectification of the environmental impacts of agriculture, social awareness linked to media coverage of it, and redefinition of the objectives of agriculture due to agricultural policies have

J.-E. Bergez
AGIR, Université de Toulouse, INRA, Castanet-Tolosan, France
e-mail: jacques-eric.bergez@inra.fr

O. Therond (✉)
LAE, Université de Lorraine, INRA, Colmar, France
e-mail: olivier.therond@inra.fr

J.-E. Bergez et al. (eds.), *Agroecological Transitions: From Theory to Practice in Local Participatory Design*, https://doi.org/10.1007/978-3-030-01953-2_1

been the sources of two forms of ecological modernisation of agriculture (Horlings and Marsden 2011). The first one, which stems from the productivist model, corresponds to "a weak Ecological Modernisation of Agriculture" (weak-EMA). It is based on an increase in resource-use efficiency (e.g., water), the recycling of waste or by-products (Kuisma et al. 2013), and the application of good agricultural practices (Ingram 2008) and/or of precision-agriculture technologies (Rains et al. 2011). It can also correspond to new off-the-shelf technologies, such as organic inputs (Singh et al. 2011) or genetically modified organisms. Since it primarily aims to reduce the main negative environmental impacts, it is often called "ecological intensification". The other one is a real departure from the productivist model. It corresponds to "a strong Ecological Modernisation of Agriculture" (strong-EMA). Compared to weak-EMA, strong-EMA needs a paradigm shift in the conceptualisation of the link between environment and production. Along with the principles of resource recycling and flow management, it includes the use of biodiversity to produce "input services" that support production (e.g. water availability, fertility, pest control) and regulate flows (e.g. water quality, control of biogeochemical cycles) (le Roux et al. 2008). These services depend on the practices implemented at field and farm scales, as well as at the landscape scale (Kremen et al. 2012). Strong-EMA allows agricultural production and management (conservation, improvement) of natural resources (Griffon 2006) to be reconciled. This form of ecological modernisation of agriculture founded on ecological concepts is also called "ecologically intensive" (Bonny 2011). While weak-EMA is essentially based on off-the-shelf technologies and/or agricultural practices that render the environment artificial, the goal of strong-EMA is to apply agricultural practices that can capitalise on functional complementarities between organisms, or on services that agro-ecosystems can render.

Strong-EMA requires the implementation of agricultural practices that can exploit functional complementarities between diverse species and genotypes in resource use and biological regulations at multiple spatial and temporal scales (Ostergard et al. 2001; Kremen et al. 2012). Biggs et al. (2012) identify seven general key principles to maintain or increase the production of ecosystem services within an agro-ecosystem, along with their resilience to social and environmental changes. They distinguish three system properties to manage, all of which concern the biophysical and social dimensions of the system, and four attributes for its governance.

The three system properties are: diversity and redundancy; connectivity; and the state of slow dynamic variables. (i) Diversity and redundancy: diversity (taxonomic and functional), and biological (genes, species, ecosystems, spatial heterogeneity) and social (individual, social groups, strategies, institutions) equilibriums, and their levels of redundancy, define the potential for adaptations, innovations, and learning about the system. (ii) Connectivity defines the conditions and level of circulation of material and cognitive resources and actors in the system that determine the exchange capacity among system components and thus the system's performance level. (iii) The state of slow dynamic variables: the dynamics of complex systems are determined by the interaction between slow dynamic variables (e.g. farm size, soil organic matter, management agencies and social values) and fast dynamic variables (e.g. water withdrawals, authorisation to access to resources). The way of middle- or long-term management of the former determine the conditions under which the latter occur and, most often, the ecosystem services of regulation.

The four key management and governance principles are: (i) understand the system as a complex adaptive one, i.e. characterised by emergent and non-linear behaviour and a high capacity for self-organisation and adaptation based on past experiences and ontological uncertainties; and accordingly consider that governance and adaptive management are structurally necessary; (ii) encourage learning and experimentation as a process for acquiring new knowledge, behaviour, skills, or preferences at the individual or collective levels, ultimately to support decisions and actions in situations of uncertainty; experimentation, particularly in the framework of adaptive management, is a powerful tool for generating such learning; (iii) develop participation: the participation of system actors in governance and management processes facilitates collective action, as does the relevance, transparency, legitimacy, and ultimately acceptability of social organisations, decisions, and actions within the system; (iv) promote polycentric subsystems of governance that structure debate and decision-making among different types of actors, at different levels of organisation, and of different forms (e.g., bureaucratic, collective, associative, informal). The basic principle of polycentric governance is to organise governance systems at the spatial scale at which the problems to manage emerge.

The implementation of strong-EMA to ensure the expression of ecosystem services faces various difficulties (Duru et al. 2015):

(a) Strong-EMA requires a redesign of the agricultural systems (Meynard et al. 2012);
(b) Strong-EMA assumes that actors coordinate with one another, particularly for the arrangement of landscape structures, spatial crop distribution, and exchanges of matter (Brewer and Goodell 2010);
(c) The development of new cropping systems based on crop diversity (e.g. crop associations) and a decrease of inputs may cause problems for production and marketing chains (Fares et al. 2011);
(d) Incomplete information during implementation of practices (difficulty in observing ecosystem states, or difficulty in predicting the effects of actions) leads to risk-taking by farmers (Williams 2011);
(e) Given the decidedly local character of production methods to be implemented to take advantage of biological regulating services (Douthwaite et al. 2002), the process of innovation must also be localised (Klerkx and Leeuwis 2008);
(f) Steering strong-EMA at a territorial level will not happen without changes in the mode of production of knowledge and socio-technical systems (Vanloqueren and Baret 2009). An effective integration of societal concerns into scientific practice may require more fundamental changes in the nature of scientific inquiry, and a move towards truly trans-disciplinary research strongly involving external stakeholders in the research process (Pahl-Wostl et al. 2013).

Considering the key challenges of strong-EMA, three conceptual frameworks are potentially suitable to support its implementation: the farming system framework to analyse the organisation and dynamics of production systems, the social-ecological system framework to analyse the management of natural resources locally, and the socio-technical system framework to understand the dynamics of activities, especially transitions in production methods.

Farming systems (FS) range from simplified to more diversified and integrated (Hendrickson et al. 2008). The simplest ones generally have a limited number of crops and pre-planned management. Their dynamics are grounded in plant and animal genetic improvement and the acquisition of high-performance equipment. Innovation on such farms is mostly linear and top-down. In contrast, diversified systems include multiple crops or subsystems that interact dynamically in space and time, which allows them to benefit from multiple synergies emerging from interactions between components. These production systems are managed dynamically, to make the best use of opportunities by performing annual or seasonal adjustments. Innovation on such farms is generally based on the development of coordination between actors to co-produce knowledge and technologies, sometimes assisted by participatory and transdisciplinary research (Knickel et al. 2009). While this type of approach enables us to analyse the structure and dynamics of farming systems, it has three main limits: (i) it does not really consider the risks of implementing specific agroecological practices, due to knowledge gaps; (ii) the social system considered is often reduced to the farmer; and (iii) the impact of farmers' practices on the state of natural resources at the local scale is barely considered or assessed, if at all.

The Social Ecological System (SES) framework allows us to analyse interactions between a social system composed of users, managers, and institutions using technologies and infrastructures to manage resources and a complex ecological system generating these resources (Anderies et al. 2004; Sibertin-Blanc et al. 2011). Through this framework the dynamics of complex systems is analysed through the concepts of resilience, adaptation, and transformation (e.g. Folke et al. 2011). In many situations, the problems of managing natural resources are associated with a failure in governance due to an underestimation of the changing nature and complexity of the SES concerned (Pahl-Wostl et al. 2010). The challenge is therefore twofold: (i) to strengthen the adaptive capacities of governance systems for sustainable management of natural resources; and (ii) to implement adaptive management that aims for continual improvement in policies and practices for the management of natural resources. The application of management strategies is then considered as part of a system for experimentation and learning. Here management methods correspond to an adaptive, deliberative, and iterative decision-making process that is often associated with the organisation of social learning, whose main objectives are mutual understanding, sharing of viewpoints, collective development of new adaptive management strategies for resources, and the establishment of "communities of practice" (Armitage et al. 2008; Newig et al. 2008). While analysis of the social-ecological system allows us to decipher their structure and dynamics, it poorly takes into account (i) the agronomic and organisational constraints of farming systems and (ii) the necessary changes in agricultural supply chains.

The Socio Technical System (STS) framework allows us to analyse the dynamics of innovations and ways of producing goods within economic sectors or production chains as the result of interactions between three levels of organisation (Geels 2002): (i) production niches (an unstable configuration of formal and informal networks of actors in which radical innovations emerge); (ii) socio-technical regimes (a relatively stable and dominant configuration associating institutions, techniques, and artefacts, as well as regulations, standards, and norms of production, practices, and actor networks); and (iii) the global context (the set of factors outside regimes that "frame" interactions among actors: cultural values, political institutions, environmental problems, etc.). Its dynamics are addressed by analysing the adoption and dissemination of the innovations that niches bring, and the transformation of one or more dominant socio-technical regimes under the pressure of niche development and incentives, and regulatory changes from the global context (Geels 2005; Smith and Stirling 2010). Currently, the dominant socio-technical regime is the weak-EMA model (Vanloqueren and Baret 2009). Niches correspond to alternative production models of varying structure, which coexist in a complementary or competitive manner. Analysis of STS allows us to highlight both how regimes adapt when they are threatened, along with the obstacles that prevent regime changes (Schiere et al. 2012); and conditions for the emergence and stabilisation of niches or their access to the status of a regime. However, the STS approach, like the social-ecological approach, has some limitations for dealing with strong-EMA. It fails to consider the stakes and constraints of: (i) the collective management of natural resources; and (ii) farming systems.

To deal with the limitations of these three approaches to implement strong-EMA, Duru et al. (2014, 2015) built a unified framework (DTF-Framework, Fig. 1). They

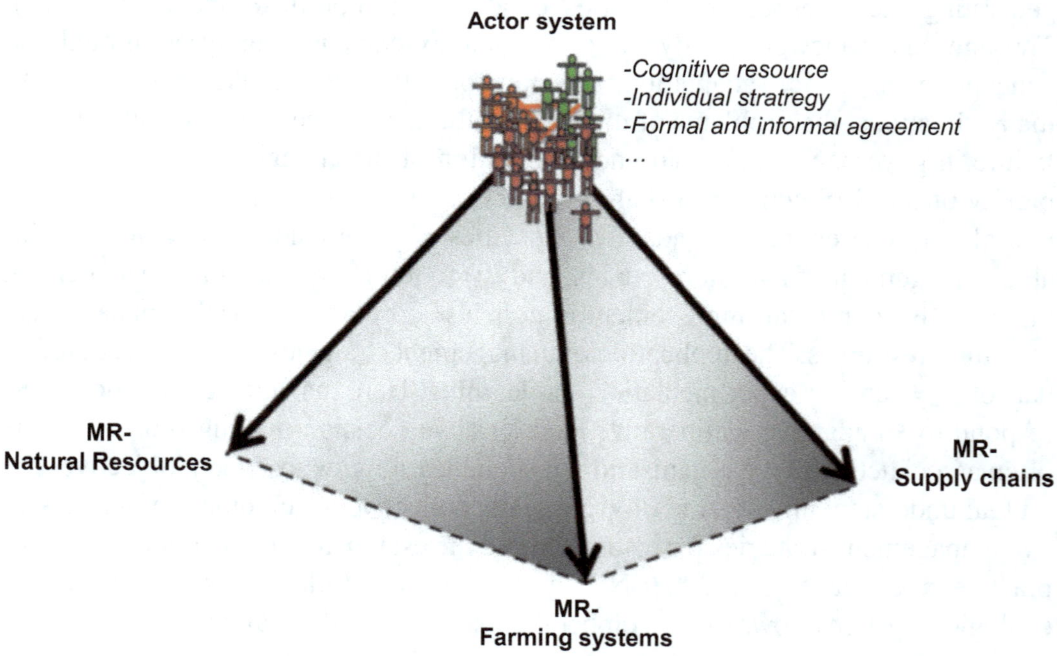

Fig. 1 The DFT framework

use this framework to describe the nature of the complex system concerned by the agroecological transition (AET) of agriculture. The framework is supposed to help thinking on and to organise the transition towards a strong-EMA of local agriculture. It is also used to assess current agricultural production methods and define the types of management and governance systems that promote a strong-EMA. Integrating key concepts of FS, SES and STS approaches, the DTF-Framework represents local agriculture as a system of various actors whose behaviour is determined by formal and informal norms and agreements that interact, via technology, with the material resources specific to farms, supply chains, and natural resource management. Two main types of managed resources are distinguished: material resources (with a biophysical dimension) and cognitive resources. The latter are intangible assets corresponding to the knowledge, beliefs, values, and procedures that actors use to define their objectives, devise their own strategies or alliances, and drive their actions. This framework distinguishes three main systems of material resources (MR) associated with the three management processes: (i) the MR system of the farm (MR-F), used by the farmer for agricultural activities; (ii) the MR system used by actors of each supply chain for collection, processing, and marketing activities (MR-PC); and (iii) the MR system used by actors for management of the natural resources of local agriculture (MR-NT). These MR systems include components that interconnect or interact, such as fields, planned biodiversity (crops, domestic animals), associated biodiversity, machinery, buildings, water resources, and labour for the MR-F system; transportation, storage, and processing equipment and roads for the MR-PC system; and water, soil, and biodiversity (including associated) resources and landscape structures (hedgerows, forests, hydrological network) for the MR-NT system. The three systems of material resources are interdependent, if not interlocked. Material resources, more particularly natural resources, are considered as a social construct and not as an intrinsic characteristic of biophysical objects that become resources for actors. The dimensions and properties that qualify a biophysical object as a resource depend directly on the management process considered. Each management process is based on, and determined by, technologies that are specific to it and used to act upon the concerned resource system. Importantly, within these technologies, information systems determine the methods for characterising resources, the knowledge that actors have about the state of material resources over time, and consequently their actions for managing them in time and space, and ultimately, their ability to meet their performance objectives. Following New Institutional Economics (Williamson 2002) and the Sociology of Organised Action (Crozier and Friedberg 1977), the DTF-Framework considers that formal norms do not completely determine the behaviour of actors. Thus, having limited rationality, actors have a certain degree of freedom and autonomy in their choices and actions.

This integrative conceptual framework can be used to analyse and characterise current forms of agriculture called “Agricultural Systems in a Territory” (ASaT) and to design a future “Territorial AgroEcological System” (TAES) corresponding to a strong-EMA of current ASaT. A key characteristic of the TAES is to organise interactions at the local level between the production systems, in order to take advantage of their complementarities whether they be biophysical (best use of differing soil and/or climate characteristics and/or of access to some natural resources of the farms) and/or production-oriented (e.g. organisation of crop-livestock interactions at the local scale) (Moraine et al. 2017).

Duru et al. (2015) also present a generic transdisciplinary methodological framework for designing, at the local scale, an AET to foster a strong-EMA and allow stakeholders to develop a Territorial AgroEcological System (TAES). This methodology is sketched in Fig. 2.

To support local stakeholders in the design of such transitions, they identified three key methodological challenges:

1. designing, developing and steering a multi-level, multi-domain participatory approach dealing explicitly with trade-off issues;
2. developing boundary objects (conceptual model, computerised-model, indicators, dashboard, etc.) used in the different participatory workshops by stakeholders and enabling trade-off analysis and multicriteria representations;
3. characterising adaptive governance and management enabling stakeholders to locally steer the AET.

The general goal of the TATA-BOX project was to deal with these three challenges through the operationalisation and application of the transdisciplinary methodological framework of Duru et al. (2015). For this, an operational participatory

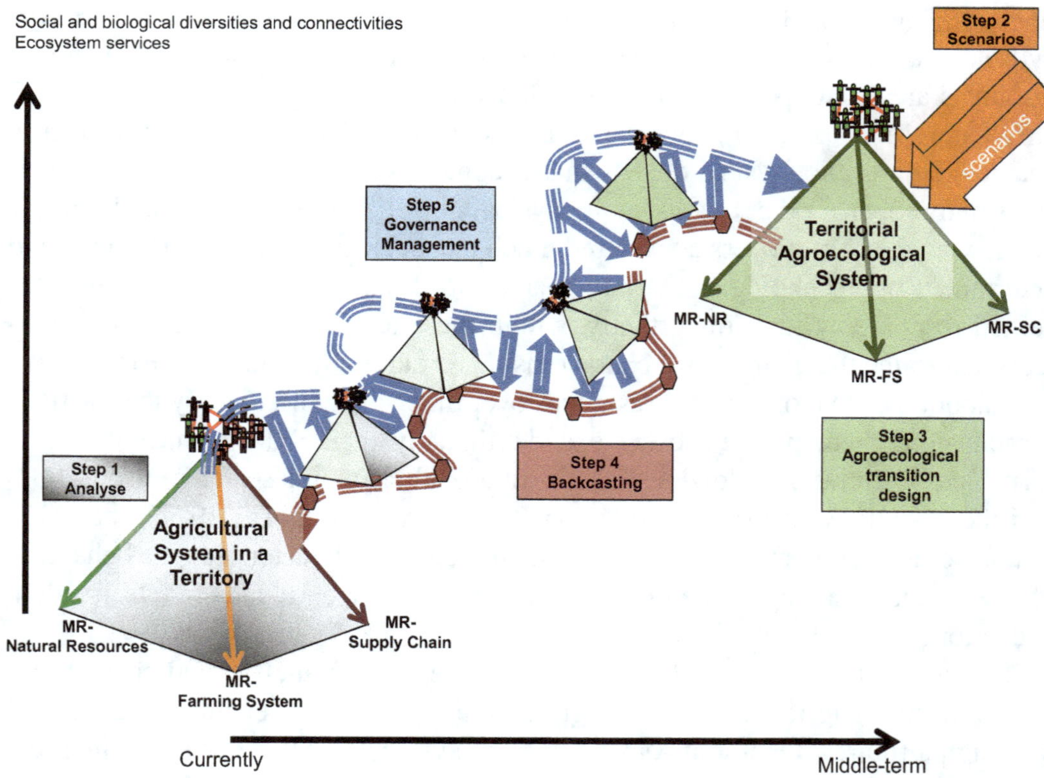

Fig. 2 The conceptual methodology for design and the agroecological transition of a territory

design approach was developed and applied in the Tarn River watershed (south-western France) where farming systems range from arable to livestock ones. In this area water and biodiversity resources are at stake and some collective dynamics toward agroecology already exist.

The TATA-BOX participatory design approach seeks to deal with the interdependences of technological, technical, social, economic and institutional innovations at the farm, supply chain and natural-resources management scales. It is based on the development and support of a "transition arena"; a relatively small group of innovation-oriented stakeholders who reached a consensus on the need and opportunity for systemic changes, and engaged in a process of social learning about future possibilities and opportunities (Foxon et al. 2009). An adhocratic organisation of this arena, based on an institutionalised dialogue involving all the partners and enhancing mutual and permanent adjustments, was organised and implemented.

This book gives some insights of the main outcomes of the TATA-BOX project. It is structured into three sections.

The first section deals with key concepts, challenges and stakes related to agriculture transition: (i) the socio-economic characterisation of the different agriculture models; (ii) the stakes of autonomies and sovereignties; (iii) the AET to a territorialised food system; (iv) the management of uncertainties in AET; (v) the governance of AET; and (vi) the role of actors in the AET.

The second section deals with methodological issues. It contains three chapters. The first describes the transdisciplinary methodology developed and the main outcomes of its application. The second chapter is an assessment of social impacts of the participatory methodology for designing AET developed during the project. The third chapter provides a reflective approach on the characteristics of a research project seeking to support stakeholders in AET design.

The third section opens the field studied during the TATA-BOX project. The first chapter is a foresight on the potential use of information and communication technologies (ICT) for an AET. In the second chapter we asked three other research groups to analyse and discuss outcomes of the TATA-BOX project.

References

Anderies JM, Janssen MA, Ostrom E (2004) A framework to analyze the robustness of social-ecological systems from an institutional perspective. Ecol Soc 9:18

Armitage D, Marschke M, Plummer R (2008) Adaptive co-management and the paradox of learning. Glob Environ Chang 18:86–98. https://doi.org/10.1016/j.gloenvcha.2007.07.002

Biggs R, Schlüter M, Biggs D et al (2012) Toward principles for enhancing the resilience of ecosystem services. Annu Rev Environ Resour 37:421–444

Bonny S (2011) L'agriculture écologiquement intensive: nature et défis. Cah Agric 20:451–462

Brewer MJ, Goodell PB (2010) Approaches and incentives to implement integrated pest management that addresses regional and environmental issues. Annu Rev Entomol 57:41–59

Crozier M, Friedberg E (1977) L'acteur et le système. Paris, Edition du

Douthwaite B, Manyong VM, Keatinge JDH, Chianu J (2002) The adoption of alley farming and Mucuna: lessons for research, development and extension. Agrofor Syst 56:193–202. https://doi.org/10.1023/A:1021319028117

Duru M, Fares M, Therond O (2014) A conceptual framework for thinking now (and organising tomorrow) the agroecological transition at the level of the territory. Cah Agric 23:84–95. https://doi.org/10.1684/agr.2014.0691

Duru M, Therond O, Fares M (2015) Designing agroecological transitions: a review. Agron Sustain Dev 35:1237–1257. https://doi.org/10.1007/s13593-015-0318-x

Fares M, Magrini M, Triboulet P (2011) Transition agroécologique, innovation et effets de verrouillage: le rôle de la structure organisationnelle des filières. Le cas de la filière blé dur française. Cah Agric 21:34–45. https://doi.org/10.1684/agr.2012.0539

Folke C, Jansson Å, Rockström J et al (2011) Reconnecting to the biosphere. Ambio 40:719–738. https://doi.org/10.1016/j.gloenvcha.2006.04.002

Foxon T, Reed M, Stringer L (2009) Governing long-term social–ecological change: what can the adaptivemanagement and transition management approaches learn from each other? Environ Policy Gov 2:3–20. https://doi.org/10.1002/eet.496

Geels FW (2005) Processes and patterns in transitions and system innovations: refining the co-evolutionary multi-level perspective. Technol Forecast Soc Change 72:681–696. https://doi.org/10.1016/j.techfore.2004.08.014

Geels FW (2002) Technological transitions as evolutionary reconfiguration processes: a multi-level perspective and a case-study. Res Policy 31:1257–1274. https://doi.org/10.1016/S0048-7333(02)00062-8

Griffon M (2006) Nourrir la planète, pour une révolution doublement verte. Odile Jaco, Paris

Hendrickson JR, Liebig MA, Sassenrath GF (2008) Environment and integrated agricultural systems. Renew Agric Food Syst 23:304–313. https://doi.org/10.1017/S1742170508002329

Horlings LGG, Marsden TKK (2011) Towards the real green revolution? Exploring the conceptual dimension of a new ecological modernisation of agriculture that couls "feed the worl". Glob Environ Chang 21:441–452. https://doi.org/10.1016/j.gloenvcha.2011.01.004

Ingram J (2008) Agronomist–farmer knowledge encounters: an analysis of knowledge exchange in the context of best management practices in England. Agric Hum Values 25:405–418. https://doi.org/10.1007/s10460-008-9134-0

Klerkx L, Leeuwis C (2008) Balancing multiple interests: Embedding innovation intermediation in the agricultural knowledge infrastructure. Technovation 28:364–378. https://doi.org/10.1016/j.technovation.2007.05.005

Knickel K, Brunori G, Rand S, Proost J (2009) Towards a better conceptual framework for innovation processes in agriculture and rural development: from linear models to systemic approaches. J Agric Educ Ext 15:131–146. https://doi.org/10.1080/13892240902909064

Kremen C, Iles A, Bacon C (2012) Diversified farming systems: an agroecological, systems-based alternative to modern industrial agriculture. Ecol Soc 17:44. https://doi.org/10.5751/ES-05103-170444

Kuisma M, Kahiluoto H, Havukainen J et al (2013) Understanding biorefining efficiency – the case of agrifood waste. Bioresour Technol 135:588–597. https://doi.org/10.1016/j.biortech.2012.11.038

Lamine C (2011) Transition pathways towards a robust ecologization of agriculture and the need for system redesign. Cases from organic farming and IPM. J Rural Stud 27:209–219

le Roux X, Barbault R, Baudry J et al (2008) Agriculture et biodiversité. Valoriser les synergies, ESCo. INRA Editions, Paris

Meynard JM, Dedieu B, Bos AP (2012) Re-design and co-design of farming systems: an overview of methods and practices. In: Darnhofer I, Gibbon D, Dedieu B (eds) Farming systems research into the 21st century: the new dynamic. Springer, Dordrecht, pp 407–431

Moraine M, Duru M, Therond O (2017) A social-ecological framework for analyzing and designing integrated crop–livestock systems from farm to territory levels. Renew Agric Food Syst 32:43–56. https://doi.org/10.1017/S1742170515000526

Newig J, Haberl H, Pahl-Wostl C, Rothman DS (2008) Formalised and non-formalised methods in resource management – knowledge and social learning in participatory processes: an introduction. Syst Pract Action Res 21:381–387. https://doi.org/10.1007/s11213-008-9112-x

Ostergard RL, Tubin M, Altman J (2001) Stealing from the past: globalisation, strategic formation and the use of indigenous intellectual property in the biotechnology industry. Third World Q 22:643–656

Pahl-Wostl C, Giupponi C, Richards K et al (2013) Transition towards a new global change science: requirements for methodologies, methods, data and knowledge. Environ Sci Policy 28:36–47. https://doi.org/10.1016/j.envsci.2012.11.009

Pahl-Wostl C, Holtz G, Kastens B, Knieper C (2010) Analyzing complex water governance regimes: the management and transition framework. Environ Sci Policy 13:571–581. https://doi.org/10.1016/j.envsci.2010.08.006

Rains G, Olson D, Lewis W (2011) Redirecting technology to support sustainable farm management practices. Agric Syst 104:365–370. https://doi.org/10.1016/j.agsy.2010.12.008

Schiere JB (Hans), Darnhofer I, Duru M (2012) Dynamics in farming systems: of changes and choices BT. In: Darnhofer I, Gibbon D, Dedieu B (eds) Farming systems research into the 21st century: the new dynamic. Springer, Dordrecht, pp 337–363

Sibertin-Blanc C, Therond O, Monteil C, Mazzega P (2011) Formal modeling of social-ecological systems. In: 7th international conference of the European Social Simulation Association (ESSA 2011), September 19–23, Montpellier, France, p 47

Singh JS, Pandey VC, Singh DP (2011) Efficient soil microorganisms: a new dimension for sustainable agriculture and environmental development. Agric Ecosyst Environ 140:339–353. https://doi.org/10.1016/j.agee.2011.01.017

Smith A, Stirling A (2010) The politics of social-ecological resilience and sustainable socio-technical transitions. Ecol Soc 15:11. https://doi.org/10.5751/ES-04565-170208

Vanloqueren G, Baret PV (2009) How agricultural research systems shape a technological regime that develops genetic engineering but locks out agroecological innovations. Res Policy 38:971–983. https://doi.org/10.1016/j.respol.2009.02.008

Williams BK (2011) Adaptive management of natural resources – framework and issues. J Environ Manag 92:1346–1353. https://doi.org/10.1016/j.jenvman.2010.10.041

Williamson OE (2002) The theory of the firm as governance structure: from choice to contract. J Econ Perspect 16:171–195. https://doi.org/10.1257/089533002760278776

TATA-BOX at a Glance

Jacques-Eric Bergez and Olivier Therond

Abstract In this chapter we present very briefly the main framework used to establish the TATA-BOX project and the general methodology developed to codesign the territorial agroecological transition.

Context

Environmental degradation, human health, energy crises and climate issues are forcing policy-makers and farmers to rethink the industrial and input-based model of agriculture. One way to deal with these issues is to promote a strong ecologisation of agricultural systems, based on diversification at field, farm and landscape levels to develop ecosystem services (Duru et al. 2015b). Designing and implementing such an approach requires profound change in the management of farming systems, natural resources and food chains, and in turn entails a wide range of environmental and societal changes. To support this change, agricultural actors and researchers require new tools.

Roots of TATA-BOX

To describe the nature of the complex system concerned by the agroecological transition of agriculture, Duru et al. (2014) proposed a new conceptual framework. This Framework represents local agriculture as a system of various actors whose

J.-E. Bergez (✉)
AGIR, Université de Toulouse, INRA, Castanet-Tolosan, France
e-mail: jacques-eric.bergez@inra.fr

O. Therond
LAE, Université de Lorraine, INRA, Colmar, France
e-mail: olivier.therond@inra.fr

J.-E. Bergez et al. (eds.), *Agroecological Transitions: From Theory to Practice in Local Participatory Design*, https://doi.org/10.1007/978-3-030-01953-2_2

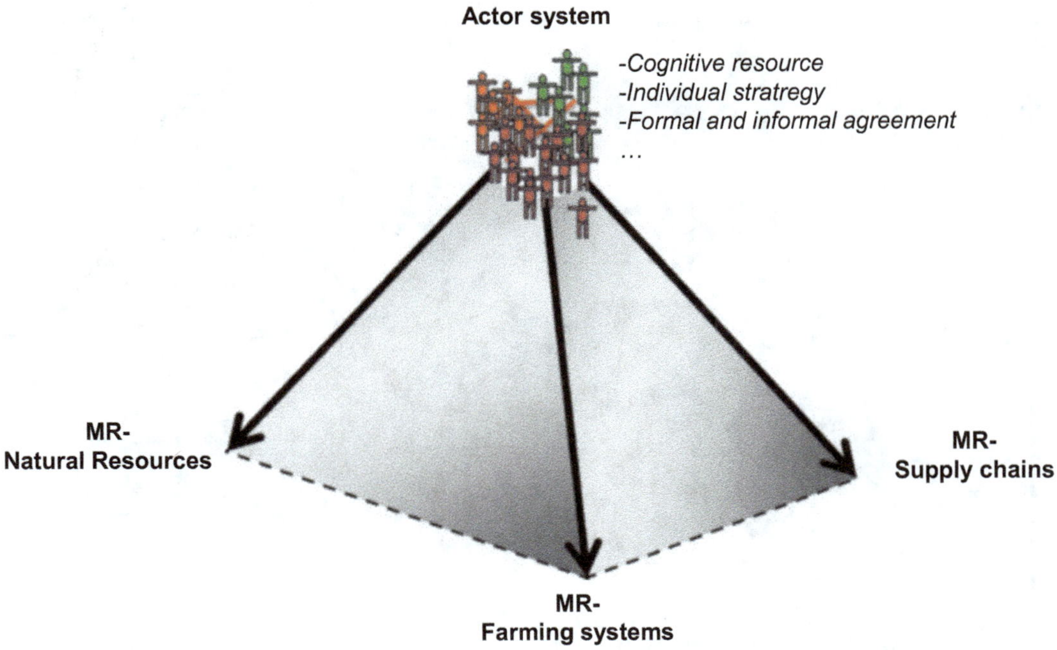

Fig. 1 The DFT framework

behaviour is determined by formal and informal norms and agreements that interact, via technology, with material resources specific to farms, as well as with supply chains and natural resources (cf. Fig. 1). Two main types of resources are managed: material resources, and cognitive resources. This framework distinguishes three systems of material resources (MR) associated with the three management processes: (i) the MR system of the farm (MR-F), used by the farmer for agricultural activities; (ii) the MR system used by actors of each supply chain for collection, processing, and marketing activities (MR-PC); and (iii) the MR system used by actors for management of the natural resources of local agriculture (MR-NT). These MR systems include components that interconnect or interact, such as fields, planned biodiversity (crops, domestic animals), associated biodiversity, machinery, buildings, water resources, and labour for the MR-F system; transportation, storage, and processing equipment and roads for the MR-PC system; and water, soil, and biodiversity (including associated) resources and landscape structures (hedgerows, forests, hydrological network) for the MR-NT system. The three MR systems are interdependent. Each management process is based on, and determined by, technologies that are specific to it and used to act upon the concerned resource system. Actors with limited rationality have a certain degree of freedom and autonomy in their choices and actions.

This conceptual framework (cf. Fig. 2) can be used to analyse and characterise current forms of agriculture called "Agricultural Systems in a Territory", and to design a future "Territorial AgroEcological System" (TAES) corresponding to a strong ecologisation of current Agricultural Systems in a Territory. A key character-

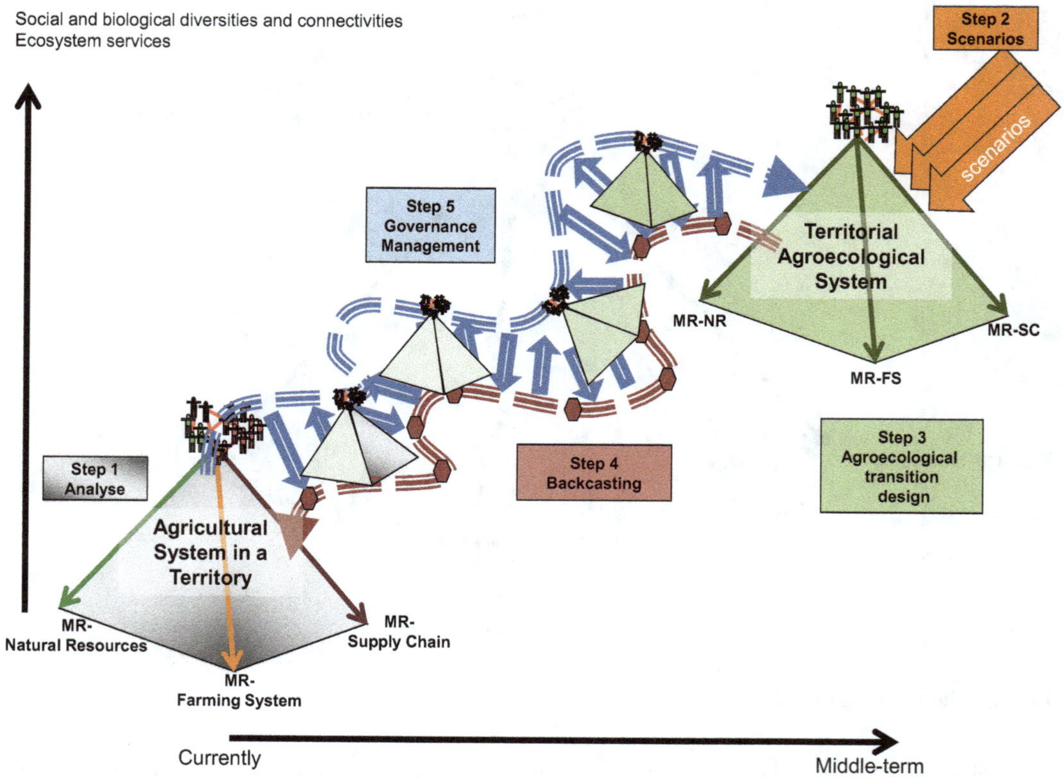

Fig. 2 The conceptual methodology for design and the agroecological transition of a territory

istic of the TAES is that it organizes interactions locally between production systems, to take advantage of their complementarities, whether these be biophysical (best use of differing soil and/or climate characteristics and/or of access to some natural resources of the farms) and/or production-oriented (e.g. organisation of crop-livestock interactions at the local scale) (Moraine et al. 2012, 2014). Duru et al. (2015a) also present a generic methodological framework to support local stakeholders in designing transition to Territorial AgroEcological Systems (tTAES). The TATA-BOX project is designed for testing and adapting a methodology to help local agricultural stakeholders to develop a vision of the desirable transition of local agricultural systems and to steer that process. The methodology is based on 5 steps: (i) characterisation of the current local agriculture; (ii) definition of the exogenous forces that will impact local agriculture in a near future; (iii) design of a Territorial Agroecological system (TAES) based on ecological principles (Biggs et al. 2012); (iv) definition of steps to attain such a system i.e. the tTAES; and (v) proposal of local governance and management to steer this transition. Importantly, each step must be performed by considering and integrating characteristics of and interactions between farming systems, and food chains, and natural resource management.

The TATA-BOX Project

The TATA-BOX project is rooted in post-normal science (Funtowicz and Ravetz 1993) and the participatory integrated assessment paradigm (Rotmans 1998). It is a transdisciplinary and participatory project with interaction between labs and on-fields arenas. To meet the requirements of the French National Research Agency which funded it, it was structured as a set of workpackages, but destructured to follow fields' requirements. Starting from the conceptual and methodological proposals of Duru et al. (2014, 2015a), the TATA-BOX project developed an operational set of articulated methods for supporting stakeholders to design a transition to Territorial AgroEcological Systems at local level. This methodology was structured into three workshops: (i) construction of a shared diagnosis of current issues in local agriculture; (ii) identification of the exogenous and endogenous drivers of change in the territory that determine the future of local agriculture and the co-design of a shared vision of the forms of agriculture to be developed locally to respond to current and future challenges; and (iii) co-design of the adaptive action plan to develop these forms of agriculture by specifying (a) the actions to be implemented, considering local impediments and resources and (b) the polycentric governance to be developed. To evaluate the efficiency of this operational methodology it was applied to two neighbouring study territories in south-western France, downstream and upstream of the Aveyron Valley, in partnership with the PETR (Territorial and Rural Balance Pole) Midi-Quercy Country (48 municipalities, 1192 km^2) and PETR of Centre Ouest Aveyron (129 municipalities, 2998 km^2).

Each of the three workshops of the methodology resulted in a one-day workshop in each of the two study territories. In each of them, the actors were invited to work in sub-working groups on each of the three key sub-domains of local agriculture (farming systems, natural resource management systems, and food chains) and in plenary sessions on interactions between sub-domains. Various artefacts (boundary objects) were developed and used to formalise and integrate knowledge and proven collaborative methodologies (e.g. meta-plan, participatory mapping, rich picture, mind-map, icebreaker), and customised methodologies were applied. One of original products was the development of a method and associated artefacts for determining the transition pathway (sequence of actions and objectives to be achieved) and the governance to steer it.

Some Key Figs

- 4-year project
- €600,000 in funding
- 4 participatory workshops
- 57 participants at the different workshops
- 40 researchers

References

Biggs R, Schlüter M, Biggs D et al (2012) Toward principles for enhancing the resilience of ecosystem services. Annu Rev Environ Resour 37:421–444

Duru M, Fares M, Therond O (2014) A conceptual framework for thinking now (and organising tomorrow) the agroecological transition at the level of the territory. Cah Agric 23:84–95. https://doi.org/10.1684/agr.2014.0691

Duru M, Therond O, Fares M (2015a) Designing agroecological transitions; A review. Agron Sustain Dev 35:1237–1257. https://doi.org/10.1007/s13593-015-0318-x

Duru M, Therond O, Martin G et al (2015b) How to implement biodiversity-based agriculture to enhance ecosystem services: a review. Agron Sustain Dev 35:1259–1281. https://doi.org/10.1007/s13593-015-0306-1

Funtowicz S, Ravetz JR (1993) Science for the post-normal age. Futures 31:735–755

Moraine M, Duru M, Leterme P et al (2012) Un cadre conceptuel pour penser l'intégration agroécologique de systèmes combinant cultures et élevage. Innovations Agronomiques 22:101–115.

Moraine M, Duru M, Nicholas P et al (2014) Farming system design for innovative crop-livestock integration in Europe. Animal 8:1204–1217. https://doi.org/10.1017/S1751731114001189

Rotmans J (1998) Methods for IA: the challenges and opportunities ahead. Environ Model Assess 3:155–179

Part II
Territorial Agroecological Transition at a Concept Crossroads

Socio-economic Characterisation of Agriculture Models

Olivier Therond, Thomas Debril, Michel Duru, Marie-Benoît Magrini, Gaël Plumecocq, and Jean-Pierre Sarthou

Abstract Analyses of transition towards a more sustainable agriculture often identify two different pathways that can be linked to either strong or weak sustainability. In this interdisciplinary work, we aim at overcoming this narrow choice between these two alternatives, by offering a socio-agronomic characterisation of multiple agriculture models that currently coexist in Western economies. We use an agronomic typology of farming systems based on the role of exogenous inputs and endogenous ecosystem services in agricultural production, and on the degree of embeddedness of farming systems within local/global food systems. This typology identifies six agriculture models that we analyse in socio-economic terms. We then clarify the structuring principles that organise these models, and the social values underpinning their justification. This analysis enables us to discuss the efficiency conditions of political instruments.

This chapter is a translation of an article published in French in vol. 363(1) of Economie Rurale. *The authors kindly thank the editors of the journal for authorizing the translation and publication of that article in this book.*

O. Therond (✉)
LAE, Université de Lorraine, INRA, Colmar, France
e-mail: olivier.therond@inra.fr

T. Debril
UMR AGIR, Université de Toulouse, INRA, Castanet-Tolosan, France
e-mail: thomas.debril@inra.fr

M. Duru · M.-B. Magrini · J.-P. Sarthou
AGIR, Université de Toulouse, INRA, Castanet-Tolosan, France
e-mail: michel.duru@inra.fr; marie-benoit.magrini@inra.fr; jean-pierre.sarthou@inra.fr

G. Plumecocq
AGIR, Université de Toulouse, INRA, Castanet-Tolosan, France

LEREPS, Université de Toulouse, ENSFEA, Toulouse, France
e-mail: gael.plumecocq@inra.fr

J.-E. Bergez et al. (eds.), *Agroecological Transitions: From Theory to Practice in Local Participatory Design*, https://doi.org/10.1007/978-3-030-01953-2_3

Introduction

The Western societal model, largely grounded in the industrialisation process, has transformed the nature of agricultural activities and their role in society. Industrial agriculture is based on intensifying the use of synthetic inputs (pesticides, nitrogen fertilisers, antibiotics, etc.), mining inputs (oil, potassium, phosphates), and irrigation water. Its mass development has created significant external environmental damage (Rockström et al. 2009; Gomiero et al. 2011). Society's awareness of these impacts along with environmental regulations are driving farmers to change their relationship to nature (Horlings and Marsden 2011; Duru et al. 2015a, b). The scientific literature often identifies two agricultural evolution pathways to improve agricultural sustainability. In this sense, starting at the end of the 1990s, Hill (1998) contrasted "shallow sustainability" with "deep sustainability"; more recently, Wilson (2008) speaks of "weak versus strong multifunctionality"; Horlings and Marsden (2011) of "weak versus strong ecological modernisation of agriculture"; and Levidow et al. (2012) of "life sciences versus an agroecology vision". These conceptual dichotomies schematically oppose two different relationships to nature: one based on technological progress placed in the service of industrialising agricultural production; the other based on the protection or restoration of natural capital in order to develop associated ecosystem services.

Without denying the importance of such approaches, our research aims to more closely analyse the diversity of agricultural transformation models (hereinafter "agriculture models"). The dualisms presented above relate to oppositions that are often developed from the viewpoint of a single discipline, leading to incomplete descriptions focused on technicity and science, without exploring the social values orienting these choices. These analyses are often limited to: (i) inventorying the negative effects of the dominant model; (ii) presenting a more virtuous alternative model; and (iii) identifying the barriers and levers to transition from one to the other. In our opinion, the usefulness of frameworks that distinguish different agriculture models in more detail resides in the emphasis placed on the coexistence and co-evolution of these models, potentially demonstrating how they intertwine with one another. Yet the few publications that distinguish more than two agriculture models and include a socio-economic dimension in their analysis (ex. Gliessman 2007) tend to grant moral status only to "deep sustainability" forms (see also Wilson 2008).[1] We espouse the contrary belief that it is beneficial to develop a better understanding of the social value systems on which dominant practices rely in order to consider how to transform them. While all of this research appears to agree on certain features of the two main agriculture models, it does not question the social foundations that legitimise the choices, individual strategies, or practices in each of them. This research is consequently incapable of accounting for the variety of

[1] For example, while Sulemana and James Jr. (2014) show that environmental ethics qualify "conservationist" farmers rather than productivist farmers, this does not mean that the practices of the latter are devoid of ethical foundations.

agriculture models in our Western economies, of understanding their founding principles, and therefore of contemplating both the variety of pathways to transition toward greater agricultural sustainability, and the mechanisms of coexistence of these models.

To describe the different Western agriculture models, it is important to qualify the different moral values and social principles that legitimise and underpin coherent sets of practices, agricultural technologies, and embeddedness within food systems. While these principles are general and tacit, they nonetheless remain effective. This multidisciplinary work, which combines agronomic and socio-economic approaches, thus emphasises the technical and social rationales underlying current and emerging farming systems. It enables a reconceptualisation of the coexistence of agriculture models which, along with the multitude of institutional devices supporting them – including political devices –, are involved in the agricultural transition faced with sustainability issues.

The originality of this research lies in the fact that it combines multiple analysis frameworks to construct a detailed characterisation of the various agriculture models developing in contemporary Western economies. To this end, we first briefly present the analysis framework created by Therond et al. (2017), which defines diverse agriculture models based on: (i) the way that farming systems combine exogenous inputs (synthetic and biological) and ecosystem services provided to farmers[2]; and (ii) their level of embeddedness in globalised food systems versus territorial dynamics (circular economy, local food system, integrated-landscape management). This typology was created from an agronomic viewpoint (*sensu lato*) based on an extensive literature review. We then characterise these different agriculture models based on their main social features, drawing on the "Economies of Worth" framework (Boltanski and Thevenot 1991) which allows us to objectify the correspondence of organisation rules and social norms with the practices described in the agronomic typology. The result of this interfacing of analysis frameworks allows us to put to the test the socio-economic consistency of the agronomic description of seven agriculture models currently present in our Western societies, including six agriculture models constituting responses to the sustainability issues of industrial agriculture. By emphasising the variety of agriculture models constituting alternatives to the industrial model that emerged in the wake of World War II, this typological analysis is intended to go beyond existing frameworks that are often reduced to a dichotomy. This research thus provides categories allowing us to analyse the influence of human-technology-nature relations on modes of social organisation, on the practices and uses of nature, as well as on the institutional and political forms framing them.

Section "Economies of worth and sustainable agriculture" presents the Economies of Worth socio-economic analysis framework and its usefulness in examining human-nature relations. Section "Agriculture models at the intersection between

[2] Here, the notion of "ecosystem service"is focused on the services provided to agricultural ecosystem managers (or farmers) corresponding to ecological processes regulating the nutrient cycle, water, soil structure, and biological regulations (including pollination).

farming systems, food systems, and local dynamics" presents the agronomic typology of agriculture models. The latter are then analysed in Section "Socio-economic characterisation of agriculture models", based on the economies of worth model. Section "The usefulness of characterising sustainable agriculture models for designing public policies" discusses political support mechanisms.

Economies of Worth and Sustainable Agriculture

The economies of worth socio-economic model (Boltanski and Thevenot 1991) is a framework to analyse the formation and dynamics of collective actions. We will start by presenting its concepts and will then show how this socio-economic approach can put different typologies of agriculture models into perspective.

From Justification Principles to Organisation Principles

The Economies of Worth model emphasises the fundamental role of social values in establishing and structuring collective actions. These values serve as a basis for the justifications put forward to defend the well-founded nature of an individual choice. The justifications are collectively examined and tested, in particular during conflicts, and may eventually be accepted as legitimate. Boltanski and Thevenot (1991) use the term "higher common principles" to denote this set of collectively-accepted social values. These principles establish spaces of commensurability between individuals and objects (individuals and objects are "qualified" according to this principle) and of ranking (some individuals and objects qualified according to this principle "are acknowledged as being worth" more than others).

Boltanski and Thévenot identify six higher common principles drawn from Western political philosophy that theoretically constitute "cities", in other words, social orders (cf. columns of Table 1):

- wealth as the basis of a market city,
- efficiency as the basis of an industrial city,
- equity as the basis of a civic city,
- honesty as the basis of a domestic city,
- grace as the basis of an inspired city,
- fame as the basis of an opinion-based city.

Additional research has also sought to demonstrate the existence of other cities, particularly an ecological city based on the principle of good intentions directed at the environment or on the symmetry between humans and nonhumans (Latour 1998 – cf. last column of Table 1). Other authors argue that the theoretical requirements of the model preclude this possibility, primarily because an ecological city would imply that the biotic and abiotic elements of ecosystems can "exercise" their

Table 1 Overview of the common worlds (or cities) identified in the Economies of Worth model according to Thévenot et al. (2000)

	Common worlds (Boltanski and Thevenot 1991), according to different common principles						Additional common world (various authors)
	Merchant (wealth)	Industrial (efficiency)	Civic (equity)	Domestic (honesty)	Inspired (grace)	Opinion (fame)	Green/Ecological (good intentions)
Evaluation method (magnitude)	Price, cost	Technical efficiency	Collective well-being	Esteem, reputation	Grace, singularity, creativity	Renown, fame	Symmetry, environmental good intentions
Test	Competition	Competence, reliability, planning	Equality and solidarity	Reliability	Passion, enthusiasm	Popularity, audience, acknowledgement	Sustainability, resilience
Form of proof of admissibility	Monetary	Measurable, statistic	Formal, official	Oral, personal guarantee, exemplary nature	Expression, emotional engagement	Semiotic	Ecological ecosystem
Qualified objects	Market goods or services	Technical infrastructure, method, plan	Rules, fundamental rights	Local heritage, legacy	Body or emotionally charged object	Sign, media	Virgin, wild, natural habitat
Qualified beings	Customer, consumer, vendors, merchant	Engineer, professional, expert	Citizens, unions	Authority figure	Creative being	Celebrity	Environmentalist, ecosystem actors
Time of formation	Short-term, flexibility	Long-term, planned future	Perpetual	Customary	Eschatological, revolutionary or visionary moment	Vogue, trend	Future generations
Space of formation	Globalisation	Cartesian space	Detachment	Local, anchoring in proximity	Presence	Communication networks	Planetary ecosystem
Qualified nature (Godard 1990)	Market nature, consent to pay	Exploited, productive nature	Freely accessible nature	Domesticated nature	Exceptional natural sight, aesthetic value	Nature capable of mobilising opinion	Virgin nature, value of existence

"choice" (or their operation) in the city (Lafaye and Thévenot 1993). This impossibility leads to ecological justifications that are legitimised either by qualifying nature within existing cities (for example, as a source of spirituality or a productive resource – Godard 1990 – cf. last row of Table 1), or by combining principles from different cities (Lafaye and Thévenot 1993; Thévenot et al. 2000).

Actors refer to these principles to organise collective action, evaluate and justify the validity of their actions, criticise those of others, and/or build institutional or material devices that frame practices and collective actions. Theoretical "cities" are therefore declined in the "world" (in the sociological sense of the word), that is, in assemblies of objects (whether tangible and/or intangible) and in people.

Approaches to Sustainable Agriculture Put to the Test of Economies of Worth

Creating a typology consists of grouping individuals and/or objects together on the basis of criteria that qualify individuals and objects according to the same register. In this sense, the economies of worth model can contribute both to the creation of the typology (the establishment of equivalence within categories) and to the social relevance of these groupings.

The majority of the research that has studied the variety of sustainable agriculture models or analysed the diversity of transition pathways towards more sustainable agriculture presents certain typology problems. For example, the work of Gomiero et al. (2011), which presents a number of different "philosophical approaches to agriculture", or of Féret and Douguet (2001), who analyse various agricultural governance frameworks that they call "agricultural families", groups together types of agriculture (agroecology, organic agriculture, permaculture, intensification, etc. for the former; organic agriculture, peasant agriculture, rational agriculture, etc. for the latter) under conventional designations with varying levels of institutionalisation. These inventories are not the result of typological approaches, given that the criteria distinguishing these "philosophies", these "families", or these "reference frameworks" are not always explicit. On the contrary, our typology-based approach demonstrates that some of these "philosophies", by relating to very different agricultural practices, are qualified in different sustainable agriculture models (this is the case in particular for organic agriculture[3] or conservation agriculture).

Another line of research stems from the observation of agricultural practices and establishes distinctions based on better-defined criteria (Hill 1998; Gliessman 2007). Without overlooking the role of socio-economic context in these practices,

[3] In reality, organic agriculture relates to various practices (cf. Allaire and Bellon, 2014) described in various sustainable agriculture models (Therond et al. 2017) and justified by very different principles. Therefore, we believe it would be contradictory to acknowledge the diversity of organic agriculture while treating it as a single model (for example, cf. Benoit et al., 2017), whether to evaluate overall performance or to outline practices.

this research grants decisive importance to individuals' decision-making capabilities (see also certain ground-breaking research on agricultural multifunctionality, in particular Van der Ploeg 1996). It thus tends to consider that different agriculture models are independent of one another (alternative models are treated like contextual elements for another model). By contrast, the Economies of Worth model invites us to consider the fundamental role played by other agriculture models in the internal structure of a given model. This research also tends to attribute ethical or moral virtues to those models that most radically break with the conventional one. The Economies of Worth model teaches that even the most self-interested reasons are underpinned by powerful value (ethics) systems. In this sense, it is consistent with another body of research that more specifically addresses the problem of the transition to more sustainable agriculture (Horlings and Marsden 2011; Levidow et al. 2012). While this research emphasises the aspects structuring contexts (in particular food systems or territorial dynamics, as well as the practical aspects of farming systems, it tends to liken one to the other, considering that production practices (such as agroecological practices) go hand-in-hand with certain territorial dynamics (in this example, short circuits). However, according to the Economies of Worth model, these activities can stem from very different principles, which is what our typology aims to demonstrate.

Lastly, another set of research addresses the multifunctionality of agriculture (Laurent et al. 2003; Wilson 2008; Renting et al. 2009; van der Ploeg et al. 2009). For example, the typological approach developed by Laurent et al. (1998) highlights 11 types of agricultural activity model based on the domains in which the activity in question is embedded, the professional standard systems to which it refers, and the negotiation bodies mobilised to resolve difficulties related to the agricultural activity. This typology allows us to discuss three principal social functions of agricultural activities (providing professional income, insertion within a regime of social transfers, and own consumption and bartering). These functions nevertheless point to the principles of fairness that delimit and socially justify them. Apart from the implications for the agricultural sector, shining light on these principles allows us to assess the societal reach of agricultural activities. We believe that this focus is fundamental when it comes to considering agricultural sustainability stakes. Research on multifunctionality has therefore included the environmental function of agriculture. Some researchers establish "styles of agriculture" based on criteria comparable to those that we use (for example, Van der Ploeg 1996), but neglect the effects of supervision (or of authority) driven by productive structures, social structures, and more generally agro-food activity governance structures. Other multifunctionality research shows, on the contrary, that different types of agricultural activities (traditional agricultural logic, capitalist agriculture, agriculture as a structured profession, etc.) correspond to specific organisational structures with specific goals (agricultural income, increasing equity, subsistence and trade, etc.), the social relations of which are mediated by different forms of legitimate institutions (Laurent et al. 1998; van der Ploeg et al. 2009). However, nothing is said about the source of these institutions' legitimacy or of the principles and values underpinning the organisational forms described. The Economies of Worth model allows us to describe these social systems in detail.

Agriculture Models at the Intersection Between Farming Systems, Food Systems, and Local Dynamics

By considering the classic classifications of types of agricultural transformation (i.e. sustainability or strong versus weak multifunctionality), Therond et al. (2017) developed an analysis framework and a more detailed typology of sustainable agriculture models. These models correspond to types of farming systems that vary in their dependence on exogenous inputs or ecosystem services (y-axis of Fig. 1). They present a varying level of embeddedness in global food systems compared to the territorial dynamics that determine their biotechnical functioning (x-axis of Fig. 1). Each criterion relates to fundamental agricultural sustainability concerns (van der Ploeg 1996; Fraser et al. 2016). In this section, we elucidate these two main dimensions that differentiate sustainable agriculture models.

Sustainable Farming Systems: Exogenous Inputs and Ecosystem Services

In post-WWII Western economies the industrial development process required an increase in agricultural production. This was achieved primarily by selecting more productive plant species and animal breeds. Farmers also developed farming systems based on the use of exogenous inputs allowing them to control abiotic (water and nutrients) and biotic (the negative effects of pests) factors that could limit or reduce agricultural production (van Ittersum and Rabbinge 1997). The large-scale use of these inputs with a low relative cost enabled farmers to simplify cropping plans and rotations, and therefore led to the specialisation of systems and the standardisation of practices and products. Farmers became accustomed to adopting assurance practices consisting in intensifying the use of these inputs to limit production risks (e.g. pests). To combat the environmental damages caused by the development of these specialised farming systems, farmers can implement three agronomic strategies (Duru et al. 2015a, cf. also Hill 1998).

The first strategy consists in optimising the efficiency of input use considering the space and time needs of plants and animals, thus limiting fertiliser and pesticide use (efficiency optimisation strategy of the ESR model[4]). This strategy requires the best possible evaluation in time and space of the ecosystem services provided by the soil-plant(−animal) system in order to minimise the additional exogenous inputs necessary to reach production goals (Fig. 1, bottom of the y-axis). The development of this type of farming system is based on technological innovations, particularly so-called precision agriculture technologies, and on the use of plant and animal varieties less sensitive to limiting biotic and abiotic factors. Agricultural practices remain standardised and are therefore not well rooted in local knowledge.

[4] Efficiency, Substitution, Redesign (Hill, 1998).

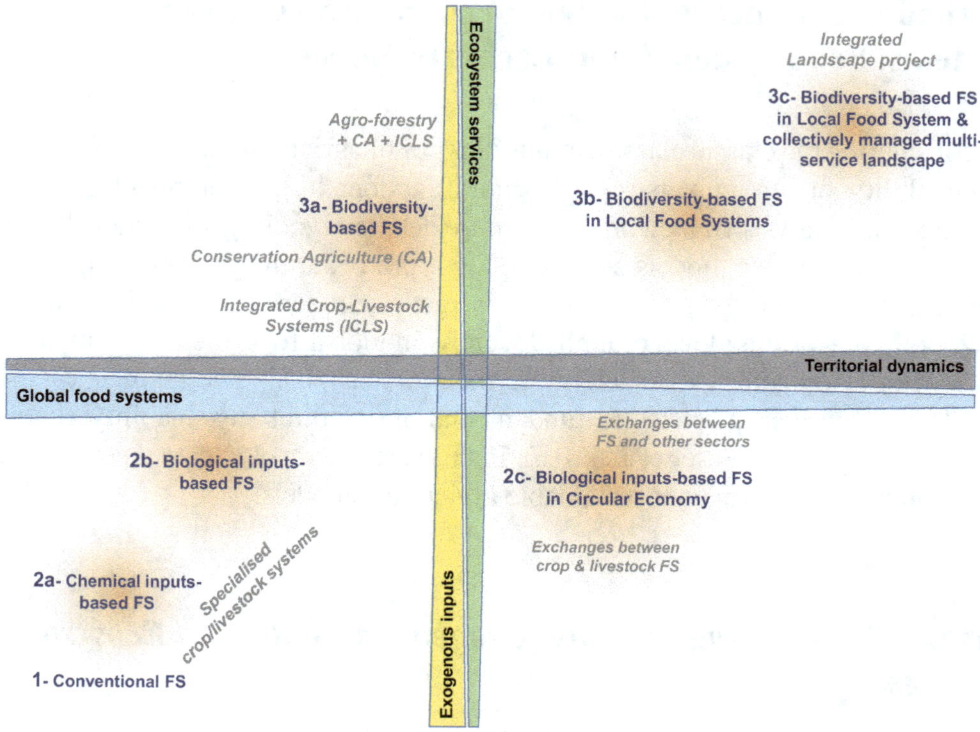

Fig. 1 Typology of agriculture models
Adapted from Therond et al. 2017. The agriculture models numbered 2 and 3 correspond to alternatives to the historical industrial agriculture model often described as "conventional", and which can also be configured differently depending on the context (numbered 1). FS means "farming system". A change from 1 and 2 to 3 indicates a profound change in biotechnical functioning in farming systems (y-axis) shifting from the use of exogenous anthropogenic inputs (1 and 2) to systems based on ecosystem services (3). The letters a, b, and c mainly refer to the relations between farming systems and global food systems or territorial dynamics (x-axis). Certain forms are already well developed; others correspond to niches or represent potential agriculture models in a given region or country. Most often, different forms coexist within a given territory, with one (or several) of them prevailing. Emblematic examples are indicated in italics; conservation agriculture – CA – is used here according to the definition of the FAO. (cf. footnote 5)

The second strategy involves farmers who are more reluctant to use synthetic pesticides and more sensitive to maintaining the health of people and ecosystems. Using a substitution strategy, they seek to develop a farming system that is based as much as possible on the use of organic fertilisers and biocontrol technologies (biopesticides, plant and soil health stimulators, addition of industrially-developed organisms to improve soil quality and biological regulations). However, even though farmers aim to reproduce the ecological operation of diversified agro-ecosystems, their farming systems often remain based on a low level of planned diversity. It is nevertheless possible that these practices (e.g. the use of biostimulants) may promote the development of ecosystem services.

The third strategy is based on managing the planned (domestic) and associated (natural) diversity of ecosystems in order to develop ecosystem services for agriculture (Zhang et al. 2007; Duru et al. 2015b). By redesigning farming systems, it consists in replacing a significant portion of exogenous inputs (whether chemical or biological) with "natural" regulation services that improve soil fertility (soil structure and nutrient cycle), water storage and restoration, pollination, and pest regulation (Fig. 1, top of the y-axis). Developing ecosystem services beneficial to the farmer requires the diversification of the species farmed/grown over time and in space (e.g. cover crops, extended crop rotations), and promotion of the diversity of semi-natural habitats on the level of the land parcel (the boundaries of fields, fallow land, hedges, and forests) (Bianchi et al. 2006), while limiting mechanical disturbances (Duru et al. 2015b). The properties thus targeted by farmers are based on the ability of ecosystems to: (i) store nutrients, energy, and water when these resources are available and to return them to the plants when necessary; (ii) regulate the dispersal and activity of biological pests; and (iii) provide an appropriate habitat for species that deliver regulation services. At field level, this requires the promotion of soil biological activity, such as through no-till practices and the insertion of cover crops during inter-crops. For animals, this may consist in using alternative livestock farming practices (e.g. low density) to ensure the health and vitality of young and adult animals throughout their life (de Goede et al. 2013). The particularity of this type of farming system is that even though its ecological principles are generic, management practices are fundamentally dependent on production/action situations (Duru et al. 2015a; Giller et al. 2015).

Sustainable Agriculture Models at the Crossroad Between Farming Systems, Food Systems, and Territorial Dynamics

Food systems consist in all of the institutions, modes of organisation, technologies, and practices that determine the modes of production, transformation, packaging, and distribution of food products. On top of influencing the nature of the products consumed and the way they are produced and traded, they also determine the conditions for accessing foods and their nutritional quality (Capone et al. 2014). Food systems have rapidly globalised over the past decades, resulting in homogenisation of initially different national food systems (Khoury et al. 2014). Farming systems overlap with these global food systems to varying degrees. They can also overlap with territorial dynamics, such as the development of circular economies, local food systems, or collective landscape management approaches (Fig. 1, x-axis).

Simplified farming systems based on the use of exogenous inputs are greatly intertwined with globalised food systems and, as such, constitute the most prevalent agriculture model in Western Europe (Levidow et al. 2012; Marsden 2013). Within these globalised food systems, power is intensely concentrated in the hands of a few companies (Marsden 2013). The economic resiliency of farming systems faced with the variability of prices and the impacts of biophysical hazards are assured by contractual or insurance devices. These assurance tools can lead farmers to maintain or develop simplified cropping systems or even monocultures that would otherwise be too risky without them (Müller and Kreuer 2016). There exist farming systems based on the use of biological inputs which, like in the previous case, are closely connected to globalised food systems for the purchase of these inputs and the sale of agricultural raw materials. Therefore, it is necessary to distinguish different ways of implementing organic agriculture given that replacing chemical inputs with biological inputs does not always result in a fundamental shift in the mode of production.[5]

In parallel, farming systems can be embedded within localised socio-economic contexts, such as circular economies (the organisation of which requires the management of input-output flows between activities, locally or regionally), alternative food systems (localised systems or those laying claim to social or environmental features), or even integrated regional development projects. In the latter case, the development of agriculture can be combined with integrated landscape management to promote the development of ecosystem services (Wu 2013; Mastrangelo et al. 2014). Most of these farming systems, which are usually diversified, are designed to address environmental or human health issues (products of biological agriculture, foods rich in omega-3s...) and satisfying consumers' demand for quality products, sometimes produced locally (Murdoch et al. 2000).

Certain farming systems may use local markets to access inputs meanwhile selling their products on global markets (or vice versa). For example, conservation agriculture,[6] agroforestry, integrated crop-livestock systems, or self-sufficient livestock systems can permit the development of ecosystem services and therefore, for certain of these, a reduction in the use of exogenous inputs while continuing to sell production via globalised food systems when no other solutions are available or when prices are appealing. Even in this case, certain raw materials can be sold off in globalised supply chains. Likewise, farming systems based on the use of biological inputs can simultaneously be connected to globalised food systems and a local circular economy. Therefore, global and local markets potentially appear to be complementary.

By cross-referencing the three biotechnical strategies of more sustainable agricultural production and the strategies for insertion in global food systems and territorial dynamics, Therond et al. (2017) obtain six sustainable agriculture models (cf. Fig. 1). The notation (2a, b, c and 3a, b, c) indicates a break between the first three forms, which are based on efficiency or substitution strate gies, and the last three forms, which require in-depth redesigning of farming systems. In the following section, we detail the socio-economic features of these forms by showing how each one is distinguished from the others.

---

[5]In particular, this type of biological agriculture does not encompass practices based on the development of ecosystem services as an essential mechanism in managing crops.

[6]Various agricultural practices can be described as conservation agriculture. Here, it corresponds to an agriculture based on three key principles: no-till, permanent soil cover, and diversified and long rotations (http://www.fao.org/ag/ca/1a.html). Debates exist surrounding the environmental performance of these practices, in particular because they may include an increase in herbicide use. In any event, to complement or replace the use of these phytosanitary products, these practices require farmers to manage ecosystem services as best they can.

Socio-economic Characterisation of Agriculture Models

Drawing on this typology, in this section we characterise the different agriculture models identified by Therond et al. (2017) in socio-economic terms according to the Economies of Worth grammar. Model by model, we detail the main socio-agronomic features of each one. These features have been selected because they correspond to the main examples of the different sustainable agriculture models. Table 2 presents a more complete overview of these features.

The Conventional Productivist Model Based on an Industrial/Market Compromise

The industrial agriculture model of Western economies, which we qualify as "conventional" (1) pertains, in the sense of the Economies of Worth model, to a system of practices that are market-based and structured to maximise productivity (in the sense of the industrial principle). These two principles are based on standardising infrastructure, production technologies (machinery, petrochemical inputs, etc.), and end products enabling mass production and distribution. Striving for efficiency and profitability come together in economies of scale and agglomeration, which concentrate production to reduce unit costs. Agricultural practices are essentially oriented at the artificialisation of the environment in order to control or even eliminate the biophysical factors of production variability. For example, the use of chemical pesticides for crop-pest control, or of antibiotics to ensure animal health, are practices minimising the effects of these factors.

This model leads humans to instrumentalise nature by reducing the farming system to a technical economic system: production strategies are rationalised over relatively short time frames (crop season, short crop rotation); the global standardisation of seeds, breeds, production technologies, and products eliminates the local particularities of ecosystems (products are commodities; inputs and technologies are generic), and so on.

The Technology-Intensive Model Based on an Industrial Efficiency/Market Profitability Compromise (2a)

This model is essentially structured around the use of new digital technologies, precision agriculture, and improved varieties or breeds economising the use of industrial inputs, which are massively used in the conventional productivist model (1). In technology-intensive agriculture, the shift in practices in farming systems is driven by the idea that technological mastery can meet environmental requirements, reduce production costs, and thus improve farmers' incomes. Beyond

Table 2 Socio-agronomic characterisation of contemporary agriculture models in Western countries

Farming systems (Section "The conventional productivist model based on an industrial/market compromise")	Historical-conventional system	Farming systems based on chemical inputs connected to global food systems	Farming systems based on biological inputs connected to global food systems	Farming systems based on biological inputs connected to local food systems	Farming systems based on biodiversity connected to global food systems	Farming systems based on biodiversity connected to local food systems	Farming systems based on biodiversity connected to local food systems and multiservices landscape
Agriculture models	**Productivist model (1)**	**Technology-intensive model (2a)**	**Techno-domestic model (2b)**	**Circular model (2c)**	**Diversified-globalised model (3a)**	**Diversified local model (3b)**	**Diversified integrated-landscape model (3c)**
Common worlds involved	**Industrial (productivity) and market (income)**	**Industrial (efficiency) and market (profitability)**	**Domestic (proximity) and industrial (biotech efficiency)**	**Industrial (system efficiency)**	**Opinion and industrial (efficiency)**	**Opinion, domestic, market…**	**Green, domestic, civic…**
Higher common principles	Independence, food security, product diversity	**Same as in 1** + Efficiency and Well-being at work	Ethics of nature and human heath	Overall efficiency at the cluster scale	Ability to put nature to work and "one health" (farm level)	**Same as in 3a** + Value creation (farmer and region level)	Systemic thought, natural and landscape resources, "one health" (ecosystem level)
Modes of evaluation (test)	Productivity of labour and surface area, farm size, balance sheet, production and exports	Production costs, structure of investments, technological intensity	Social and environmental costs, sanitary conditions, environmental indicators	Waste-production balance at the cluster level, recycling rate	Sustainable use of nature, opinion of peers, income	**Same as in 3a** + Embeddedness in the local food system, added value for the region, income	Systemic (multilevel, multicriteria, multiactor)

(continued)

Table 2 (continued)

Qualified objects	Exogenous inputs, machinery, standards	**Same as in 1** + Connected and high-tech devices	**Same as in 2a** + Biological inputs	**Same as in 2b** + Biogas production, waste recycling equipment, co-products	Natural capital and ecosystem services, petrochemical and biological inputs	**Same as in 3a** + Products sold locally	**Same as in 3a** + Quality of the landscape and state of natural resources
Qualified human beings	Productive farmer, mass consumer	High-tech farmer/ entrepreneur, mass consumer	Socially responsible farmer, family, neighbourhood	**Same as in 2b** Cluster member	Agroecologist farmer, agricultural ecosystem	**Same as in 3a** Farmer connected to local consumer, other actors in the local food system, cluster member	**Same as in 3b** Local actors and ecosystems, network member
Time of formation	Short economic term	**Same as in 1**	Short- to medium-term biological cycles	**Same as in 2b** + Cluster cycles	Short-, medium- and long-term biological cycles	**Same as in 3a** + Institutional cycles	Short–/medium–/ long-term biological, social, and institutional cycles
Key spatial levels	Linked upstream and downstream of the global industry	**Same as in 1** + Infra-parcel	**Same as in 1** + Socio-environmental neighbourhood	Local industrial clusters	Peer communities, ecological environment of the farm	**Same as in 3a** + Local market	**Same as in 3b** + Local system, social networks
Mode of regulation	Contracts, intellectual property rights, incentives and financial support, production standards	**Same as in 1**	**Same as in 1**	**Same as 1** + Local industrial partnerships	Peer associations, adaptive field management	**Same as in 3a** + Local markets	**Same as in 3b** + Local system, adaptive management of landscapes
Mode of organisation/ coordination	Global markets and food systems	**Same as in 1** + Technologies shape relationship to nature, new markets	**Same as in 2a**	Circular economy	Peer communities, global food system	**Same as in 3a** + Peer communities, local markets	**Same as in 3b** + Polycentric organisation, adaptive governance

environmental regulations, the significant economic constraints of markets, both upstream (increases in input prices) and downstream from farming systems (increases in price volatility) are encouraging farmers of the conventional model (1) to increasingly change to this technology-intensive model (2a). As a result, they often wish to increase the size of their farm to benefit from economies of scale and increase their ability to invest in Technologies. The quest for efficiency and profitability justifies using these technologies while embedding them within a compromise between the industrial and business worlds. Therefore, like in the conventional model, human-nature relations remain mediated by technology, although by increasingly sophisticated technologies that require farmers to change their way of creating and using the environmental information produced (performance maps, representation of ecological areas, etc.).

The Techno-domestic Model Based on a Local Ethics / Biotechnological Efficiency Compromise (2b)

The techno-domestic agriculture model (2b) is characterised by the use of technologies derived from the living world (e.g. microbiological treatments based on *Bacillus thuringiensis*, addition of nitrogen by spreading free nitrogen-fixing bacteria over carbonaceous residues). The reason for the adoption of these living technologies is awareness of the health and environmental effects of chemical inputs. These practices are a response to the belief[7] that they are able both to improve the productive capacity of soils and plants (e.g. bio-stimulants of soil activity and plant health) and to limit the environmental and health impacts of agriculture (less eco-toxicity among inputs of biological origin). They aim to improve the operation of the agro-ecosystem without large-scale changes (diversification) in crop or livestock farming systems and without taking on board more global environmental concerns. They also aim to reduce the impact of agricultural practices on human health and the ecosystem. The use of technologies of biological origin (e.g. biocontrol) can make it necessary to take ecological time frames into account (e.g. the population dynamics of the organisms introduced). As a result, these practices induce a relationship to nature that is not strictly instrumental. The search for efficient production remains important in the techno-domestic agriculture model, but farmers' concerns for their own health, that of their family and neighbours, and for the local ecosystem make farmers (as well as the consumers qualified in this model) receptive to environmental ethics, guided by a principle of localness and embedded within a compromise between the domestic and industrial worlds.

---

[7] Here, the term belief is related to a lack of proof for this sustainable agriculture model, in which actors do not have the means to objectify the achievement of the common good (the effectiveness of their practices). In these situations, the shared belief of doing what is right (adopting environmental ethics) can be enough to sustain this order, but the absence of proof makes it fragile.

The Circular Model Based on a Compromise Between Efficiency and Industrial Ecology (2c)

Developing the circular economy at local or regional level gives farming systems opportunities to replace chemical inputs by organic materials derived from agricultural activities and other sectors of activity (e.g. organic fertilisers rather than chemical ones), or outlets for their production (biomass for energy production). Agriculture based on the circular economy is developing in some forms of crop-livestock farming combinations on the territorial level (Moraine et al. 2016). This sustainable agriculture model follows the principles of industrial ecology. It is thus essentially based on new ways of organising farmers and other agricultural actors into productive clusters. Geographic proximity plays an important role in developing the exchange of materials and energy at local/regional level. It can also contribute to redefining urban/rural relations. These forms of organisation are possible only if the actors involved in circular economies adopt a concept of production efficiency on the local/regional scale. In this sense, the practices qualified in this form relate to a relationship to nature that is peculiar to the industrial world: natural resources, waste and scraps are seen as resources to be used efficiently in an industrial economy of the environment (both in the use of resources as well as environmental impacts).

The Diversified-Globalised Model: A Compromise Between Opinion/Bioproduction Efficiency (3a)

Diversified-globalised agriculture refers to large crop or livestock farming systems that are diversified, and in particular farming systems based on conservation agriculture based on three pillars (no-till, cover crops, and long rotations) or agroforestry. These typical examples of this sustainable agriculture model are characterised by the adoption of production principles based on the "work" of nature (biodiversity at the origin of ecosystem services), without, however, prohibiting the use of synthetic or biological inputs. These practices make it impossible to use the underlying information on which farmers base their choices in type 2 agriculture models, in particular the technical benchmarks for standardised production associated with specialised and artificialised farming systems. To better understand the uncertainty of nature and the effects of biodiversity management practices, farmers organise into peer groups. These groups allow them to communicate and share experiences around nature and the effects of agricultural practices, thus activating social networks as a production resource. These forms of organisation, the main goal of which is to share knowledge and learn situated practices, also have the effect of re-drawing the boundaries of agronomic "standard practices". This entails redefining what constitutes good cropping practices, what a field is in a good state is like, what acceptable production levels are, or even the criteria for judging efficiency, and so on

(Cristofari et al. 2017). This way of organising knowledge circulation contributes to "valuing" others' viewpoints while simultaneously enabling the construction of a shared representation of collective values. These peer groups thus establish a test based on opinions, which makes the set of production practices stable and coherent. These practices are supported by the effects of reputation, with a principle of legitimacy resulting from a compromise between the industrial world and the opinion world. Two key characteristics set apart this agriculture model from the three previous models: (i) it institutes new social models for organising and validating practices, and (ii) it leads stakeholders to perceive nature as a place of life and as the main factor in production.

The Diversified Local Model Based on Opinion/Domestic/ Market Elements (3b)

As with the diversified-globalised model (3a), the environmental sustainability of this model's agricultural production is based on the development of ecosystem services. However, while the production of the former is essentially sold on global markets, the second distributes agricultural products locally. This enables farmers to sell the products of diversified crops that are more difficult to sell in global food systems (unappealing prices), and to participate in the local economy and development. Two organisational forms exist, supported by two different value sets. The first consists of communities of farmers within which agronomic practices are put to the test (in the sense of the economies of worth model) and socially validated. The other concerns the sale of products within local food systems. It exposes farmers (and their production practices) to consumers' judgements when evaluating the environmental, organoleptic, and sanitary quality of products (even if a portion of outlets are provided by global food systems). This market test is combined with that of the world of opinion (peer groups). By bringing consumers and producers closer together, this form of organisation answers the needs of the former to reconnect with nature. This requires them to be capable of recognising the specific qualities of the products of this system. In France, for example, *Associations pour le Maintien de l'Agriculture Paysanne*, (AMAP, associations for the maintenance of peasant agriculture) are the most visible alternative food systems today, but other distribution forms also exist (direct producer stores, farmers' markets, and different forms of "short-circuits", such as selling along the road, outdoor markets, local producers supplying retail stores, etc.; cf. Deverre and Lamine 2010). This production world thus broadens the relationship between society and nature, drawing from elements of the worlds of opinion and industry as well as the market, in a relatively loose compromise.

Diversified Integrated-Landscape Agriculture Based on Green/Domestic/Civic Elements (3c)

For the time being, the existence of diversified farming systems embedded in integrated-landscape approaches is mainly theoretical in Western Europe, even though it is possible to spot its building blocks in reality. It is characterised by local stakeholders (farmers, consumers, local supply chain actors, citizens, etc.) sharing systemic thought and evaluating the effects of their actions at the territorial level. Beyond the adoption of production practices based on the work of nature (biodiversity at the origin of ecosystem services), this form of sustainable agriculture requires agricultural activities to be embedded within integrated approaches of landscape conceptualised as a social-ecological system. In certain respects, in France, the Biovallée project, based on the preservation and valorisation of natural resources in the service of the population's drinking water, food, habitat, health, energy, and leisure needs, is a relevant example of this form (even though some of its aspects borrow much from other forms – cf. Lamine 2012). In the agriculture model, the agricultural practices that contribute to territorial development, such as organising the spatial distribution of cropping systems and semi-natural habitats to develop ecosystem service at the landscape level, are considered legitimate. This is an extreme form of embedding agriculture in the socio-economic context, insofar as its development requires a participatory "landscape design" approach (cf. Nassauer and Opdam 2008) and the collective governance of land use and semi-natural habitats. Consequently, this form of agriculture borrows legitimising elements from the domestic world (a locally-based regime), the civic world (fair treatment of stakeholders within the territorial system), and the ecological world (nature is treated as an organised whole of living beings, whose intrinsic value is recognised). The preferred social organisation in this agriculture model is the network. It establishes the fair treatment of all the members, specific to the civic world and potentially extended to landscape's biotic elements, and has the particularity of not establishing a hierarchy of individuals within a social order. In this sense, but also because it is based on ecological justifications (in the sense of the ecological city), this model lies outside of the axiomatic system of the Economies of Worth model.[8] It more clearly appears to break with other sustainable agriculture models.

[8] The Economies of Worth model is based on an axiomatic system that postulates: (i) the common humanity of beings belonging to cities (which excludes from the outset, for example, considering the possibility of dialogue with beneficial organisms), and (ii) a principle of the ranking of individuals, which defines, for varying durations of time, states of worth (with "worth beings" constituting the individuals who are legitimate to take responsibility for the common good according to certain principles), which the "network" form, in this context, does not do.

The Usefulness of Characterising Sustainable Agriculture Models for Designing Public Policies

Research on the multifunctionality of agriculture (Wilson 2008; van der Ploeg et al. 2009) emphasises the multiple functions of agriculture not only as an economic and social activity, but also in view of its environmental impacts (Laurent et al. 2003). Drawing from this research, characterising the different sustainable agriculture models presented above illustrates how multifunctionality mechanisms can operate on different levels, and what the role of the interconnection between various locally-embedded stakeholders, as well as of consumers and citizens, is (cf. Renting et al. 2009).

By closely describing the specific consistency of the sustainable agriculture models that coexist, our framework provides paths to an improved consideration of the adjustment of policies to the models that they address. For example, the ecological conditionality of Common Agricultural Policy subsidies can shift the practices of conventional farmers towards technology-intensive and techno-domestic sustainable agriculture models (2a and 2b), because this public policy device is compatible with the principle underpinning them. Concretely, farmers primarily motivated by increasing their income are sensitive to monetary signals, including incentives, subsidies, or compensation. On the other hand, these devices do not trigger a transition toward diversified (3a), diversified local (3b), or diversified integrated-landscape (3c) models, because practices that take place in these diversified agriculture models are justified by a will to restore natural capital or to maintain local/regional economic activities. The stakeholders of these diversified models may be more sensitive to political devices seeking to socially animate the local/regional territory or to develop more sustainable agriculture. The effectiveness and efficiency of these public policies thus depend on their adjustment to the sustainable agriculture model that they address, which implies taking into consideration the value system underlying them.

Moreover, considering that the agriculture models are embedded in one (or multiple) specific world(s) and disqualified in others, by supporting certain agriculture models and not others, public policies reveal the extent of their contribution to the reproduction of relations of domination. For example, in France, the maintenance of the intellectual property devices underpinning the technology-intensive model (2a) prohibits seed exchange practices, which constitute an institutional and organisational device (despite operating via informal rules) that addresses the technical problems encountered in models based on adapting crops to local production situations. Likewise, international trade agreements promote the access of national products to global food systems and encourage models that are essentially dependent on outlets on these types of markets, but which discourage the production of products that do not meet their standards. Lastly, environmental services' payment devices, which compensate farmers for maintaining practices that respect the environment, can change the reference frameworks for judging farmers not motivated primarily by financial compensation, and can requalify technological practices (such as in models 2a and 2b) as producers of environmental services (Froger et al. 2016). Our socio-economic analysis of the agronomic typology thus allows us to show the extent to which a public policy, even when adjusted to the sustainable agriculture model that it addresses, can have harmful effects on other types of practices. In this sense, our analysis draws attention to the variety of lock-in mechanisms – not only those that are technical or cognitive, but also those that are normative and political – driving the stability of technology-intensive (2a) and techno-domestic (2b) sustainable agriculture models. While this stability draws from the ability of public policies to legitimise the practices of these agriculture models, it also resides in the ability to exclude the criticism directed at them, for example, originating from agriculture models with production based on ecosystem services (3a, b, and c).

Ultimately, the success of public policies depends on the agriculture models whose principles are consistent with those they convey, or on their way of arranging the complementarity (potentially on different levels) between various agriculture models. They are therefore potentially unsupportive of, or even antagonistic to, the development of other models, either because they appear to be illegitimate with respect to the principles and values underpinning these policies, or because they produce perverse effects (prohibition or discouragement of practices alternative to the target model), or because they disqualify the criticism that the most radical sustainable agriculture models level at the most conventional ones.

Conclusion

Our analysis presents the socio-agronomic features of seven agriculture models that currently coexist in Western economies. We have insisted on oppositions between the historical "conventional" model underpinned by industrial and market organisation principles, and six alternative models that provide answers to environmental and social sustainability issues. We have explained how these agriculture models are based on different ways of implementing practices and technologies to organise and regulate agricultural production. We have qualified the ways of doing and acting, with varying levels of incompatibility between models, based on the value system that socially justifies them in terms of sustainability. This research shines light on the complexity of agricultural territories that are composed of different models that coexist and co-evolve to differing degrees and at different levels, and on the multiplicity of transitions to more sustainable agriculture.

This analysis allows us to clarify the conditions under which public policy instruments are effective. First, the mechanisms for implementing policies must be consistent with the agriculture models that they address; in other words, they have to take into account the reasons why the stakeholders of these models act as they do. It is therefore necessary to properly design these policies based on the features of agriculture models and the multiple possible configurations of coexistence and interweaving of these models.

Faced with this complexity, an extension of this research would consist in two complementary directions. The first avenue would specify the socio-economic reality of this typological characterisation by providing quantified data (in terms of agricultural surface area, labour time, agricultural employment, added value, product volume, etc.). The objectives would be to evaluate the representativeness of each of these agriculture models and to better characterise configurations of coexistence and hybridisation between these models in a few developed countries. The second line of research would evaluate, in these countries, the sustainability of these configurations through multi-criteria assessments. These analyses would provide useful information to public decision-makers by allowing them to better adjust the instruments and targets of their policies. For the moment, we believe that adopting a precautionary principle appears to be necessary in order not to hinder the most marginal models' development, especially considering the likely porosity between different agriculture models. While the technical or organisational innovations that develop in minority models can be passed on to the most prevalent agriculture models and thus improve their sustainability, the systemic nature of these transition necessitates, beyond technical changes, profound moral and philosophical shifts in the way that we conceive of our relationship to nature and our food. The different agriculture models present distinct particularities in this respect.

References

Allaire G, Bellon S (2014) L'AB en 3D : diversité, dynamiques et design de l'agriculture biologique. Agron Environ sociétés 4:79–90

Benoît M, Tchamitchian M, Penvern S et al (2017) Potentialités, questionnements et besoins de recherche de l'Agriculture Biologique face aux enjeux sociétaux. Économie Rural 361:49–69

Bianchi FJJA, Booij CJH, Tscharntke T (2006) Sustainable pest regulation in agricultural landscapes: a review on landscape composition, biodiversity and natural pest control. Proc R Soc B Biol Sci 273:1715–1727. https://doi.org/10.1098/rspb.2006.3530

Boltanski L, Thevenot L (1991) De la justification. Les économies de la grandeur. NRF Essais, Paris

Capone R, El Bilali H, Debs P et al (2014) Food system sustainability and food security: connecting the dots. J Food Secur 2:13–22. https://doi.org/10.12691/jfs-2-1-2

Cristofari H, Girard N, Magda D (2017) Supporting transition toward conservation agriculture: a framework to analyze the learning processes of farmers. Hungarian Geogr Bull 66:65–76. https://doi.org/10.15201/hungeobull.66.1.7

de Goede D, Gremmen B, Rodenburg TB et al (2013) Reducing damaging behaviour in robust livestock farming. NJAS – Wageningen J Life Sci 66:49–53. https://doi.org/10.1016/j.njas.2013.05.006

Deverre C, Lamine C (2010) Les systèmes agroalimentaires alternatifs. Une revue de travaux anglophones en sciences sociales. Économie Rural Agric Aliment Territ 317:57–73. https://doi.org/10.4000/economierurale.2676

Duru M, Therond O, Fares M (2015a) Designing agroecological transitions; a review. Agron Sustain Dev 35(4):1237–1257. https://doi.org/10.1007/s13593-015-0318-x

Duru M, Therond O, Martin G et al (2015b) How to implement biodiversity-based agriculture to enhance ecosystem services: a review. Agron Sustain Dev 35:1259–1281. https://doi.org/10.1007/s13593-015-0306-1

Féret S, Douguet JM (2001) Agriculture durable et agriculture raisonnée: quels principes et quelles pratiques pour la soutenabilité du développement en agriculture? Nat Sci Sociétés 9:58–64

Fraser E, Legwegoh A, Kc K et al (2016) Biotechnology or organic? Extensive or intensive? Global or local? A critical review of potential pathways to resolve the global food crisis. Trends Food Sci Technol 48:78–87. https://doi.org/10.1016/j.tifs.2015.11.006

Froger G, Méral P, Muradian R (2016) Controverses autour des services écosystémiques. L'Économie Polit 69:36–47. https://doi.org/10.3917/leco.069.0036

Giller KE, Andersson JA, Corbeels M et al (2015) Beyond conservation agriculture Front Plant Sci 6:870. https://doi.org/10.3389/fpls.2015.00870

Gliessman SR (2007) Agroecology: ecological processes in sustainable agriculture, Second edn. Lewis Publishers (CRC Press, Second edition), Boca Raton

Godard O (1990) Environnement, modes de coordination et systèmes de légitimité : analyse de la catégorie de patrimoine naturel. Rev économique 41:215–242. https://doi.org/10.3917/reco.p1990.41n2.0215

Gomiero T, Pimentel D, Paoletti MG (2011) Is there a need for a more sustainable agriculture? CRC Crit Rev Plant Sci 30:6–23. https://doi.org/10.1080/07352689.2011.553515

Hill SB (1998) Redesigning agroecosystems for environmental sustainability: a deep systems approach. Syst Res Behav Sci 15:391–402. https://doi.org/10.1002/(SICI)1099-1743(1998090)15:5<391::AID-SRES266>3.0.CO;2-0

Horlings LGG, Marsden TKK (2011) Towards the real green revolution? Exploring the conceptual dimension of a new ecological modernisation of agriculture that couls “feed the worl.”. Glob Environ Chang 21:441–452. https://doi.org/10.1016/j.gloenvcha.2011.01.004

Khoury CK, Bjorkman AD, Dempewolf H et al (2014) Increasing homogeneity in global food supplies and the implications for food security. Proc Natl Acad Sci 111:4001–4006. https://doi.org/10.1073/pnas.1313490111

Lafaye C, Thévenot L (1993) Une justification écologique ? Conflits dans l'aménagement de la nature. Rev française Sociol 34:495–524

Lamine C (2012) Changer de système: une analyse des transitions vers l'agriculture biologique à l'échelle des systèmes agri-alimentaires territoriaux. Terrains Trav 20:139–156

Latour B (1998) To modernize or to ecologize? That's the question. In: Castree N, Willems-Braun B (eds) Remaking reality: nature at the millenium. Routledge, London/New York, pp 221–242

Laurent C, Cartier S, Fabre C et al (1998) L'activité agricole des ménages ruraux et la cohésion économique et sociale. Économie Rural CN – Z5 244:12–21. https://doi.org/10.3406/ecoru.1998.4996

Laurent C, Maxime F, Maze A, Tichit M (2003) Multifonctionnalité de l'agriculture et modèles de l'exploitation agricole. Economie Rurale 273(1):134–152

Levidow L, Birch K, Papaioannou T (2012) Divergent paradigms of European agro-food innovation: the knowledge-based bio-economy (KBBE) as an R&D agenda. Sci Technol Hum Values 38:94–125. https://doi.org/10.1177/0162243912438143

Marsden T (2013) From post-productionism to reflexive governance: contested transitions in securing more sustainable food futures. J Rural Stud 29:123–134. https://doi.org/10.1016/j.jrurstud.2011.10.001

Mastrangelo ME, Weyland F, Villarino SH et al (2014) Concepts and methods for landscape multifunctionality and a unifying framework based on ecosystem services. Landsc Ecol 29:345–358. https://doi.org/10.1007/s10980-013-9959-9

Moraine M, Grimaldi J, Murgue C et al (2016) Co-design and assessment of cropping systems for developing crop-livestock integration at the territory level. Agric Syst 147:87–97. https://doi.org/10.1016/j.agsy.2016.06.002

Müller B, Kreuer D (2016) Ecologists should care about insurance, too. Trends Ecol Evol 31:1–2. https://doi.org/10.1016/j.tree.2015.10.006

Murdoch J, Marsden T, Banks J (2000) Quality, nature, and embeddedness: some theoretical considerations in the context of the food sector*. Econ Geogr 76:107–125. https://doi.org/10.1111/j.1944-8287.2000.tb00136.x

Nassauer JI, Opdam P (2008) Design in science: extending the landscape ecology paradigm. Landsc Ecol 23:633–644. https://doi.org/10.1007/s10980-008-9226-7

Renting H, Rossing WAH, Groot JCJ et al (2009) Exploring multifunctional agriculture. A review of conceptual approaches and prospects for an integrative transitional framework. J Environ Manag 90:S112–S123

Rockström J, Steffen W, Noone K et al (2009) A safe operating space for humanity. Nature 461:472

Sulemana I, James HS (2014) Farmer identity, ethical attitudes and environmental practices. Ecol Econ 98:49–61. https://doi.org/10.1016/j.ecolecon.2013.12.011

Therond O, Duru M, Roger-Estrade J, Richard G (2017) A new analytical framework of farming system and agriculture model diversities. A review. Agron Sustain Dev 37:21. https://doi.org/10.1007/s13593-017-0429-7

Thévenot L, Moody M, Lafaye C (2000) Forms of valuing nature: arguments and modes of justification in French and American environmental disputes. In: Lamont M, Thévenot L (eds) Rethinking comparative cultural sociology: repertoires of evaluation in France and the United States. Cambridge University Press, Cambridge, pp 229–272

van der Ploeg JD (1996) Styles of farming: an introductory note on concepts and methodology. In: van der Ploeg JD, Long A (eds) Born from within. Practices and perspectives in endogenous rural development. Van Gorcum, Assen, pp 7–30

van der Ploeg JDD, Laurent C, Blondeau F, Bonnafous P (2009) Farm diversity, classification schemes and multifunctionality. J Environ Manag 90:S124–S131. https://doi.org/10.1016/j.jenvman.2008.11.022

van Ittersum MK, Rabbinge R (1997) Concepts in production ecology for analysis and quantification of agricultural input-output combinations. F Crop Res 52:197–208. https://doi.org/10.1016/S0378-4290(97)00037-3

Wilson GA (2008) From 'weak' to 'strong' multifunctionality: conceptualising farm-level multifunctional transitional pathways. J Rural Stud 24:367–383. https://doi.org/10.1016/j.jrurstud.2007.12.010

Wu J (2013) Landscape sustainability science: ecosystem services and human well-being in changing landscapes. Landsc Ecol 28:999–1023. https://doi.org/10.1007/s10980-013-9894-9

Zhang W, Ricketts TH, Kremen C et al (2007) Ecosystem services and dis-services to agriculture. Ecol Econ 64:253–260. https://doi.org/10.1016/j.ecolecon.2007.02.024

An Integrated Approach to Livestock Farming Systems' Autonomy to Design and Manage Agroecological Transition at the Farm and Territorial Levels

Marie-Angélina Magne, Guillaume Martin, Marc Moraine, Julie Ryschawy, Vincent Thenard, Pierre Triboulet, and Jean-Philippe Choisis

Abstract In agroecological approaches, autonomy emerges as a central concept. It is also meaningful for farmers, for whom implementing the agroecological transition of livestock farming systems (LFS) requires greater autonomy with respect to inputs and the dominant socio-economic and technical regime. How does this concept of autonomy encompass the complexity of the agroecological transition? This chapter provides an answer through an overview of the various approaches used to analyse the autonomy of LFS, as well as a conceptual framework that can serve to comprehensively examine it. Three approaches to LFSs' autonomy are presented, based on whether they are focused on the flows of material between system components, on the functioning and management of the system, or on the socio-economic organisation and the values underpinning it. Each of these addresses autonomy in its biotechnical or decisional dimension, as well as in terms of three analysis components: embeddedness, dependency, and footprint. The conceptual

M.-A. Magne (✉)
AGIR, Université de Toulouse, INRA, ENSFEA, Castanet-Tolosan, France
e-mail: marie-angelina.magne@inra.fr

G. Martin · J. Ryschawy · V. Thenard
AGIR, Université de Toulouse, INRA, Castanet-Tolosan, France
e-mail: guillaume.martin@inra.fr; julie.ryschawy@inra.fr; vincent.thenard@inra.fr

M. Moraine
Innovation, INRA, CIRAD, Montpellier SupAgro, Univ Montpellier, Montpellier, France
e-mail: marc.moraine@inra.fr

P. Triboulet
AGIR, Université de Toulouse, INRA, Castanet-Tolosan, France

LEREPS, Université de Toulouse, ENSFEA, Toulouse, France
e-mail: pierre.triboulet@inra.fr

J.-P. Choisis
DYNAFOR, Université de Toulouse, INRA, Castanet-Tolosan, France
e-mail: jean-philippe.choisis@inra.fr

J.-E. Bergez et al. (eds.), *Agroecological Transitions: From Theory to Practice in Local Participatory Design*, https://doi.org/10.1007/978-3-030-01953-2_4

framework inter-relates these two dimensions and three components, thus providing an integrated approach to LFSs' autonomy. Its application to two case studies, one on the farm level and the other on the farm and territorial levels, demonstrates its relevance to design and implement the agroecological transition of LFSs.

Introduction

Over the past decades, the industrial livestock farming model has enabled a massive increase in agricultural production through: (i) animals and plants selected on the basis of their high production potential; (ii) the use of synthetic inputs that minimise the effect of production limiting factors and environmental heterogeneity; and (iii) the standardisation of modes of production and the specialisation of farms and regions. Today, the limits of this model are well-documented (Brussaard et al. 2010; Duru and Therond 2015). Among them are a loss of biodiversity, including agrobiodiversity (i.e. crops and livestock), negative impacts on the environment (pollution, climate change, exhaustion of fossil fuels and water resources), and ethical issues related to the lack of consideration of animal well-being on livestock farms (Clark et al. 2016). All these elements call into question the relevance of the industrial animal production model for the future. In this context, a major challenge for livestock farmers is to simultaneously contribute to the food and nutritional security of humanity, based on limited resources, all the while reducing the negative impacts of agriculture on human health and the environment, and maintaining decent living conditions. Many researchers believe that agroecology is a promising way to overcome all of these challenges (Dumont et al. 2014; Altieri et al. 2017).

As a scientific discipline, agroecology is defined as "the application of ecological concepts and principles to the design and management of sustainable agroecosystems" (Altieri 1987; Wezel and Soldat 2009; Duru et al. 2015). This definition emphasises the fact that natural processes, and in particular biodiversity and the interactions between biotic and abiotic elements, can support the sustainability of livestock farming systems (LFSs) and enable production at adequate levels while simultaneously reducing dependency on agricultural and agrochemical inputs, as well as negative impacts on human health and on the environment, even under sub-optimal conditions. Francis et al. (2003) define agroecology on the level of the food system as a whole as the integrative study of the operation of the entire food system encompassing ecological, economic, and social dimensions. This definition highlights the transdisciplinary nature of agroecology and the fact that transformations on the farm level are the result of or trigger transformations upstream and downstream of the farm. In line with this, some authors stress the need for farmers to rediscover the sovereignty of their food production, technological, and even energy system (Rosset and Martínez-torres 2012; Koohafkan et al. 2012; Altieri et al. 2017). Within these different perspectives of agroecology, a common and central concept emerges: that of autonomy. The agroecological transition (AET) of farming systems, and in our case, LFSs, would thus take place through a quest for autonomy in terms of inputs, as well as the reconfiguration of the decisional autonomy of livestock farmers with respect to the socio-economic and technical regime within which they evolve. Addressing the AET of LFSs through the concept of autonomy makes sense for actors in the field (livestock farmers, advisers, etc.). Many livestock farmer networks are seeking to develop LFSs that are self-sufficient in terms of inputs or which use them in small amounts only (Brocard et al. 2016).

In research on livestock farming autonomy, studies focus on improving the feed self-sufficiency of herds, defined as the ratio between the feed produced on the farm and the feed consumed by the animals of this farm. It is expressed in terms of mass autonomy (based on the amounts of dry feed materials, fodder, and concentrates), energy autonomy (based on the amount of energy provided by these foods, expressed in feed units for milk or meat production), or protein autonomy (based on the amount of protein provided by these foods, expressed in total nitrogen). Other studies focus on integrating crop and livestock farming as a pathway to designing livestock farms that are more self-sufficient in terms of feed and use fewer agrochemical products. These studies inter-relate different spatial levels – the farm and the territory – to improve LFSs' autonomy (e.g. Moraine et al. 2016; Ryschawy et al. 2017). Focused on the biotechnical dimension of autonomy, they show that the individual and collective decisional dimension constitutes an impediment to integrating crop and livestock farming, and draw support from participatory processes aiming at overcoming this. Therefore, in the literature, it is clear that LFSs' autonomy: (i) can be understood in its biotechnical or decisional dimensions; (ii) is achieved through the use of local resources and would require the cooperation of actors in the sociotechnical system; and (iii) can be analysed according to different approaches focusing on flows of materials between system components, the functioning and management of LFS, or the organisation of activities around it. Autonomy is therefore a complex topic that it is necessary to understand comprehensively in order to support the AET of LFSs. The goal of this chapter is to give a brief overview of the different approaches to the autonomy of LFSs and to develop a framework to comprehensively analyse it (Section "Framework to analyse the autonomy of farming systems"). We apply this conceptual framework to two LFS case studies, one on the farm level and the other on the farm and territorial levels. The intention is not to demonstrate that the case studies encompass all elements of the framework, but rather to show the utility of the framework for critically analysing studies on LFSs' autonomy, and for identifying lines of research to complete them (Section "Case study 1: a methodology to analyse the overall autonomy of dairy sheep farms in Aveyron").

Framework to Analyse the Autonomy of Farming Systems

Based on the framework proposed by Madelrieux et al. (2017) to analyse agricultural activity as a function of its interactions with the territory, we are developing a framework to analyse LFSs' autonomy that considers biotechnical and decisional dimensions through three analysis components (Fig. 1):

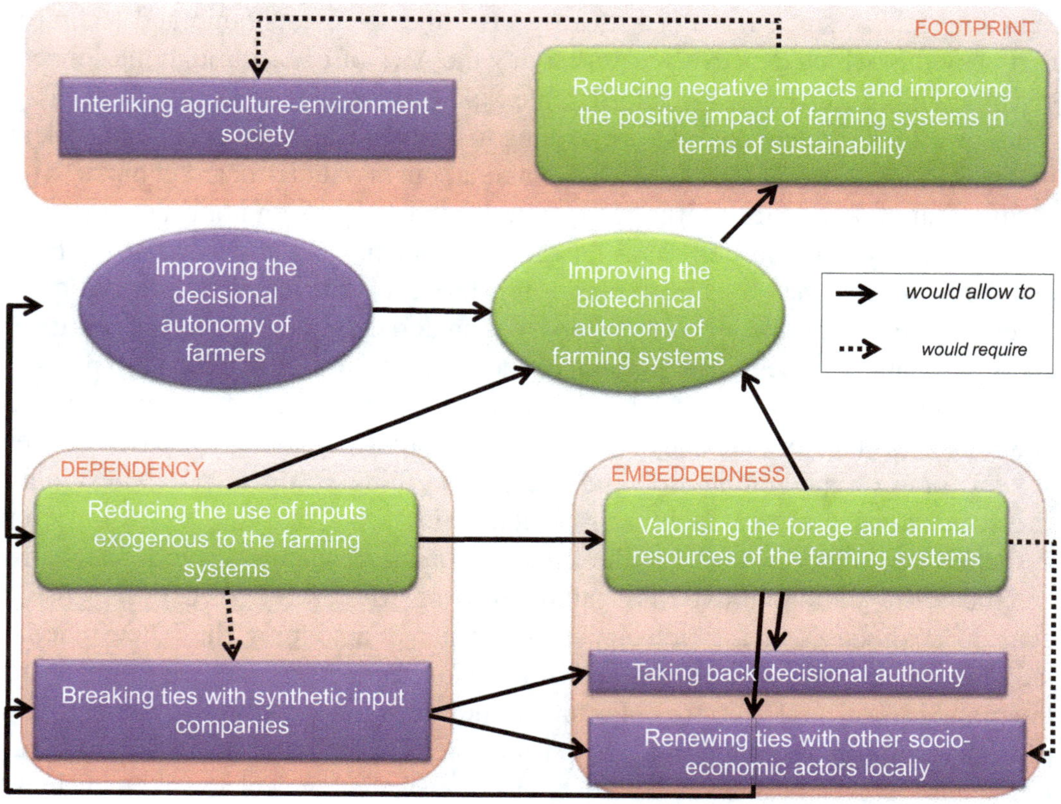

Fig. 1 Conceptual framework for analysing the autonomy of LFSs. It integrates the biotechnical (in green) and the decisional (in purple) dimensions of autonomy and distinguishes the three components that are useful to analyse the overall autonomy of farming systems: embeddedness, dependency, and footprint

- the **forms of embeddedness** of farms, groups of farmers, or agricultural supply chains within territories, that enable to understand how these entities support themselves by using local resources (both natural and socio-economic resources) and valorise these resources. The embeddedness of agricultural activity within a territory thus constitutes a means to increase the autonomy of LFSs.
- the **forms of dependency** of farms, groups of farmers, or territories with respect to inputs, technologies, and the actors that provide them. Increasing the autonomy of LFSs in terms of inputs and technologies calls into question and reconfigures their dependency on socio-economic actors.
- the **forms of the footprint** of farms, groups of farmers, or agricultural supply chains on territories, in terms of their social, economic, and environmental impact. Increasing the autonomy of LFSs has to be assessed in view of their sustainability.

Three approaches grounded on different disciplines were reviewed to analyse the forms of embeddedness, dependency, and footprint, and consequently the autonomy of farming systems and territories. We present these three approaches, their advantages and limitations, and the opportunity to hybridise them in order to get an integrated view of LFSs' autonomy.

Closing Cycles: A Material Flows-Based Approach

Autonomy can be understood as the flows of material existing between components within a farming system and between farming systems and other environmental and socio-economic components of territories and supply chains. Upstream of production, natural resources are used to produce the inputs (e.g. energy) necessary for systems to function. Downstream, beyond the sale of agricultural products, the "storage" compartments of the biophysical environment (the biosphere, the atmosphere, oceans) absorb, accumulate, and sometimes recycle the elements rejected by production systems. In these studies, the solution to improve the autonomy of farming systems and to reduce their negative impact on the environment is to promote intra- or inter-system internal recycling and thus to reduce the use of resources upstream and waste downstream of the production process.

The analysis of territorial metabolism (Bonaudo et al. 2016) has given a toolset to describe which products or by-products of an activity, considered its wastes, could be valorised as resources for another activity. Likewise, life cycle assessment (LCA)-based eco-design or assessment approaches follow the same logic, which aims at considering a "material" form of autonomy by organising optimal recycling of flows of materials. On the territorial level, this ideal state can be achieved by combining systems into a complex organisation, which is not, however, taken into consideration when the only thing measured is flows, in other words, that which is consumed, produced, reused, transformed, and ultimately rejected. In this case, the system is thus considered a "black box".

Flows between crop and livestock farming components in LFSs at the farm or the territory levels can be analysed from this point of view: crops provide the energy, protein, minerals, and vitamins to animals, which in return provide fertilisers that are beneficial for plant growth through their excrement. Several authors have used this approach to show that beyond the "apparent" autonomy in the complementarities between LFS subsystems, the whole can remain heavily dependent on exogenous resources. For example, Nesme et al. (2016) show that in the exchange of materials between organic cereal farms and livestock farms, the production of crops used as animal feed is also heavily dependent on manure fertilisation from conventional livestock farms, as such importations are allowed in organic crop farming. However, these manures themselves come from conventional livestock farms that may use feed from conventional cereal farms. Likewise, Regan et al. (2017) demonstrate that increasing the exchanges between cereal and livestock farms to close the biogeochemical cycles can sometimes lead to increasing the local fodder supply of livestock farms. To balance animal rations (energy and protein), livestock farmers had to buy protein concentrates, and so increase their dependence on nitrogen inputs. Last of all, depending on the geographical level considered, the energy costs of transporting materials can be very high (Asai et al. 2018) and can call into question the relevance of exchanges in economic and environmental terms.

The material flows-based approach focuses on the biotechnical dimension of the autonomy of farming systems, and more particularly on their footprint. It is often limited in terms of dealing with topics related to the embeddedness of production systems, because it only allows to establish an assessment of flows, which has led many authors to broaden their perspective, in particular in the direction of territorial ecology (Buclet 2015; cf. § 2.3).

Managing Agroecosystems: A Functional Approach

The autonomy of LFSs can also be analysed by looking at the technical management of the biological resources of agroecosystems on different levels of space and time, and the performances resulting from this management. It is therefore necessary to study the structure and functioning of agroecosystems and to identify levers for action that increase "biotechnical" autonomy (Fig. 1). The valorisation of local plant resources and organic fertilisers (embeddedness) in LFSs reduces the use of feed inputs, synthetic fertilisers, and fossil energy exogenous to the system (dependency). The assessment of the impact of such practices on the LFSs in terms of sustainability and resilience (footprint) is required.

As for the flows-based approach (cf. § 2.1), the functional approach based on LFSs' autonomy focuses on farming practices that increase the local embeddedness of animal and plant productions by matching them (Hendrickson et al. 2008; Lemaire et al. 2014). However, the latter aims at integrating plants and animals to offer a balanced ratio of energy and protein to animals, and in return, for crops (e.g. legumes) or livestock manures to allow soil fertility to be maintained rather than reducing material losses as a whole. In particular, ruminant LFSs that are self-sufficient in inputs are mainly systems that combine several crops with livestock farming, and in which grass makes up a significant part (Grolleau et al. 2014; Coquil et al. 2014). Grass has multiple advantages: it has a good balance between energy and protein for ruminants, provides permanent ground coverage, and is an inexpensive resource. Legumes also have advantages owing to their symbiotic fixation of atmospheric nitrogen and the provision of high-protein animal feed. Last of all, the insertion of by-products into monogastric animal rations or dairy cattle farming, as well as inter-cropped meslins, also promote a reconnection between animals-plants-soil, all the while allowing for waste recycling, to limit the footprint (Dumont et al. 2017).

The levers for action based on the functional management of agrobiodiversity are concretised in the form of the animal or plant component of the agroecosystem, with the goal being to maintain consistency between these components in view of promoting embeddedness and limiting dependency of LFSs. In mixed crop-livestock farming systems, diversification of the cropping plan and the extension of rotations provide feed that is more balanced in terms of energy and protein for animals (Russelle et al. 2007), and therefore limits the use of external feed inputs. It also promotes synergy between species or functional types of fodder and/or farmed plants (e.g. the combination of grasses and legumes) in time and space. Thus, it induces a better management of plant health by minimising the use of phytosanitary products (Martin et al. 2016), and ensures soil fertility by minimising the use of mineral fertilisers (Lemaire et al. 2014). To improve LFSs' autonomy, levers also concern the diversification of animals themselves (Magne et al. 2017). This consists in: (i) choosing genotypes best suited to local soil-climate conditions and, in particular, local fodder resources, such as local breeds (Lauvie et al. 2011), crossbred animals (Lopez-Villalobos et al. 2000), and/or breeds with a good feed conversion efficiency (Delaby et al. 2009); (ii) combining animals to take advantage of the complementarity of their features, such as combining breeds in dairy herds to produce milk with low feed inputs (Magne et al. 2016), or cattle and small ruminants during grazing to make the best use of fodder resources and to achieve better overall animal productivity and parasite management (Dumont et al. 2013); and (iii) using the diversity of the physiological stages of animals within the herd to match animal needs with the fodder offering, and to deal with the risks of limited fodder resources during certain periods of the year (Blanc et al. 2006).

Some studies carried out on the assessment of autonomous LFSs (systems with little dependency on inputs) showed that these systems were a win-win situation for all three dimensions of sustainability (the economic, environmental, and social dimensions). From the economic point of view, they prioritise high added value per hectare by decreasing input consumption (and therefore dependency) and by mobilising ecosystem services (Garambois and Devienne 2012). They are less dependent on market fluctuations (Benoit and Laignel 2009). From an environmental point of view, they have a smaller footprint in terms of nitrogen, pesticides, and wild biodiversity (Le Rohellec et al. 2009). Last of all, from the social point of view, they allow for more decisional autonomy (Coquil et al. 2014).

The literature reports on multiple limits to this functional approach to LFSs' autonomy. First, few studies address the input autonomy of farming systems while integrating all components of the system. Specifically, the study of LFSs and crop systems has long been carried out separately by livestock production researchers and agronomists, respectively. Studying mixed crop-livestock farming systems requires animal production to be associated again with plant production. It is therefore necessary to analyse the complementarities and the flows between these productions, as well as the recycling of by-products, alternative crops, and inter-crops for animal feed. In addition, few studies combine an analysis of farmers' practices with an analysis of the forms of organisation of the socio-economic and sociotechnical actors involved in the management of autonomous LFSs. This functional approach to LFSs' autonomy is therefore focused on its biotechnical dimension, and is useful for studying the embeddedness, dependency, and footprint of LFSs. However, it is not particularly relevant for studying the decisional dimension of autonomy and its variants in terms of these three components.

Coordinating Actors: An Approach Based on Organisation and Values

The LFSs' autonomy can also be examined by looking at the actors' forms of organisation and the values that they share within farms, groups of farmers, or agricultural supply chains in territories. These forms of organisation can either impede or promote LFSs' autonomy, as they determine the nature and the extent of the coordination between actors and material, economic, and potentially labour flows at the different levels of action (within farms, farm networks, agricultural supply chain, etc.). Sharing values helps farmers build the necessary bond for effective cooperation to develop farming systems' self-sufficiency in terms of inputs (Asai et al. 2018). For that, these farmers draw support from self-organised networks of actors and the experience-based knowledge that they acquire along the way (Coquil et al. 2014). In this sense, they are autonomous in establishing their own technical guidelines and resource portfolios, partially independent of the dominant sociotechnical regime (Rosset and Martínez-Torres 2012; cf. chapter "The Key Role of Actors in the Agroecological Transition of Farmers: A Case-Study in the Tarn-Aveyron Basin"). The issues and determinants of LFSs' autonomy can therefore be addressed as comprehensive research or intervention-research problems focusing on a system of socio-ecological interactions (McGinnis and Ostrom 2014). Comprehensive research seeks to understand how flows of materials and the social, political, and economic organisation of human societies are structured. Intervention-research seeks to participate in designing a collective organisation aimed at achieving a territorialised system of actors, such as a group of crop and livestock farmers to collectively integrate crop and livestock production.

On the territorial level, comprehensive research analyses a variety of actors and issues – whether industrial, urban, or agricultural – from a multidisciplinary viewpoint. Territorial ecology (Barles 2011; Buclet 2015) is an example of an approach offering a combined analysis of territorial resources, systems of activities, and the forms of governance of these resources and activities. It encourages the adoption of a perspective on the interactions between farms, groups of farmers, agricultural supply chains, and territories that takes the organisational and identity dimensions into account. It requires the interplay between actors (capacity for action, negotiation, etc.), the values of these actors, and their impact on forms of territorial embeddedness to be described by identifying what resources and activities they will prioritise. This ranking of priorities is based on their power of action and their vision of the system's autonomy. Different focuses can be adopted, depending on the goals pursued: a business strategy, the values of actors, the qualification of resources, or the relation to consumption. Analysing business strategies (Saives 2002; Hannachi et al. 2010) allows one to distinguish between two types of spatial behaviours of companies: localisation behaviours and territorialisation behaviours. The analysis of values and in particular the vision of autonomy enables one to understand farmers' relations (or the absence thereof) with their ecological, economic, and social environment (Stock and Forney 2014). Autonomy as a value determines actors' strategies to better valorise the resources available in their territory (embeddedness). These strategies can be manifest in the search for and the sharing of knowledge and technologies through local networks of farmers and advisors. They can also aim at bringing together a broader diversity of actors, in particular around the development of local food systems (Bellows and Hamm 2001). The analysis of the valorisation of agricultural products, and in particular the territorial qualification processes for food products (Ilbery et al. 2005), affords insight into the process of constructing territorial resources jointly between farmers or within supply chains and territories. This clarification could benefit from an analysis of the relations between production and consumption within supply chains, specifically in terms of the socio-spatial proximity between the producers and consumers of a territorial resource (Deverre and Lamine 2010).

In a territory, intervention research, such as that carried out during the TATA-BOX project, aims at supporting the design of a collective organisation oriented towards a territorialised system of actors, on the basis of a transdisciplinary viewpoint. It can draw support from the result of comprehensive research in order to understand the interactions between farms, groups of farmers, agricultural supply chains, and territories from the organisational and identity perspectives. It subsequently requires the organisation of a debate around the notions of autonomy and the motives behind collective organisation (Ryschawy et al. 2017). This phase should establish common values between actors who wish to engage in the collective organisation, or alternatively, allow them to exit the process. The following stage consists in applying tools to design and assess scenarios that enable the actors involved to analyse the advantages and limits of diverse forms of collective organisation, choosing one to ultimately implement (Moraine et al. 2016). The scenario assessment phase can partially use a flows-based and/or functional approach (via the associated practices and performances), in particular to balance the material, economic, or labour flows between actors (Barnaud and Van Paassen 2013). It is necessary to ensure that the scenario retained does not contribute to increasing power disparities between actors. The assessment must also consider actors' degree of satisfaction with respect to their decisional autonomy (Ryschawy et al. 2017).

The organisation and values approach to LFSs' autonomy thus proves to be appropriate for addressing its decisional dimension of autonomy, based on the components of embeddedness, dependency, and footprint. However, it is not suited to addressing its biotechnical dimension and its variants in terms of these three components.

An Integrated Approach to Autonomy

This brief literature review shows that the three approaches implemented to analyse the autonomy of LFSs put emphasis on either one or two of its components (i.e. embeddedness and/or dependency and/or footprint), as well as integrating one or both of its dimensions (i.e. biotechnical and/or decisional autonomy (Table 1).

Table 1 Contributions of the three research approaches to analysing the autonomy of LFSs. The "X"s indicate the dimensions and components of autonomy to which each approach contributes. The coloured rectangles indicate the dimensions, approaches, and components addressed in each of the case studies (presented in section "Case study 1: a methodology to analyse the overall autonomy of dairy sheep farms in Aveyron"): case study 1 is in green; case study 2 is in orange. The continuous/dotted lines refer to the spatial level taken into account in each case study: the farm level is indicated with a continuous line; the territorial level with a dotted line

		DIMENSIONS					
		Biotechnical			Decisional		
	COMPONENTS	Embeddedness	Dependency	Footprint	Embeddedness	Dependency	Footprint
APPROACHES	Flows-based		X	X			
	Functional	X	X	X			
	Organisation and values				X	X	X

Our analysis framework (Fig. 1) structures and hybridises these three components and these two dimensions of the autonomy of LFSs. To illustrate this integrated approach, we apply this analysis framework *a posteriori* to two case studies carried out as a part of the ANR TATA-BOX project. The goal is to show how these two case studies address the different dimensions and components of the analysis framework that we are developing here, and the limits of these studies with respect to the framework. The first case study aims at producing a methodology to analyse the global autonomy of dairy sheep farming systems in the Roquefort region (Thenard et al. 2014, 2016). The level of analysis is that of the farm. The second case study explores the design of LFSs' autonomy through integrating crop and livestock farms on the level of a small territory in the Occitanie region (Ryschawy et al. 2017).

Case Study 1: A Methodology to Analyse the Overall Autonomy of Dairy Sheep Farms in Aveyron

This study used a functional approach and focused on the biotechnical dimension of LFSs' autonomy (Table 1). It consisted in analysing farmers' management and assessing the multiple performances of the LFSs of a group of dairy sheep farms in south-western France (territory of the Roquefort PDO), that were seeking to become more autonomous through better use of the territory's fodder resources. To do so, we developed a three-step methodology: (i) collectively defining what autonomy encompasses in these LFSs; (ii) describing and characterising LFSs, based on the combinations of levers of action implemented by livestock farmers to increase their autonomy; and (iii) assessing the multiple performances, including autonomy, of LFSs. The methodology implemented enabled us to address the decisional dimension of these farms' autonomy without, however, studying it.

Step 1. Participatory Workshops to Comprehensively Describe Autonomy in Sheep Farming Systems

This first step was carried out as a participatory workshop with sheep farmers and some of their advisers. It aimed at building a common framework for LFSs' autonomy. The workshop consisted of an individual "post-it" session, followed by the drawing of a collective cognitive map to establish common ground (Fig. 2). The map showed that LFSs' autonomy related to three main categories of goals for the farmers and their advisers. The first goal was to valorise local resources to feed sheep, and in particular the fodder and pastoral resources of the Roquefort territory (in green, Fig. 2), which expressed the embeddedness of production systems in the "*terroir*" (term used by the farmers). The second goal was to reduce input

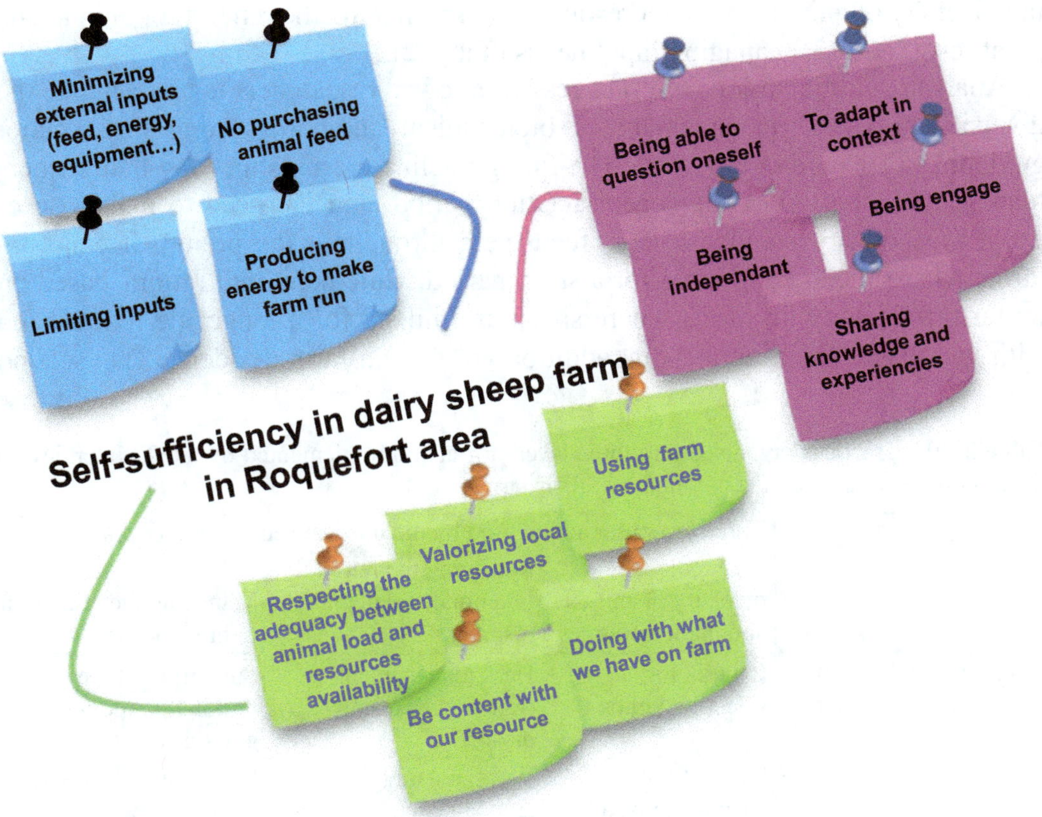

Fig. 2 Cognitive map built in collaboration with the group of sheep farmers and their advisers to define the meaning of autonomy for them in Roquefort territory: valorising local resources (in green); limiting the use of inputs and purchases (in blue); and the ability to make their own decisions (in pink)

use – whether feed inputs for animals, agrochemical inputs for crops, or equipment inputs for livestock or crop management (in blue, Fig. 2) – and thus for farmers to remove themselves from a situation of dependency on suppliers. The last goal was for farmers to be able to make their own decisions, to adapt to the soil-climate and economic contexts, and to share their experiences within peer groups (in pink, Fig. 2). It thus represented the decisional dimension of autonomy.

Step 2. Characterisation of the Operation of Dairy Sheep Farming Systems from the Angle of Biotechnical Autonomy

The second step was based on analysing the data collected from the 27 dairy sheep farmers in the Roquefort area, expertly selected based on the criteria of "seeking autonomy of sheep farming systems". This step was based on an approach focusing on the flock and fodder practices managed by farmers to increase the embeddedness and reduce the dependency of sheep farms. Ten kinds of practices categorised were identified as levers for action implemented by farmers to increase their "biotechnical" autonomy. They were organised into three types of levers for action: (1) managing the diversity of animal and plant resources; (2) managing the renewal of animal and plant resources; (3) managing input needs (Table 2).

Analysing combinations of practices has allowed researchers to characterise the diversity of LFSs' management of biotechnical autonomy along three major guidelines (Thénard et al. 2014). The first guideline presents the way that farmers manage the duration of the sheep lactation period, and the need to make use of exogenous dietary supplements to feed them throughout the period. It contrasts farms where sheep are milked for a short period while being fed rations based on on-farm fodders, with farms where sheep are milked for a longer period and fed with purchased concentrates in addition to the on-farm fodders. The second

Table 2 Ten practices organised into three levers for action implemented by the 27 interviewed sheep farmers to increase the autonomy of their farm

	Three levers for action used by the interviewed sheep farmers		
	Managing the diversity of animal and plant resources	Managing the renewal of animal and plant resources	Managing the reduction in input needs
Livestock and fodder management practices	Criteria for selecting lambs	Herd management (reproduction and dry-off)	Suitability of the milking period with grass growing
	Diversity of pastures grazed in the spring	Ways of using the animal genetic progress	Origin of concentrates for sheep feed
	Fodder resources used in summer	Diversity of the fodder and/or pastoral resources of the farm	Outdoor or indoor management of lambs
			Supplementary feeding of sheep in summer

guideline describes the way that farmers manage the diversity of fodder resources and sheep reproduction. It thus contrasts farms based on diversified fodder systems and natural animal reproduction, with farms based on more intensive fodder systems with limited diversity and artificial reproduction. Last of all, the third guideline presents the types of females desired and selected at the farm. It contrasts farmers who use "milk yield" as the only criterion for selecting females and raising lambs in a sheep pen, with farmers who use selection criteria other than milk yield and raise lambs outside for some months of the year.

Four types of livestock farms are therefore distinguished based on their strategy to increase their autonomy.

Type 1: Producing the Milk Permitted by the Territory's Resources These farms adapt the duration of the sheep-milk production period to the on-farm forage and pastoral resources. The milking period overlaps with the grazing period, including during the summer, thanks to the use of pastoral resources. These farmers try to combine significant embeddedness with low dependency on feed inputs, even if it means producing less than the average in the region. Biotechnical autonomy is also closely tied to a desire for decisional autonomy and to the values promoted by these sheep farmers.

Type 2: Producing Milk by Optimising Fodder Stocks to Provide for Significant Sheep Needs These farms are based on the use of "intensified" and low-diversified seeded grasslands with grasses or grass/alfalfa mixtures. These grasslands are used to produce fodder stocks and are also grazed during the spring. The farms are autonomous in terms of energy supply of animals but not protein supply. Farmers therefore use nitrogen concentrates to provide for the significant nutritional needs of their sheep, which are selected based on their milk yield. They also use mineral fertilisers to ensure the production of the fodder necessary for milk production, which is mainly carried out in a sheep barn during the winter and for a short time in the spring. Therefore, feed self-sufficiency indicates a strong desire for embeddedness in the *terroir*, but follows an efficiency approach that results in high nitrogen dependency of farms.

Type 3: Producing Milk through Organic Farming These farms use a wide variety of forage resources, including pastoral resources, native grasslands and highly diversified seeded grasslands. They do not use agrochemical fertilisers, which are prohibited in organic farming. On these farms, milk production is managed in accordance with the grass-growing season, beginning in the spring and often lasting until the autumn. Taking into account the lower quality of fodder resources, in particular due to the fact that they do not receive mineral fertilisers, sheep farmers use nitrogen and energy concentrates to provide sheep rations. Autonomy is based on a low level of dependency on synthetic inputs. Yet the high degree of embeddedness of these sheep farms in the local resources of the terroir leads to their dependency on animal feed inputs. Reducing their environmental footprint is one of the ultimate goals of this type of sheep farm.

Type 4: Producing Milk by Diversifying the Fodder System and by Integrating Crop and Livestock Farming These farms are based on a wide diversity of cultivated plant resources which enable them to establish stocks and ensure grazing by alternating types of seeded grasslands – or crops – over the year, including summer crops (e.g. intercropping, sorghum, etc.). Using a wide diversity of crops allows farmers to limit purchases not only of feed concentrates but also of agrochemical fertilisers because legumes and intercropped cover crops are used extensively, and conservation agriculture practices are sometimes implemented. This diversified fodder and crop system enables sustainable milk production both in the sheep barn and during the grazing period, including in the summer. The autonomy of these sheep farms is thus based on a significant embeddedness in the local soil-climate context, along with a low dependency on inputs.

Step 3. Assessing the Performance Profiles of the Different Types of Sheep Farms

This third step aimed at assessing the technical-economic and environmental footprint of the four types of sheep farm. The technical-economic performances of sheep farms were therefore assessed based on three categories of performance: herd productivity (ewe milk production, lambing rate and prolificacy), economic efficiency, and the feed self-sufficiency of the herd. For each category, multiple indicators were defined and aggregated (Fig. 3). Environmental performances were assessed during a second series of interviews with farmers, based on their agronomic practices. These were categorised into different criteria, depending on whether they related to practices to conserve soil fertility, to limit agrochemical inputs, or to manage plant diversity.

The analysis of technical-economic performances shows that the four types of sheep farm present different trade-offs between herd productivity, economic efficiency, and feed self-sufficiency (Fig. 3). It appears that type 4 farms present the most balanced profile. Type 2 farms have the least feed self-sufficiency, the highest herd productivity and the lowest economic efficiency. This proves that increasing animal production does not systematically entail best economic performances. Compared to type 4 farms, type 1 farms have the same level of feed self-sufficiency, slightly lower economic efficiency and significantly lower herd productivity. One of the main reasons is that these farmers seek to minimise all kinds of purchases and use local natural resources without seeking to better use the agronomic potentialities of the environment to diversify the fodder system. Last of all, type 3 farms have the least balanced performance profiles. They have the same herd productivity as type 1 farms, with a slightly lower economic efficiency, but they have the lowest feed self-sufficiency of the four types of farm identified. The added value of organic milk production therefore allows them to have an economic efficiency that is not too strongly impacted by the low feed self-sufficiency of the herd.

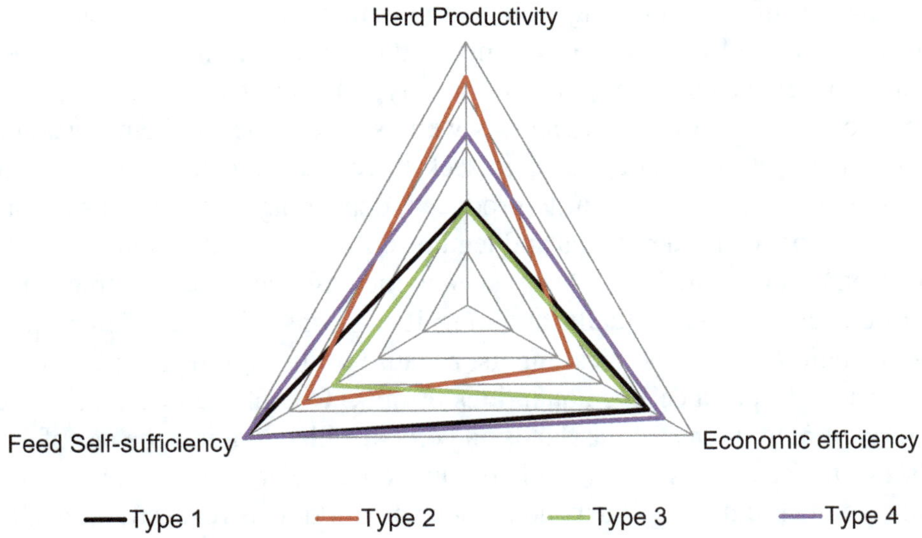

Fig. 3 Technical-economic performance profiles of the four types of sheep farm characterised above

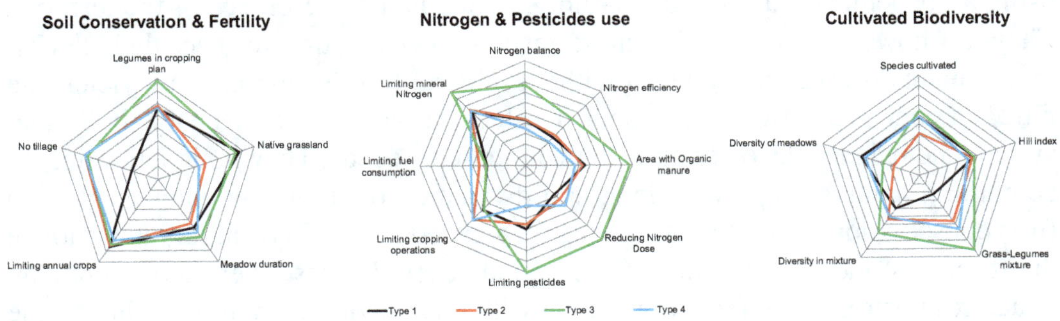

Fig. 4 Assessment of the environmental performances of the four types sheep farming system characterised

The assessment of sheep farms' environmental performance (Fig. 4) showed that type 4 farms had the best scores for the preservation of soil fertility. These farms practice no-tilling and simplified cropping techniques. Inversely, type 1 farms have poorer performance around the maintenance of soil fertility, due to the use of tilling. As for type 3 farms, their performance is good, owing to the extensive use of legumes (sainfoin, alfalfa, clover, etc.). With respect to the "use of chemical inputs", type 3 farms present the best performance in terms of indicators related to: (i) the risks of pesticide use, because it is the only type that does not use pesticides; and (ii) the nitrogen use due to planting legumes and not purchasing mineral fertilisers. The other three types farm have equivalent performances. Last of all, with respect to the "valorisation of crop diversity", the strengths of type 3 concern the management of species diversity, in particular legumes, whereas for types 1 and 4, their advantages are around managing types of grasslands.

The analysis of sheep farming systems shows that the forms of autonomy sought by sheep farmers differ. The components of the analysis framework that we offer allow us to observe these forms. For example, type 1 is built around autonomy based on the valorisation of local resources as well as independence from upstream and downstream structures. Sheep farmers seek to reduce herd feed inputs even if it means reducing the volume of milk produced, depending on the farm's agronomic potential to produce fodder, crops and legumes. In this way, they reduce the farm's environmental footprint. Type 2 farms seek to increase their flocks' forage and feed self-sufficiency, without actually attaining it in terms of either dietary nitrogen supplementation or agrochemical inputs, because of the significant pressure on the selected sheep to produce large quantities of milk. One of the levers mobilised by these farmers is to intensify grasslands and crops, which requires the use of mineral fertilisers and pesticides. They seek production efficiency over valorising local natural resources and being independent of upstream/downstream structures. On the other hand, they limit tilling to reduce the workload or soil erosion. Yet they cannot go without pesticides (glyphosate in particular), which causes the farm to have a larger environmental footprint. Type 3 farmers, who have organic farming management, naturally seek to reduce their farms' environmental footprint by not using agrochemical inputs and by valorising the natural resources of the territory. They are however forced to purchase feed supplementation to meet their flocks' requirements, as fodder produced without mineral fertilisation has low yields and nitrogen contents. Last of all, type 4 livestock farmers act to diversify the fodder system and balance the offering with their animals' needs. They limit soil tilling by combining legume crops with long and diversified rotations, drawing inspiration from conservation agriculture. By doing so, they reduce synthetic inputs and valorise the resources and potentialities of the region. They decrease their dependency on structures upstream from the farm, along with their environmental footprint, all the while maintaining the best profile in terms of productivity/economic efficiency and feed self-sufficiency.

Case Study 2: Co-Design of Scenarios of Exchanges Between Crop and Livestock Farmers to Improve Autonomy on the Level of a Small Territory

Coordination Between Farmers to Strengthen Autonomy on the Collective Level

This case study presents an attempt to integrate crop and livestock farming in territories in collaboration with a group of crop and livestock farmers. The implementation of this coordination between farmers has a twofold impact: on each farm, and collectively. Various approaches were therefore used to address autonomy, according to the organisational level considered and in view of the components and priority dimensions (Table 1).

Scenarios of exchanges between crop and livestock farmers were jointly designed in collaboration with organic farmers in Tarn-et-Garonne wishing to increase their embeddedness and decrease their dependency on feed inputs and fertilisers. In this group, livestock farmers wished to develop a local supply of concentrates, whereas crop farmers wished to diversify their cropping plans by inserting legumes into them, and to collect manure to enrich their soils. A functional analysis was carried out from the biotechnical perspective to estimate the demand and offering of concentrates and manure, i.e., the dependency, and the potential to increase the embeddedness of each crop or livestock farm. Subsequently, an analysis of biotechnical flows allowed for a comparison of the overall supply and demand on the collective level, in order to estimate the dependency of the group. This first analysis involved 24 livestock and crop farmers belonging to the Bio82 collective and their facilitator (Fig. 5).

Several scenarios of crop-livestock integration were designed, depending on the form of organisation of biotechnical exchanges between farms. The scenario chosen by the livestock farmers was based on the insertion of grasslands (mainly alfalfa) and cereal-legume mixtures into rotations and manure exchanges (Moraine et al. 2016). It increased the embeddedness and limited the dependency of all the farms. In such a scenario, annual exchanges amounted to 341 tonnes of alfalfa, 125 tonnes of a barley/peas-type cereal-legumes mixtures, and 88 tonnes of hay provided by the crop farmers. In return, 1059 tonnes of composted manure were available to restore the soil organic matter exported by crop farms. This scenario was very promising in terms of closing the mineral cycles, managing the agroecosystems, and bringing together actors around common values. However, across an area of 1655 ha, the distances between crop and livestock farmers were very large, and logistic constraints proved to be too complex to manage.

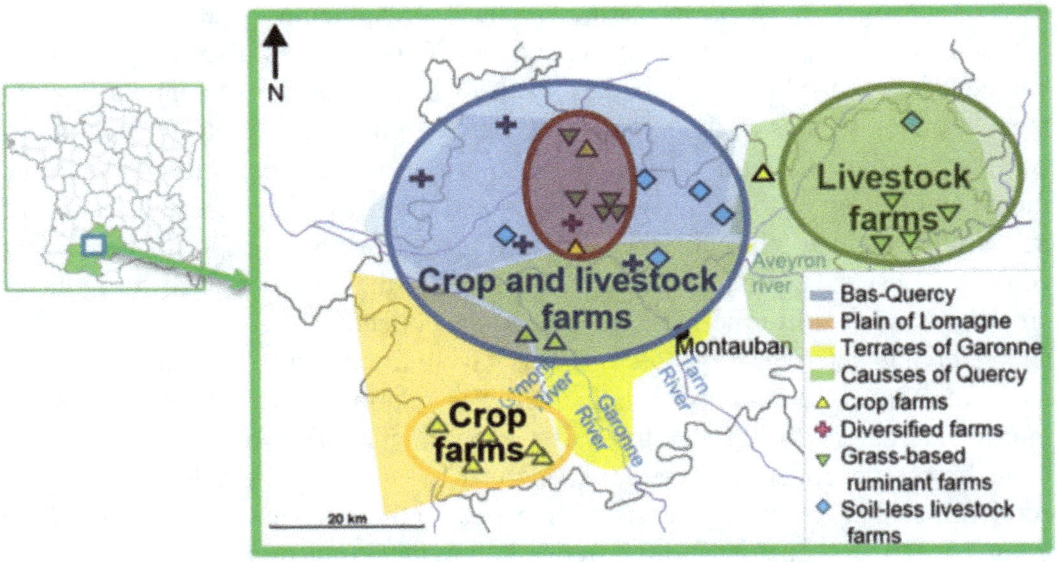

Fig. 5 Location of the crop and livestock farmer groups involved in the process. The 24 farmers initially involved in the research, were located in three contrasting soil-climate areas represented in yellow, green, and blue. The seven farmers selected for the final crop-livestock integration scenario were located in a single soil climate area (in blue) and were delimited by the red circle

As a result, the approach was repeated with seven livestock and crop farmers (Fig. 5), who knew one another very well and were very close geographically (less than 20 km apart). The scenario favouring the maximum synergy was based on exchanges of 40 tonnes of harvested barley-pea cereal-legumes mixture, 18 tonnes of maize for grains, 8 tonnes of alfalfa hay, and 4 tonnes of sunflower for grains provided by the crop farmers. In return, the livestock farmers supplied 105 tonnes of manure. The exchanges within this subgroup were less ambitious in terms of volume, but they enabled the redesign of livestock and crop farming systems to close mineral cycles, with a view to achieving a smaller footprint and to coordinating actors in moving towards more decisional embeddedness. The scenario also appeared to be more feasible in terms of coordination and logistics (transportation, storage, etc.), and it limited the decisional dependency on other actors (transportation or storage companies, etc.). To assess these scenarios, we simultaneously considered footprint and dependency from the biotechnical perspective, by carrying out a functional analysis on the farm level as well as a flows-based analysis on the level of the group of farms. We also assessed embeddedness, dependency, and footprint from the decisional perspective on the level of the group of farms.

Sustainability and Performance of the Crop-Livestock Integration Scenarios

The crop-livestock integration design scenarios allowed farmers to collectively increase autonomy with respect to inputs, and thereby to reduce dependency on exogenous supplies by increasing the territorial embeddedness of farms (Fig. 6). On the collective level, livestock farmers became completely autonomous thanks to local exchanges with crop farmers, thus strengthening their biotechnical embeddedness and reducing their dependency. Crop farmers also improved their embeddedness and limited their dependency on organic nitrogen inputs (feather meal, etc.) exogenous to the territory, by introducing legumes into their rotations and through the contribution of organic manure from livestock farms. In addition, the diversification of rotations allegedly limited the risks of disease and the use of irrigation water by limiting the surfaces planted with crops with significant consumption needs, such as maize, thus reducing the environmental footprint of crop farms. The multi-criteria assessment on the collective level showed that spatial and temporal heterogeneity of crops was favoured, and that autonomy (energy, mass, and protein autonomy) was increased in relation to the decrease in use of inputs external to the group, thus also increasing embeddedness and reducing biotechnical dependency. Therefore, as Asai et al. (2018) emphasised, higher economic and environmental costs with respect to fuel use should be estimated for groups of farmers, due to the more frequent individual transportation of crops, fodder, and manure.

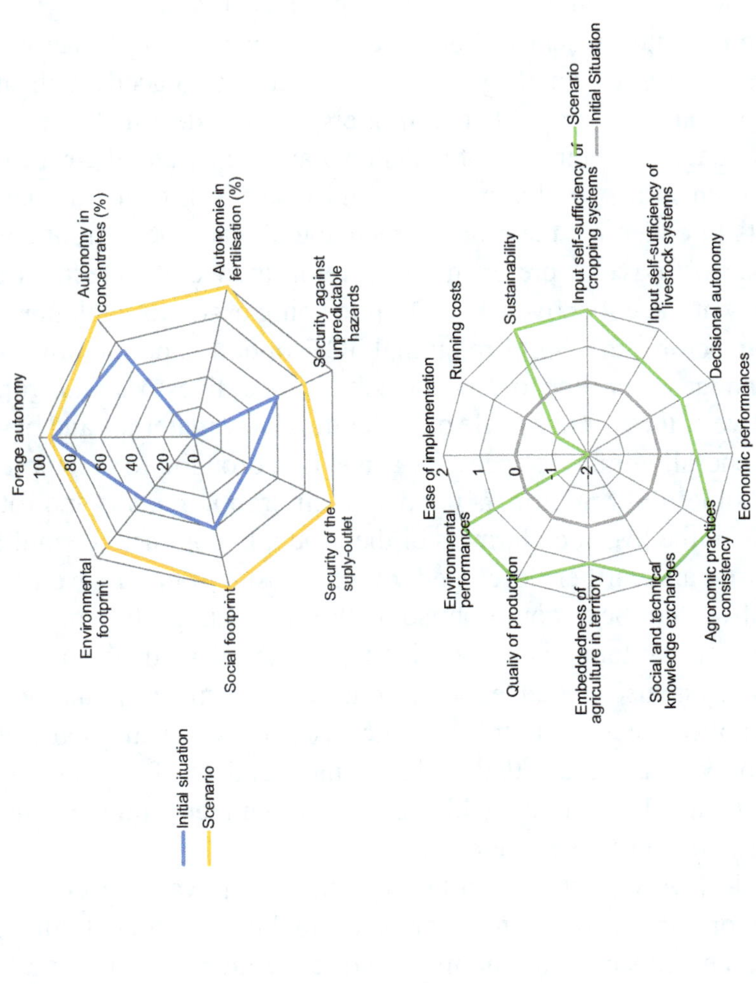

Fig. 6 Radar chart assessing the scenarios of crop-livestock integration designed at the collective level compared to the initial situations (to the left: group of 24 farmers; to the right, group of 7 farmers)

A Participatory Process to Examine the Decisional Dimension of Autonomy

Beyond the purely technical aspect, the exchanges numerically represented through a biotechnical analysis of flows required complex coordination between the actors. We considered this coordination and its impact on embeddedness, dependency, and footprint from the decisional perspective in the context of the participatory design process implemented. In this case, for the first analysis of 24 farmers, we proposed three types of organisation: cooperative-type centralised organisation; multi-relational organisation of the purchase/sale platform type based on ICTs (cf. chapter "Information and Communication Technology (ICT) and the Agroecological Transition"); and an intermediate option called multi-centred organisation, in which multiple, more localised small groups self-organised. Faced with the three organisational scenarios proposed, the farmers clearly declined the centralised option, as it went against their goal of decisional autonomy and their idea of direct exchanges between farmers. This option would result in higher transaction costs and investments in collective materials, which sounds like the current cooperative model they wished to avoid, preferring to develop their embeddedness and limit their dependency at the collective level. Despite being easier to implement in terms of coordinating actors, the multi-relational ICT option did not offer sufficient stability of exchanges over time to permit the redesign of livestock and crop farming systems. Effectively, the purchase-sale of agricultural raw materials and by-products was therefore generally limited to occasional needs (a drought, for example) and did not make it possible for crop farmers to adapt their cropping plans and rotations to match the needs of the livestock farmers of the group. In contrast, the multi-centred option was chosen and further developed within the subgroup of seven farmers. It appeared to offer the best compromise between closing the mineral cycles, redesigning farming systems, and coordinating actors around common values to develop farming systems' embeddedness, limit their dependency, and reduce their footprint at the collective level. In this specific case, collective autonomy was based more on autonomy as a value – that is, being independent of suppliers – than on decisional and financial autonomy, which can help to understand the compromises made by the farmers within the group.

In the scenario involving the seven farmers, the group was expected to manage exchanges in coordination with one another by making reciprocal commitments. The increase in the decisional autonomy of farmers with respect to input suppliers was replaced by a high degree of dependency on the farmers in the group. Even though all the farmers in the group were able to improve their overall gross margins as well as their environmental footprint in the exchange scenario designed, they had to invest time in coordinating exchanges, and money in storage materials, and consequently had to agree to reduce their individual decisional autonomy to increase it on the level of the group of farmers. Moreover, compromises between the collective level (with a clear improvement in input autonomy as well as decisional autonomy with respect to suppliers) and the individual level had to be made, with trade-offs that were different depending on the farmer. The question of sharing materials for storage and potentially transportation was addressed and required investments. To address these questions, an Economic and Environmental Interest Group was set up, also enabling skills exchange and institutional acknowledgement of the agroecological process, as well as increased dependency on one another in the case of collective materials purchases.

During the operationalisation of the process, new locks appeared. For the livestock farmers, modifying rations constituted a large risk in the absence of crop farmers' guarantees around the quality of the feed provided. Moreover, the farm-based manufacture of foods implied more work for them than when they purchased finished feed. For the crop farmers, the utility of using legumes to start their rotation or as an intercrop was high, but it did not always offset the risk of not valorising these crops if the livestock farmers did not purchase them. As Asai et al. (2018) mention, the dependency between farmers is thus reinforced, and the sustainability of exchanges is contingent on the monitoring and facilitation of these exchanges, which require agreements and individual and collective learning processes. Monitoring product orders and deliveries and the implementation of contracts appeared essential. In the context of Bio82, the group leader's departure resulted in a lack of follow-up and disagreements between farmers around schedules and crop exchange commitments, endangering the organisation implemented. Therefore, even though the scenarios designed promoted autonomy in terms of quantity, energy, and protein content, and were in line with the decisional autonomy values of farmers in the group, in terms of reducing the dependency on suppliers, the need for the process to be facilitated during its implementation appeared to be a key factor determining the operationalisation of the scenarios considered.

Conclusion

The conceptual framework developed here enables one to comprehensively analyse the biotechnical and decisional dimensions of the autonomy of LFSs. It is based on three main components for analysing relations between LFSs and their territory: embeddedness, dependency, and footprint. This framework, applied to two case studies carried out under the ANR TATA-BOX project, shows that it is initially the biotechnical autonomy of LFSs that is addressed in this research, with the decisional autonomy dimension being taken into account subsequently and to varying degrees. For example, in the first case study, at farm level, the decisional autonomy of LFSs was not studied as such, even though it was taken into account during the first step of the research process aimed at collectively defining what the notion of autonomy encompassed for the farmers. On the other hand, in the second case study, at the territorial level, it was studied as such, because it constituted a compulsory step to design crop-livestock integration at the territorial level. At the farm level, the two case studies focus on the biotechnical dimension. Developing the decisional autonomy of LFSs would be interesting to understand the factors influencing farmers' choices. Switching to the territorial level requires articulating the functional approach of LFSs with flows-based and organisational approaches, as demonstrated by case study 2. The tools, methods, and concepts for studying this decisional dimension of autonomy are addressed in other chapters in this book, in particular those on the governance and adaptive management of the AET (*cf.* Chaps. 7 and 6 respectively), as well as the analysis of farmers' networks and information systems for the AET (*cf.* Chap. 8). Ultimately, applying the conceptual framework to our case studies clearly illustrates that to support the AET of LFSs, it is important to integrate the three components of analysis constituted by embeddedness, dependency, and footprint into the biotechnical dimension. It furthermore shows that research efforts should be made to better integrate the biotechnical and decisional dimensions of autonomy.

References

Altieri MA (1987) Agroecology: the scientific basis of alternative agriculture. Westview Press, Boulder

Altieri MA, Nicholls CI, Montalba R (2017) Technological approaches to sustainable agriculture at a crossroads: an agroecological perspective. Sustain 9:1–13. https://doi.org/10.3390/su9030349

Asai M, Moraine M, Ryschawy J et al (2018) Critical factors for crop-livestock integration beyond the farm level: a cross-analysis of worldwide case studies. Land Use Policy 73:184–194. https://doi.org/10.1016/j.landusepol.2017.12.010

Barles S (2011) L'écologie territoriale : qu'est-ce que c'est ? In: Ecotech&tool Conference, 30 novembre–2 décembre 2011, Montpellie

Barnaud C, Van Paassen A (2013) Equity, power games, and legitimacy: dilemmas of participatory natural resource management. Ecol Soc 18:21. https://doi.org/10.5751/ES-05459-180221

Bellows AC, Hamm MW (2001) Local autonomy and sustainable development: testing import substitution in more localized food systems. Agric Hum Values 18:271–284. https://doi.org/10.1023/A:1011967021585

Benoit M, Laignel G (2009) Performances techniques et économiques en élevage ovin viande biologique: observations en réseaux d'élevage et fermes expérimentales. Innov Agron 4:151–163

Blanc F, Bocquier F, Agabriel J et al (2006) Adaptive abilities of the females and sustainability of ruminant livestock systems. A review. Anim Res 55:489–510

Bonaudo T, Domingues JP, Tichit M, Gameiro A (2016) Intérêts et limites de la méthode du métabolisme territorial pour analyser les flux de matière et d'énergie dans les territoires d'élevage, 217–220

Brocard V, Jost J, Rouillé B et al (2016) Feeding self-sufficiency levels in dairy cow and goat farms in Western France: current situation and ways of improvement. Grassl Sci Eur 21:53–55

Brussaard L, Caron P, Campbell B et al (2010) Reconciling biodiversity conservation and food security: scientific challenges for a new agriculture. Curr Opin Environ Sustain 2:34–42. https://doi.org/10.1016/j.cosust.2010.03.007

Buclet N (2015) Essai d'écologie territoriale. L'exemple d'Aussois en Savoie. CNRS, Paris

Clark B, Stewart GB, Panzone LA et al (2016) A systematic review of public attitudes, perceptions and Behaviours towards production diseases associated with farm animal welfare. J Agric Environ Ethics 29:455–478. https://doi.org/10.1007/s10806-016-9615-x

Coquil X, Béguin P, Dedieu B (2014) Transition to self-sufficient mixed crop–dairy farming systems. Renew Agric Food Syst 29:195–205. https://doi.org/10.1017/S1742170513000458

Delaby L, Faverdin P, Michel G et al (2009) Effect of different feeding strategies on lactation performance of Holstein and Normande dairy cows. Animal 3:891–905. https://doi.org/10.1017/S1751731109004212

Deverre C, Lamine C (2010) Les systèmes agroalimentaires alternatifs. Une revue de travaux anglophones en sciences sociales. Économie Rural Agric Aliment Territ:57–73. https://doi.org/10.4000/economierurale.2676

Dumont B, Fortun-Lamothe L, Jouven M et al (2013) Prospects from agroecology and industrial ecology for animal production in the 21st century. Animal 7:1028–1043. https://doi.org/10.1017/S1751731112002418

Dumont B, González-García E, Thomas M et al (2014) Forty research issues for the redesign of animal production systems in the 21st century. Animal 8:1382–1393. https://doi.org/10.1017/S1751731114001281

Dumont B, Ryschawy J, Duru M et al (2017) Les bouquets de services, un concept clé pour raisonner l'avenir des territoires d'élevage. INRA Prod Anim 30:407–422

Duru M, Therond O (2015) Livestock system sustainability and resilience in intensive production zones : which form of ecological modernization ? Reg Environ Chang 15:1436–3798. https://doi.org/10.1007/s10113-014-0722-9

Duru M, Therond O, Martin G et al (2015) How to implement biodiversity-based agriculture to enhance ecosystem services: a review. Agron Sustain Dev 35:1259–1281. https://doi.org/10.1007/s13593-015-0306-1

Francis CA, Lieblein G, Gliessman SR et al (2003) Agroecology: the ecology of food systems. J Sustain Agric 22:99–118. https://doi.org/10.1300/J064v22n03_10

Garambois N, Devienne S (2012) Les systèmes herbagers économes du Bocage vendéen: une alternative pour un développement agricole durable? Innov Agron 22:117–134

Grolleau L, Falaise D, Moreau J, et al (2014) Autonomie et productivité : évaluation en élevages de ruminants grâce à trois indicateurs complémentaires Résumé 1. L ' autonomie des systèmes de production Autonomie alimentaire quantitative et azotée, 17–24

Hannachi M, Coléno F-C, Assens C (2010) La collaboration entre concurrents pour gérer le bien commun: le cas des entreprises de collecte et de stockage de céréales d'Alsace. Ann des Mines-Gérer Compr 3:16–25. https://doi.org/10.3917/geco.101.0016

Hendrickson JR, Hanson JD, Tanaka DL, Sassenrath G (2008) Principles of integrated agricultural systems: introduction to processes and definition. Renew Agric Food Syst 23:265–271. https://doi.org/10.1017/S1742170507001718

Ilbery B, Morris C, Buller H et al (2005) Product, process and place: an examination of food marketing and labelling schemes in Europe and North America. Eur Urban Reg Stud 12:116–132. https://doi.org/10.1177/0969776405048499

Koohafkan P, Altieri MA, Gimenez EH (2012) Green agriculture: foundations for biodiverse, resilient and productive agricultural systems. Int J Agric Sustain 10:61–75. https://doi.org/10.1080/14735903.2011.610206

Lauvie A, Audiot A, Couix N et al (2011) Diversity of rare breed management programs: between conservation and development. Livest Sci 140:161–170. https://doi.org/10.1016/j.livsci.2011.03.025

Le Rohellec C, Falaise D, Mouchet C et al (2009) Analyse de l'efficacité environnementale et énergétique de la mesure agri-environnementale «Système fourrager économe en intrants»(SFEI), à partir de l'analyse de pratiques de quarante quatre signataires. Campagne culturale 2006/2007:109–112

Lemaire G, Franzluebbers A, de Faccio Carvalho PC, Dedieu B (2014) Integrated crop–livestock systems: strategies to achieve synergy between agricultural production and environmental quality. Agric Ecosyst Environ 190:4–8. https://doi.org/10.1016/j.agee.2013.08.009

Lopez-Villalobos N, Garrick DJ, Blair HT, Holmes CW (2000) Possible effects of 25 years of selection and crossbreeding on the genetic merit and productivity of New Zealand dairy cattle. J Dairy Sci 83:154–163. https://doi.org/10.3168/jds.S0022-0302(00)74866-1

Madelrieux S, Buclet N, Lescoat P, Moraine M (2017) Écologie et économie des interactions entre filières agricoles et territoire: quels concepts et cadre d'analyse? Cahiers Agricultures 26:24001. https://doi.org/10.1051/cagri/2017013

Magne MA, Thénard V, Mihout S (2016) Initial insights on the performances and management of dairy cattle herds combining two breeds with contrasting features. Animal 10:892–901. https://doi.org/10.1017/S1751731115002840

Magne MA, Ollion E, Cournut S, et al (2017) Some key research questions about the interest of animal diversity for the agroecological transition of livestock farming systems. In: First agroecology Europe forum fostering synergies between movement, science and practice, 25–27 October 2017, Lyon, France

Martin G, Moraine M, Ryschawy J et al (2016) Crop–livestock integration beyond the farm level: a review. Agron Sustain Dev 36:53. https://doi.org/10.1007/s13593-016-0390-x

McGinnis M, Ostrom E (2014) Social-ecological system framework: initial changes and continuing challenges. Ecol Soc 19:30. https://doi.org/10.5751/ES-06387-190230

Moraine M, Grimaldi J, Murgue C et al (2016) Co-design and assessment of cropping systems for developing crop-livestock integration at the territory level. Agric Syst 147:87–97. https://doi.org/10.1016/j.agsy.2016.06.002

Nesme T, Nowak B, David C, Pellerin S (2016) L'Agriculture Biologique peut-elle se développer sans abandonner son principe d'écologie? Le cas de la gestion des éléments minéraux fertilisants. Innov Agron 51:57–66. https://doi.org/10.15454/1.4721176631543018E12

Regan JT, Marton S, Barrantes O et al (2017) Does the recoupling of dairy and crop production via cooperation between farms generate environmental benefits? A case-study approach in Europe. Eur J Agron 82:342–356. https://doi.org/10.1016/j.eja.2016.08.005

Rosset PM, Martínez-Torres ME (2012) Rural social movements and agroecology: context, theory, and process. Ecol Soc 17(3):17. https://doi.org/10.5751/ES-05000-170317

Russelle MP, Entz MH, Franzluebbers AJ (2007) Reconsidering integrated crop–livestock systems in North America. Agron J 99:325–334. https://doi.org/10.2134/agronj2006.0139

Ryschawy J, Martin G, Moraine M et al (2017) Designing crop-livestock integration at different levels: toward new agroecological models? Nutr Cycl Agroecosyst 108:5–20. https://doi.org/10.1007/s10705-016-9815-9

Saives A-L (2002) Territoire et compétitivité de l'entreprise: territorialisation des entreprises industrielles agroalimentaires des pays de la Loire. L'Harmattan, Paris, p 492

Stock PV, Forney J (2014) Farmer autonomy and the farming self. J Rural Stud 36:160–171. https://doi.org/10.1016/j.jrurstud.2014.07.004

Thénard V, Jost J, Choisis JP, Magne MA (2014) Applying agroecological principles redesign and to assess dairy sheep farming systems. In: Options Méditerranéennes, Série A: Séminaires Méditerranéens, vol 109, pp 785–789

Thenard V, Choisis JP, Pages Y et al (2016) Towards sustainable dairy sheep farms based on self-sufficiency: patterns and environmental issues. Options Méditerranéennes Série A 116:81–85

Wezel A, Soldat V (2009) A quantitative and qualitative historical analysis of the scientific discipline of agroecology. Int J Agric Sustain 7:3–18. https://doi.org/10.3763/ijas.2009.0400

Agroecological Transition from Farms to Territorialised Agri-Food Systems: Issues and Drivers

Marie-Benoît Magrini, Guillaume Martin, Marie-Angélina Magne, Michel Duru, Nathalie Couix, Laurent Hazard, and Gaël Plumecocq

Abstract Agroecological transition corresponds to a systemic transformation consisting in the ecologisation of agriculture and food. It concerns multiple stakeholders (farmers, supply chains, natural resource managers, etc.) and is characterised by a deliberate political intention to bring about change. This chapter highlights a set of determinants of agroecological transition at play in transforming the techniques and the values underpinning both agricultural production and food consumption choices – both of which can lead to various new agri-food systems. Based on the literature on transition studies, we focus on several considerations that could help stakeholders to better engage in such a process: (i) transition takes place over time intervals that vary, depending on the analysis scale (the farm or the agri-food system as a whole); (ii) transition is complex, systemic and requires changes of the whole sociotechnical regime; (iii) transition implies strong connections between niche-innovations and the dominant sociotechnical regime; and (iv) changes in values and individuals' abilities are fundamental drivers. Hence, by focusing on the plurality of factors and stakeholders at work, we unpack the complexity of this transition, and in this way help the stakeholders to design and execute it. To conclude, we examine specific issues around the governance of agroecological transition.

M.-B. Magrini (✉)· G. Martin · M. Duru · N. Couix · L. Hazard
AGIR, Université de Toulouse, INRA, Castanet-Tolosan, France
e-mail: marie-benoit.magrini@inra.fr; guillaume.martin@inra.fr; michel.duru@inra.fr; nathalie.couix@inra.fr; laurent.hazard@inra.fr

M.-A. Magne
AGIR, Université de Toulouse, INRA, ENSFEA, Castanet-Tolosan, France
e-mail: marie-angelina.magne@inra.fr

G. Plumecocq
AGIR, Université de Toulouse, INRA, Castanet-Tolosan, France

LEREPS, Université de Toulouse, ENSFEA, Toulouse, France
e-mail: gael.plumecocq@inra.fr

J.-E. Bergez et al. (eds.), *Agroecological Transitions: From Theory to Practice in Local Participatory Design*, https://doi.org/10.1007/978-3-030-01953-2_5

Introduction: What Agroecological Transition Are We Talking About?

Faced with the urgency of sustainable development and the crises that the agricultural sector is experiencing, successive governments in France have strengthened measures to make agriculture more ecological. Since the "Ecophyto 2018" plan adopted by the Ministry of Agriculture in 2008, and up until the new agricultural framework law (the *Loi d'avenir* of 13 October 2014) that makes explicit reference to agroecology, institutional measures have been strengthened to encourage farmers to adopt more sustainable practices. Farmers are urged to implement alternative production strategies to reduce synthetic input use and to combine economic, environmental, and social performance. This political injunction to adopt agroecology is situated beyond the reference framework of organic agriculture, which today remains the only alternative framework officially recognised by government-approved labelling.

This agricultural transformation also calls food into question (Francis et al. 2003; Barbier and Elzen 2012; Hinrichs 2014; Gliessman 2015). The FAO (2012: 8) defines the sustainability of our food as being closely related to that of our agriculture, according to the following five criteria[1]: (i) protects ecosystem biodiversity; (ii) is accessible and culturally acceptable; (iii) is economically fair and affordable; (iv) is safe, nutritionally adequate, and healthy; and v) optimises natural and human resource use. The sustainability of agriculture and food systems thus simultaneously involves technical changes and the values that govern them: it requires the implementation of "non-technological changes such as those in consumer behaviour, social norms, cultural values, and formal institutional frameworks" (OECD 2010: 32). This is even more relevant, given that our "agricultural practices are not primarily determined by agronomic or ecological science, but by markets, regulations and agricultural support programs" (Weiner 2017: 869).

This systemic transformation consisting in ecologising our agriculture and food, which concerns multiple stakeholders (farmers, supply chains, or natural resource managers) and which is marked with a deliberate political will to change, is qualified as an agroecological transition (Duru et al. 2015a). Note that it is a transition and not a revolution, because it does not explicitly entail the need for other changes relative to the capitalist foundations of the societal model underpinning our agriculture and food (cf. Hinrichs 2014 or Brown et al. 2012 on this point).[2] It is a transition

[1] The FAO adopted this definition during the International Scientific Symposium on Biodiversity and Sustainable Diets in 2010.

[2] "processes of *transition* may contain weighty seeds of ambition; but typically do not anticipate a wholesale shift in the future economic mode of production" (Brown et al. 2012). Research discussing a profound transformation in the capitalist model, in particular through degrowth theories, does not, to our knowledge, cover the subject of the agroecological transition. Degrowth is nevertheless embodied by social movements supported by alternative agriculture models (cf. for example D'Alisa et al. 2014).

in the making within our capitalist regime, to move towards a more sustainable agricultural and food system.

As this transition is currently underway, it is characterised by relative uncertainty because we cannot predict the end result (Lubello et al. 2017). The literature moreover refutes the idea that this agroecological transition is based on a single model positioned as the archetype of a new agriculture, instead defending the idea of the coexistence of a plurality of possible models that can contribute to greater agricultural ecologisation (Plumecocq et al. 2018). A major distinguishing feature of models supporting the agroecological transition is the representations and place granted to nature in the design of new solutions. The value attributed to nature underpins an ecologisation of agriculture that varies, depending on whether the new system aims at reducing its environmental impact or developing ecosystem services (Therond et al. 2017; Plumecocq et al. 2018).

The first route, which we called "weak ecological modernisation", aims at increasing the efficiency of synthetic input use (Horlings and Marsden 2011) through the implementation of standardised management practices (Ingram 2008) and the adoption of precision agriculture (Buman 2013) or genetic engineering technologies (Vanloqueren and Baret 2009). It can also be based on the replacement of chemical inputs by biological ones that are less harmful to the environment (Singh et al. 2011).

The second route, which we called "strong ecological modernisation", is based on a more radical redesign and significant biological diversification of agricultural systems (Kremen et al. 2012). It is characterised by intensified interactions with components of the biophysical system in order to substitute synthetic inputs by ecosystem services, and requires locally adapted agricultural practices and cropping or livestock farming systems (Duru et al. 2015b). This redesign of agricultural production systems is part of a broader change downstream to ensure sustainability across the entire agri-food system (Gliessman 2015). The recent literature also shows that food products strongly rooted in local production systems are those that address a broader range of sustainability concerns, whether in terms of biodiversity or ethics (Schmitt et al. 2017).

In this chapter, we focus on the agroecological transition following the route of strong ecological modernisation, broadened to include the question of food sustainability. To highlight the issues of this agroecological transition, two principles will structure our reflection:

(i) engagement of farmers and their advisers in building agroecological knowledge and techniques *in situ* to strengthen their capacities to change and to adapt;
(ii) territorialisation of agriculture, promoting a reconnection between agricultural production and local food, and enabling fair compensation of farmers for their activities.

However, as with any structural change, the stakeholders driving this transition are faced with the entrenchment of the incumbent model. Agriculture has progressively established itself as a coherent set of production and sales practices tied to the food industry (Meynard et al. 2015). For example, adopting a new pulse crop in crop rotations in a territory to reduce the use of synthetic fertilisers, strengthen crop biodiversity, and reduce greenhouse gas emissions, but can be hindered if consumers are not accustomed to eating pulses, or by a lack of transformation infrastructure, or of suitable supply circuits, ultimately leading to low pay for farmers. Those economic non-incentives do not encourage pulse farming (Magrini et al. 2016, 2018).

The aim of this chapter is to identify the broad range of concerns in transforming the values and techniques underpinning agricultural and consumption choices, in order to allow the stakeholders supporting this transition to build a governance approach adapted to these concerns. We draw on the literature on transition studies to identify the salient aspects and topics of interest regarding this agroecological transition.

This analysis is important, considering that little research on transitions has focused on the agricultural and food sectors (Picard and Tanguy 2016; Elzen et al. 2017). Moreover, most of the analysis scales used are still the territory, the production basin (Bui et al. 2016), or even an entire sector of agricultural activity (Elzen et al. 2011; Magrini et al. 2016, 2018), whereas the scale of the farm, which is nonetheless the central and essential link in any agroecological transition, tends to be overlooked (Chantre and Cardona 2014). This reflection will thus show that it is necessary to analyse the different variants of this agroecological transition on these different levels, from the agri-food system down to the farm scale, and that new conceptual and methodological developments in the fields of agronomy and system zootechnics are needed.

The first section goes over the theoretical foundations of transition analysis, in particular co-evolutionary approaches to sociotechnical changes, as well as their contribution to the analysis of the agro-ecological transition. Drawing on this theoretical clarification, the second section develops multiple topics of interest to explain the multi-dimensional nature of the agro-ecological transition, insofar as its implementation requires that the coherency of changes among a large number of stakeholders in the agri-food supply chain be taken into consideration. The third section focuses on several main issues on the farm level. The conclusion opens different paths for deeper analysis with respect to the governance of this agro-ecological transition.

The Theoretical Foundations of Transition Processes

The use of the concept of transition is relatively recent. It dates back to the nineteenth century, when it was used in different disciplines of the life sciences or social sciences and the humanities (Lachman 2013). For instance, Tocqueville used it to talk about the end of slavery; in political science, it initially designated the shift from socialist economies to capitalist economies; and it was used as the basis of some biology and population demographics publications. However, independent of its use, the word "transition" denotes a radical change of a systemic nature. Hence, transition approaches are focus primarily on deep-seated changes that very often affect both social and technical values, as opposed to incremental changes or innovations. The particularity of transition studies is that they highlight these interdependencies between social and technical values, justifying the central use of the concept of "sociotechnical" regime to describe a coherent set of stakeholders, knowledge, rules, values, and artefacts governing an incumbent production model (Elzen et al. 2004; Geels 2004).

Among the different heuristic frameworks of the transition towards sustainability developed in this literature, the Multi-Level-Perspective (MLP) framework is one of the most cited (Brauch et al. 2016; Chang et al. 2017). Drawing on several social science disciplines (including economics, sociology, political science), the MLP approach is multidisciplinary and integrative by nature, allowing one to address societal change through its multiple components and to convey the complexity of this change. This research focuses primarily on the conditions of the transition, from a society based on the intensive use of fossil fuels to one based on the use of renewable resources to satisfy various societal functions such as food, energy production, or transportation (Foxon 2011). Any societal function can be the subject of a transition. Lachman (2013) thus defines a transition as when "*[t]he dominant way in which a societal need (e.g. the need for transportation, energy, or agriculture) is satisfied, changes fundamentally*".

This systemic approach describes the mechanisms through which the target to achieve sustainability is confronted with a lock-in situation (section "Transitions are embedded in lock-in situations"), and the resources for the unlocking process that will initiate a transition (section "Unlocking in transition approaches"). The empirical literature on transitions specifically shows the importance of the conditions for the spreading of innovation niches in these processes (section "Scales and scopes of transition analysis: the major role of networks of stakeholders").

Transitions Are Embedded in Lock-In Situations

Sustainability transition approaches stem from the idea that the dominant production system (for example, in the sense of a supply chain, sector, or food system) is locked in (Geels 2004, 2011). The only changes within the incumbent system aim at improving it and therefore strengthening the technological trajectory initially chosen. Because they remain incremental, these changes do not permit a radical change (that is, a change in technological paradigm). This lock-in is strengthened by the fact that the routines and standards within which stakeholders operate hinder their creative capacity, and because the multiple dependencies between the technical and social components of the system have become reinforced over time. Several studies adopting this co-evolutionary approach enable us to understand how these lock-ins are constructed in the agricultural sector (Cowan and Gunby 1996; Vanloqueren and Baret 2009). For example, the work of Magrini et al. (2016; 2018) shows how the political drive following the Second World War, based on the conventional paradigm,[3] discouraged the development of production alternatives with less use of mineral fertilisers, such as pulses farming. Instead, the specialisation of production by country (European cereals versus American soy), region, and farm, supported by the use of synthetic inputs and specific genetic changes, were promoted. Combined with market dynamics favouring certain species, the conventional regime strengthened the economic competitiveness of a few main crops as well as geographically-concentrated industrial livestock farming, to the detriment of agriculture based on agrobiodiversity and the integration of cropping and livestock (Horlings and Marsden 2011; Duru et al. 2015b).

Evolutionary economists explain this lock-in through the concept of "increasing returns to adoption". This key concept, coined by Arthur (1989), explains how one technology progressively "prevails over" the alternatives due to the fact that its performance improves by being increasingly adopted. Five main types of mechanism (called "self-reinforcing") feed this adoption practice and highlight the role of collective action and knowledge. These mechanisms are illustrated for the conventional agricultural paradigm in (Magrini et al. 2017; 2018). We give a brief overview of them below:

(i) learning by using: the production performance of synthetic inputs and selected varieties and breeds increases with user experience;
(ii) network externalities: the greater the number of adopters, the more beneficial it is for users to adopt the system to benefit from other products or services developed, such as services to support crop management, storage, and sale;
(iii) scale economies and learning by doing: the unit cost of production is reduced over time by the volume effect and the improvement of the techniques and materials developed, such as agricultural mechanisation;
(iv) informational increasing returns: the more this production paradigm develops, the more widely known and understood it becomes, thus incentivising others to adopt and develop it;
(v) technological interrelatedness: other production technologies and standards are established in the food sector in relation to this agricultural production, such as seed quality criteria for food transformation (e.g. grain protein content).

These returns to adoption are said to be "increasing" because, as a system develops more users, the utility for each user is increased compared to alternative solutions. Knowledge on the dominant system is progressively consolidated compared to alternatives, which are more uncertain because they receive less investment. Stakeholders in the agricultural sector in particular highlight this problem of uncertainty surrounding alternative crops (cf. Chap. 6). Hence, uncertainty surrounding alternative solutions, which have benefited from less investment and knowledge, as

[3] The conventional paradigm is often called the agri-industrial and agri-chemical paradigm due to the combined logic of a high degree of standardisation of agricultural production enabled by the accumulated use of synthetic inputs.

well as the inherent cost of the change, reinforces the initial choice over time, that is, the conventional paradigm.

To unlock such a lock-in, it is necessary to understand all the components of the system that determine the triggering of a new trajectory which can be consolidated over time through these same mechanisms of increasing returns to adoption. The multilevel approach developed by Geels (e.g. 2004) offers a framework to understand how a new trajectory can begin.

Unlocking in Transition Approaches

This theory of technological lock-in has allowed for the renewal of approaches to change by highlighting the co-evolutionary nature of trajectories. Its analysis framework is however strongly focused on the role of technological innovation. The specificity of transition approaches has consisted in expanding this framework to consider the roles of a multitude of stakeholders, including civil society, and not only those using the technology (e.g. a way of producing). For example, new technological innovations can simultaneously trigger changes in scientific knowledge as well as in factors tied to the demand from civil society, in particular ethical factors. Many authors thus insist on the fundamental role of societal values, which legitimise production decisions (Plumecocq et al. 2018). The incumbent production system (e.g. the dominant sociotechnical regime) has built its coherency over time as a function of the progress of scientific knowledge, technologies and infrastructure, and networks of companies and markets. It has also based itself on values tied to consumer preferences and different institutions, defining a set of rules and standards structuring collective action (Elzen et al. 2004). As Geels (2012: 474) demonstrates, "*an important implication is that the MLP does away with simple causality [...] there is no single 'cause' or driver. Instead, there are processes on multiple dimensions and at different levels which link up and reinforce each other*".

The second specificity of the MLP approach is that it offers three main levels of analysis that influence one another and steer the evolution of the sociotechnical regime (i.e. trajectories), presented as a diagram in Fig. 1. Major evolutionary factors, such as demographic shifts or environmental problems affecting all societal functions, can place the sociotechnical regime in a situation of crisis if the principles governing it do not provide a solution to these problems. These societal contextual elements, which constitute the 1st level of analysis (called the "landscape" according to MLP terminology), open up windows of opportunity for radical changes. However, because the dominant regime (2nd level) does not constitute a space of radical innovation, the keys of the change operate on another level: that of innovation niches (3rd level of analysis). These innovation niches are built by stakeholders outside of the dominant regime. They enable the development of new ways of producing, transforming, or consuming in order to more radically address contemporary pressures on the landscape. When these niches reach a certain stage of development and internal structure, they can spread in the dominant regime. They

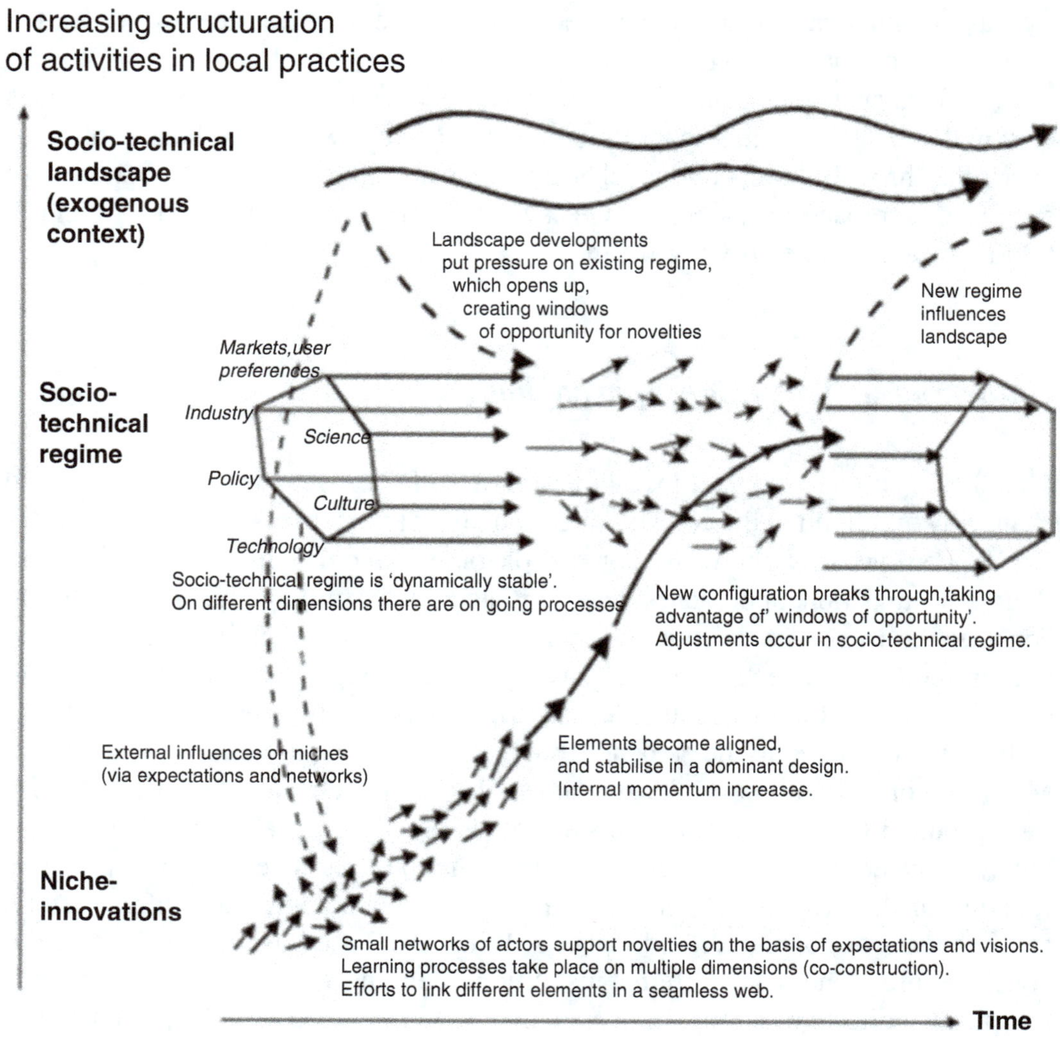

Fig. 1 The MLP approach to transitions (in Geels 2011)
The MLP approach is based on 3 levels of analysis: the landscape, the dominant sociotechnical regime, and innovation niches representing networks of stakeholders oriented towards radical innovation. The transition process encompasses all interdependencies that are woven between these 3 levels over time

can then choose between two main strategies: submitting to the selection factors of the dominant regime ("fit and conform") or trying to modify them ("stretch and transform") (cf. Smith and Raven 2012 for more details).

Organic agriculture is an emblematic example of this process: its network of stakeholders established itself progressively and is now spreading within the dominant regime. For example, organic products are now sold in large retail chains and the rate of conversion to organic has increased over the past few years in France. Yet organic has not managed to reverse or replace the conventional regime, which remains dominant. It is by obtaining a quality marking that differentiates it on the market, that organic products are able to economically develop under the conditions of the dominant regime. The specific aid for conversion provided by Europe is also evidence of a change starting in the dominant regime. Hence, the organic innovation niche has progressively become a market niche and continues in itself to be an incubator for new practices prohibiting synthetic inputs that influence the dominant regime.

Many hypotheses support the role of niches as incubators of radical innovations: for instance, whereas their development outside of the dominant regime promotes the emergence and expression of new creative capabilities not limited by the routines and standards of the dominant regime, within that regime the initiation of economic activities by new entities is less hindered when taking risks (which are often prohibitive for established players that want to secure margins or pay off specific investments tied to already-established activities). In the MLP approach, the process of the change in the dominant regime starts, strictly speaking, when niches manage to spread within the regime and to influence its evolution around its major components (the stage of the "empowerment" innovation niche, in the sense of Smith and Raven 2012. This leads to a new alignment of the trajectories of different components of the regime (details in Fig. 2). The MLP framework is thus a fundamental co-evolutionary and diachronic approach.

This heuristic framework for transitions, which was developed in the 2000s, has been extended many times. This highlights the complexity of the processes of spreading innovation niches within the dominant regime to initiate a change in trajectory, drawing in particular on the work of Smith et al. (2005), Smith and Raven (2012), and Raven et al. (2016). Various configurations are possible, depending on the type of ties maintained between niche and dominant regime stakeholders. There is often an overlap of networks of niche and regime stakeholders, as certain stakeholders are present in both systems. This is particularly manifest in research on the agricultural sector, in which the sale of agricultural products does not necessarily benefit from alternative transformation and distribution networks. In other words, differentiated products can be distributed by dominant networks, with alternative networks remaining on the sidelines. For new practices to spread there must therefore be a minimum level of adaptive capacity in the dominant regime – and

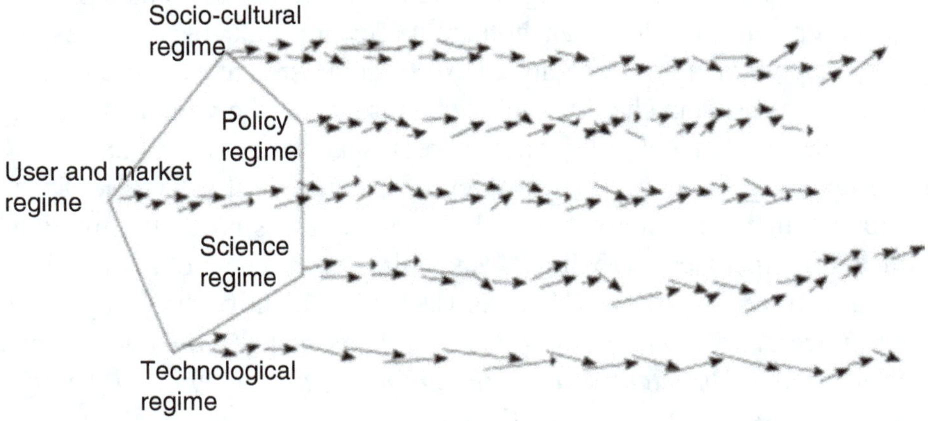

Fig. 2 The alignment of trajectories in socio-technical regimes in Geels (2004)
The sociotechnical regime is composed of multiple subsystems whose trajectories align with one another, providing coherency to the whole regime

vice versa in the niche, in relation to the dominant regime. These two levels of analysis (regime versus niche) of the MLP approach must therefore be analysed together. This is why (Ingram 2015, 2018) proposes an "overlapping niche-regime space" rather than separating these two entities, as the MLP approach suggests.

Elzen et al. (2017) more particularly focus on the role of stakeholders that are intermediaries between the dominant regime and innovation niches. These intermediary stakeholders, called hybrids, are described as participating in an innovation niche while simultaneously having direct access to the dominant regime (for example, through their participation in debate arenas or by holding positions at dominant regime stakeholders). They can also be "innovation brokers" that promote interconnection between the dominant regime and niches, and which are highlighted as essential stakeholders in these transition processes (Klerkx et al. 2012). For example, the research of Bouttes et al. (2018) reveals the major role of these stakeholders in the case of conversion to organic agriculture: for livestock farmers, the fact that the agricultural adviser supporting them was previously a part of the conventional system (at a chamber of agriculture) is a guarantee of credibility and legitimacy.

Scales and Scopes of Transition Analysis: The Major Role of Networks of Stakeholders

Transition processes require us to consider multiple scales of analysis – whether the temporal (start and duration), geographic, or sectoral scale – in relation to the scope of the network of stakeholders observed.

The majority of research insists on the emergence and initial stages of these processes, granting particular attention to the structuring of innovation niches. Some focuses more particularly on the emergence of the niche, which can be based on different strategies varying in their distance from the dominant regime. These configurations of ties between niches and the regime vary, as indicated above (Ingram 2015). For example, conservation agriculture is outside of the domain of the dominant regime in order to allow stakeholders to develop their ideas and experiment freely (Vankeerberghen and Stassart 2016). In contrast, stakeholders in the Bleu-Blanc Coeur supply chain chose to include themselves in the dominant regime from the emergence of the niche, to quickly access financial resources and allow it to spread faster (Magrini and Duru 2015; cf. also Elzen et al. (2008) or Diaz et al. (2013) concerning anchoring strategies). Different stages are often highlighted in these transition processes. In their analysis of the development of niches directed at structuring local food systems, Bui et al. (2016: 99) identify three key stages in a transition process: "*the emergence of the initiative (Stage I); the construction of a sociotechnical niche through the enrolment of new stakeholders into the initiative, leading to the diversification of objectives and activities (Stage II); the construction of an alternative model impacting various components of the agri-food regime (Stage III). The increasing diversity of stakeholders involved in the niche progressively leads to the construction of an alternative model, and the enrolment of local authorities, through spill-over effects, then triggers deep changes in practices, strategies and alliances of some regime stakeholders at the local scale"* (page 99). The evolution of the network of stakeholders structuring the innovation niche allows them to access resources and, to a certain extent, determines their development up until the empowerment stage.

These publications show that an increase in the number of stakeholders involved in the process is a necessary condition of the transition. Increasing the size of the network of stakeholders makes it possible to progressively establish ties with certain stakeholders in the dominant regime. By participating in exchanges with niche operators, the dominant regime will be able to understand the stakes and opportunities of the development of the niche, and will push towards its own reconfiguration. These intermediary stakeholders constitute relays between the niche and the dominant regime. They often enable access to public policy resources to help to develop the niche and spread innovative practices. For example, the structuring of certain locally-based food systems can involve local officials in the development of these niches in order to access new support devices (cf. the example of Bui et al. 2016). Another example of the role of political stakeholders is the policy of the city of Mouans-Sartoux in France. By organising the supply of school cafeterias with local organically-produced products, the elected officials of this commune have been successful in structuring and perpetuating a network of stakeholders around this innovation niche (Pérole 2017).

The social interactions that take place within niches or in relation to the dominant regime are thus at the core of transition processes. It is therefore useful to grant particular attention to devices aiming at establishing a consensus between stakeholders regarding visions of changes to carry out and possible routes for achieving common goals. This is precisely the goal of the TATA-BOX project, through carrying out a territorial diagnosis followed by phases for designing, evaluating, and selecting agro-ecological transition scenarios (cf. Chap. 9). Among these devices, "transition arenas", defined as spaces of shared dialogue between stakeholders engaged in moving towards a change in the system (Boulanger 2008), constitute an essential structure in the transition process (Duru et al. 2015b). Networks of stakeholders built through these arenas are fundamental in the transition process, because by "building up a broadening network of diverse actors that share the debate, thinking and experimenting, conditions are created for up-scaling of innovation and breakthrough of innovations" (Loorbach and Rotmans 2010: 238). These authors thus emphasise that in a transition process, "*we need both pioneers operating outside and inside the incumbent power structures*" (*ibid.*: 243). For example, the RIO project in the Netherlands, based on the creation of a reflexive arena to rethink livestock farming systems, has enabled significant progress in establishing a consensus of new ideas among livestock farmers, supply chain stakeholders, and consumers (Bremmer and Bos 2017 in Elzen et al. 2017).

The scope of the transition process also depends on the sector in question. In agriculture and agri-food, two main situations can be distinguished, depending on whether the agro-ecological transition is mainly based on a departure from the dominant regime in terms of production methods with little engagement of the downstream, or whether it is embedded within a larger transition of the food system (Therond et al. 2017). In the first case, the agricultural products are not distinguished from those of the dominant regime; they are often sold at global market prices. There is no distinctive quality marking, even if production methods are more respectful of the environment (e.g. grass-fed livestock, conservation agriculture). It is the relationship to the upstream that is changed through decreased input consumption or equipment needs. In the second case, beyond production methods, the transition includes a diversification of production (crops, animals, etc.), a modification in input supply, both in the choice of inputs and in the ways of accessing these resources, and new organisations in product collection, storage, or transformation, as well as in consumers' food habits.

From the time perspective, transition researchers agree that processes underway on the societal or industry level are staggered across a 25 to 50 year period and therefore involve one to two generations (Elzen et al. 2004, 2011; Geels 2004; *inter alia*). However, this period can be longer, depending on the extent of the change considered. Sovacool (2016) thus believes that while studies analysing the adoption of a new technology intended for consumers (such as the refrigerator or digital technologies) fit within short timeframes (around 25 years), transitions concerning in-depth changes (such as large energy or transportation infrastructure) take place over longer periods (from 50 to 100 years). While authors analyse transitions ex-post by retracing the history of these processes over several decades, the majority of studies on transitions in the making present the state of these processes over shorter time periods and, generally speaking, over 5- to 15-year intervals, such as in the work of Elzen et al. (2011) on pig farming, Diaz et al. (2013) on green algae in Brittany, or Lascialfari et al. (Forthcoming 2019) on product innovation in the agri-food sector. These long intervals concern situations in which the transition requires getting a large number of stakeholders in the food supply chain on-board. By contrast, independent of the timeframes observed on the societal or industry level, a farm can implement the transition over much shorter intervals of only a few years when changing its production system or the way that it sells its products.

This review thus allows us to propose that the combination of these multiple changes (or their alignment, in the words of the MLP framework) *a fortiori* prefigures the different possible trajectories of the agroecological transition. While different trajectories are possible, we continue to use the term "agroecological transition" (AET) to refer to these change processes as a whole.

What Are the Determinants of the Agroecological Transition on the Scale of the Agri-Food System?

Considering the multidimensional nature of the AET and the fact that it involves a large number of stakeholders upstream and downstream of agriculture, below, we highlight a few of the noteworthy features of changes underway that predetermine (or will predetermine) agroecological transition trajectories. We start by considering the importance of shared values, in particular those of farmers and consumers, so necessary to supporting the AET (section "What are the values underpinning the AET?"). We also show that various innovations related to transformation and sale infrastructure for agricultural productions (section "What new market infrastructure provides the basis for AET?") have a significant influence on the AET. Lastly, we analyse the role of technical norms and standards in institutionalising the AET, or in the making of new collective action rules that can support the AET (section "On what new collective rules is AET based?").

What Are the Values Underpinning the AET?

Social values are one of the bases of collective judgement governing the acceptability of individual or collective choices and decisions. For an individual, adhering to social values means recognising oneself as a member of a greater community based on moral principles. In this sense, our food choices pertain to worldviews varying in their degree of tacitness or implicitness, and in the extent to which they are owned or asserted. These worldviews relate to ethical or moral motives, amongst others, in such a way that food systems, which constitute interfaces between farmers and consumers, as well as markers of cultural identity, are excellent observation posts to understand the role of values in the agroecological transition.

Consumers' food choices are based on various lifestyle-related types of values. In Western countries, the topic of diet is currently crystallising around protein consumption concerns. Abandoning a diet mainly based on animal protein derives from a variety of overlapping values today – environmental concerns, ethics in terms of animal well-being, nutritional or economic values: Amongst these, environmental or health-related values appear not to be foremost in the minds of consumers, despite the fact that they are stressed by scientific experts (Campbell 2009; Hartmann and Siegrist 2017; de Boer and Aiking 2018). Scientists now agree that it is necessary, in terms of the health of individuals and the planet, to balance animal and vegetable protein sources in our diet: "*there is broad consensus that reduction of meat consumption will be crucial for a transition towards more sustainable food consumption*" (Hartmann and Siegrist 2017: 12). This is driving experts to highlight food information and education to change the values underpinning consumption choices: "*in order to move towards more sustainable food behaviour, consumers and citizens need to have better knowledge about the environmental consequences of their food behaviour. Otherwise, it will be unlikely that consumers will be motivated to change their food behaviour*" (*ibid.*: 22). In France, decreasing animal product consumption was recently included in new food guides (Anses 2016), but remains to be implemented at the level of food education programmes.

Lastly, the stakes involved in the AET require us to rethink the ways that interests and values can be combined to continue to ensure compensation for sector stakeholders, all the while meeting new social standards. To this end, an increasing number of consumers are willing to pay an additional price to assert their values by supporting agricultural activities that embody them (for instance buying products with "organic agriculture" labels or supporting small-scale family farming). Opinion polls tend to indicate that civil society recognises the difficulties inherent to farming (strenuousness of the work, low pay, etc.), which explains why, in 2016, "*despite a context of economic crisis, which is usually restrictive for the consumption of households, two thirds of French people said that they were willing to pay more for products to ensure fair payment to farmers*" (Gomant 2017: 15). A set of values (expressed in terms of social justice or solidarity) is thus driving a transition to a societal model in which the living and working conditions of farmers are expected to be better. The perception of deep-rooted imbalances in the distribution of added value in supply chains, which increase as supply chains get longer (Brown et al. 2013; Schmitt et al. 2017), has encouraged the organisation of multiple business initiatives driven by the values of fair trade (Therond et al. 2017). This need to overhaul the agricultural social model was stressed by the *Etats Généraux de l'Alimentation*[4] in France, in 2017.

The modalities of food distribution in supply chains, as well as ways of indicating product qualities, closely reflect the value systems underpinning them. The increase in consumers' preferences for local production places value on regional products (cf., for example, the BVA-INRA survey on the durum wheat supply chain, Triboulet et al. 2018). This leads us to consider the implications of changes in the organisation of the transformation and sale of agricultural products (section "What new market infrastructure provides the basis for AET?").

What New Market Infrastructure Provides the Basis for AET?

The theory of increasing returns to adoption shows that the logic of economies of scale has supported the development of large-scale collection, transformation, and distribution infrastructure (Magrini et al. 2016, 2018). Even when these economies of scale are not "economically" achieved, stakeholders' belief in this principle has encouraged the geographic concentration of infrastructure over time. For example, in the meat sector, the research of Soufflet (1990) shows that over twenty years (1967–1987), slaughterhouses for cattle were not profitable even though they had been concentrated and increased in size. Considering a spatial redistribution of storage and transformation infrastructure does not necessarily imply a financial loss,

[4]The *États Généraux de l'Alimentation* was a convention initiated by the State President in 2017 with the purpose of collectively building new, sustainable, food and agricultural systems, on a "win-win" basis. It was based on broad, nation-wide public consultation, as well as over 20 workshops where experts came together to draw up proposals. This work was continued in 2018, primarily through the enactment of a new Agriculture and Food law.

because economies of scale are often decreasing, starting from a certain production and concentration threshold. Moreover, the industrial concentration of certain groups often leads to distancing from production basins, which "become anonymous". Distributing infrastructure across the territory could promote territorial reconnection between agriculture and food in order to more directly valorise the diversity of crops farmed in territories.

As mentioned above, French consumers tend to want more local food rooted in their region. To meet this demand, large and medium-sized supermarkets, which are still the main place of purchase in Western countries (90% of food purchases in France), must develop their offering of regional products and therefore need to reorganise their purchase centres and logistics circuits. The development of short circuits, such as AMAP (small-farmer associations), direct sale, or local markets also aims at meeting this demand. These alternative consumption relays are making significant progress in France. According to the results of a recent survey, 42% of respondents had purchased a product from a short circuit over the past month. Once again, the shift in consumer values can orient the evolution of these distribution networks and lead farmers to rethink their ties to consumers.

The catering industry is an ideal place for promoting these connections. Recent experiments show that it is possible to develop a diet based on local production, but that this territorial transition takes place over a long period of time. In his text on Mouans-Sartoux (in the south-west of France), Pérole (2017) explains how this city, which decided "*to reconquer its food sovereignty*", undertook a number of measures starting in 1998 that allowed it to become, in 2012, "*the first commune in France with more than 10,000 inhabitants to switch to 100% organic in its cafeterias*" based on quasi-local production. In 15 years, owing to the determination of the local council, this city has implemented a number of changes, such as buying back agricultural lands, aid for converting farmers to organic agriculture, developing a municipal public company to prepare meals itself, and training cooks on how to cook with local products. The learning carried out and the internal reorganisation of production has ultimately made possible an agroecological meal price that implies no extra costs. Consumption values and habits have also changed in the targeted population, since the surveys carried out by this city show that in 2014, 66% of parents surveyed "*believe that their own food habits have changed towards more organic and local*", whereas this number increased to 85% in 2016, almost 20 years after the initiative to engage this city in an agri-food transition.

This example also shows that stakeholders can decide to include a new activity, to achieve their goals, or to develop new hybrid methods for organising exchanges between the market and the hierarchy. Production contracts are organisational bases that facilitate these change processes, particularly to secure new investments (Cholez et al. 2017). They also promote the structuring of new spaces for dialogue between operators, encouraging knowledge exchange to reduce the uncertainty underlying the change (cf. Chap. 6).

On What New Collective Rules Is AET Based?

The evolution of the norms and standards structuring collective action is just as important on the consumer side of things as on the farmer side. Standards or certifications remain an essential tool for disseminating new rules and information, especially in the agri-food sector (Mazé 2017).

On the consumer side, the recognition of the agroecological practices associated with a product can allow consumers to orientate their consumption choices. However, confusion tends to prevail today in the information directly accessible to consumers or not. Magrini and Duru et al. (2015b) recall that the absence of specific labelling for livestock farming conditions in France makes it impossible for consumers to make choices based on virtuous practices such as grazing. This is despite the fact that in the United States a label certifying farms where animals are exclusively grass-fed (Grass Fed Label) has existed since 2009. Even if new certifications are implemented, their quality has to be assured. Hoibian (2010) reports that "*only 31% of people believe that the information on 'green' products has a scientific basis*" and only "*25% find it clear". The increase in ecological claims surrounding massively-consumed products ("100% natural", "respectful of the environment", etc.) reinforce suspicions of "greenwashing*".

This calls to mind the current strategy of supply chains that are demonstrating a tendency to increase "agroecology" certifications. A recent study by Lhoste (2017) shows that many existing labels (e.g. "Nature et Progrès", "Bleu-Blanc-Cœur") currently tend to claim that they are agroecological, based on different principles in each specific case. This proliferation of food standards increases consumers' confusion and "*risks resulting in additional costs to producers and limiting access to markets*" through competition effects (Meybeck and Gitz 2017).

On the farmers' side, labelling based on agri-environmental criteria can also help to change the focus of their decision-making by granting more priority to environmental criteria than to yield criteria. This institutional framework is even more important, given that conversion to agroecology can result in a decrease in yields over the short term (Weiner 2017). Therefore, the development of standards or new practice measurement indicators is also a concern for farmers. Today, many initiatives exist to promote the development of new performance measurement indicators. For example, beyond organic agriculture standards based on method requirements, multiple attempts to develop new certifications are currently being tried out in France (such as HVE – *Haute Valeur Environnementale* (high environmental value) – or the IDAE method – *Indicateurs de Durabilité des Exploitations agricoles* (farm sustainability indicators)). Nonetheless, confusion can be just as pervasive among farmers and their advisers. It is therefore essential to find a relevant balance between recommendations and minimum obligations to promote improved environmental performance and better product composition. More recently, the *Etats-Généraux de l'Alimentation* (cf. above) carried out by the French government suggested generalising an HVE certification label that would be more comprehensive and integrate a degree of agricultural territorialisation. Ultimately, the new label or labels that will be introduced could accelerate the transition process if they earn consumers' trust.

Finally, note that the institutional framework can also be based on other regulation devices that also merit analysis, such as public payments to farmers, either to offset lower agricultural income, as in Switzerland (Schmitt et al. 2017) or to compensate for agri-environmental services (Reed et al. 2017). Through this system, consumers compensate farmers primarily through taxation rather than through prices, thus promoting food consumption that is less dependent on consumers' incomes.

The Determinants of the AET on the Farm Level

As mentioned above, few studies have carried out an in-depth analysis of an AET processes on the farm scale, whereas this is a key organisational level in the AET process. In this research, the farm is considered to be embedded within an agri-food system and an agricultural development system making up the incumbent socio-technical regime. The AET of the farm therefore takes place through the reconfiguration of the interactions between the different components of the regime: "*changes in farming practices are contingent on a profound reconfiguration of the whole agri-food system, i.e. change in the practices and modes of coordination of all incumbent actors – farmers, processors, distributors, consumers, public policies, research and extension services*" (Bui et al. 2016: 92). This research shows the need to design coupled innovations that promote interactions between the farm and supply chain operators (Meynard et al. 2015, 2017). Yet the literature on transition in agriculture has seldom analysed the transition processes at work on the farm and among farmers.

In particular, it is very interesting (and necessary) to analyse farmers' adaptive capacities. These are defined as being aptitudes with respect to designing and implementing adaptations or changes and managing new situations without compromising future options (Nelson et al. 2007; Marshall et al. 2014). They largely depend not only on farmers' personal traits (in terms of risk perception, values and goals, attitudes and beliefs, integration within collective dynamics) (Moser and Ekstrom 2010), but also on the prospects of the agricultural system with respect to the natural, physical, human, financial, and social capital of the farm (Nazari et al. 2015). In this section we emphasise the features of these adaptive capacities in the context of the AET of the farm. We start by focusing on farmers' values, aims, and attitudes (section "Triggers of the AET for farmers: values, aims, and attitudes"), later turning to their perception of the risks associated with the AET (section "Farmers' perception of risks and uncertainty with regard to the transition"). We then address their learning dynamics for and during the AET (section "Farmers' learning for and during the AET") and how innovative support devices can contribute to this (section "Reconfiguring exchange networks and "advisory" devices: is a shift towards a new regime of agricultural knowledge taking place?"). Finally, we discuss how the technological innovations (section "Using technological innovations to aid the AET of farms") may promote the AET of the farm.

Triggers of the AET for Farmers: Values, Aims, and Attitudes

Changing values around food consumption and diet are driving agricultural production systems to change. Organic agriculture, for instance, has made considerable progress over the past years, and in 2015 and 2016 in France, increases in consumption and certified organic surface areas were 21.7% and 17%, respectively. Analysing farmers' AET processes reveals that the values or motives behind these processes vary. Recent surveys show farmers' growing interest in reducing the use of phytosanitary products.[5] Their motivations are both extrinsic (opportunities related to the development of a market and a growing demand for these products) and intrinsic (return to more "agronomic" practices, desire to preserve ecosystem resources, limit pollution, respect nature, etc.) (Plumecocq et al. 2018), and potentially support the AET. Certain motivations also relate to expectations surrounding working conditions. For example, the transition of cattle farmers to grazing systems is partially related (Lusson et al. 2014; Cayre et al. 2018) to the amount and nature of the work (e.g. "*doing something other than riding a tractor*"). The motivations behind the change can also be related to a renewal in the societal function (environmental, health, or ethics) of livestock farming (e.g. "*producing healthily to eat healthy*") and to technico-economic motivations (savings on inputs). Four main triggers were identified that encouraged dairy farmers to transition towards agroecology: evidence that it was possible to earn a living after adopting breeding practices that generate less milk yield; finding that they were stuck in technical ruts; differences between farmers' values and practices; and external incentives or requirements (Coquil et al. 2017). The transition allowed these livestock farmers to overcome difficulties while more closely following personal and societal aspirations (ethical values), even if this meant not following the standards recognised by the profession in the dominant regime. For some farmers, the AET thus represented a hope of working conditions more satisfying on the professional level, allowing them to better align their aspirations and daily reality with society's standards (Dessein and Nevens 2007; Barbier et al. 2015).

Farmers' attitudes regarding change management (aversion vs. appeal, positive vs. negative viewpoint) were an important component in the AET. On the one hand, the ability to face the changes was related to their psychological disposition (in particular risk aversion) and their ability to manage the impact of those changes

[5] The findings of a survey by BVA (2015) show that 76% of farmers implement initiatives to reduce inputs, 45% say they have heard of agroecology and are interested in this initiative, and 40% implement innovative initiatives defined as practices "that are not commonly used", such as biocontrol, crop associations, or the insertion of legume crops.

(Marshall et al. 2014). A key factor in the success of farm transitions was the farmer having a proactive approach to the change while remaining attentive to global changes, the alternatives offered, and the various possible consequences (Coquil et al. 2014; Chantre et al. 2015). It was also beneficial to engage actively in the change rather than perceiving it as being costly or risky and thus enduring it. The situation prior to the transition was another determinant of the ease with which farmers would implement this transition. In the case of conversion to organic agriculture for dairy farming, Bouttes et al. (2018) emphasise the psychological tensions experienced by farmers at this stage: they were facing a very low price for conventional milk and they had a large workload following the increase in farm size (by seeking economies of scale to amortise investments).

Farmers' Perception of Risks and Uncertainty with Regard to the Transition

Farmers' perception of risks also affects their capacity to change. This dimension is even more important considering that the changes to be implemented in an AET are complex and uncertain for multiple reasons (Duru et al. 2015a, b): (i) they are systemic; (ii) they are based on capacities that farmers may have lost (for example, efficient grazing management); (iii) they are highly dependent on local conditions, which means that farmers must adjust the nature and extent of these changes based on their production situation; (iv) they are uncertain because the response of the agroecosystem to new management practices is poorly known in advance; (v) they must correspond to local socio-economic opportunities (for example, a dairy that wants to purchase organic milk) and threats; and (vi) they must adjust dynamically to changes in the production context.

Even if farmers perceive the AET as a risky and uncertain process, their capacity to perceive other risks can engage them in a shift in practices. For example, one reason for dairy farmers to convert to organic agriculture is the perception that there are fewer risks and less uncertainty in the future of organic dairy farming than in that of conventional dairy. This is because farmers believe that over the short and medium term, organic milk prices will be more stable and consumer preferences more favourable (Bouttes et al. 2018). Another reason for change is related to controlling market risks. Conversion to organic agriculture increases farm autonomy, which reduces exposure to the volatility of input prices. On the other hand, it can constitute a significant risk, often due to a lack of technical knowledge (Padel 2001). When deciding to undertake such a transition, farmers evaluate compromises between external factors such as product quality requirements, regulations, and prices, and internal requirements, such as risks related to new production techniques (Lamine 2011; Chantre and Cardona 2014; Bouttes et al. 2018).

Farmers' Learning for and During the AET

AET requires farmers to develop their learning capacity for practices that differ from those of the dominant system and with which they are mostly unfamiliar (Darnhofer et al. 2010; Marshall et al. 2014). This relates to their capacity to design/create alternative management methods in situ, to test and experiment with them, and to evaluate these experiments in order to derive lessons from them (Chantre et al. 2015). Furthermore and more generally, it relates to a variety of processes that can lead farmers to develop pragmatic judgements (Cristofari et al. 2018). Individual creativity in managing farm resources (the soil, plants, animals) relates to the way that this management integrates local pedoclimatic potential as well as the biological potential of animal and plant resources as perceived by farmers. For example, certain dairy farmers crossbreed within their herd to create genetic types that are more suitable for local feed resources and to meet their objectives (Ollion et al. 2018). These changes in practices involve at least three new management entities: (i) the management of plant and animal agrobiodiversity (Martin and Magne 2015; Magne et al. 2016, 2017; e.g. crossbreeding rotation types, the aptitudes and complementarities of breeds); (ii) managing integration between plant and animal production to reconnect the plant-soil-animal triad by closing biogeochemical loops (Bonaudo et al. 2014); and (iii) managing contact between domestic and wild fauna (Charrier et al. 2017) and production and sale (Nozières et al. 2014). Farmers are then led to move away from the professional standards and references of the dominant regime that regulated their initial practices (Meynard 2017). In particular, this includes revising the concept of performance in agriculture (Caron et al. 2014)). For example, accepting a decrease in the volume of milk produced per cow below a certain level (approximately 5000 to 6000 litres of milk) is not easy for many farmers converting to organic agriculture, whereas this is nonetheless an essential factor in the success of the conversion (Bize 2017). This departure from prevailing professional standards and references implies the creation of other evaluation and steering benchmarks for farmers. It is a key phase in their learning process as a part of the AET (Chantre et al. 2015). In addition to the intentional experiments that they carry out, farmers can also make use of unforeseen events by transforming them into a learning opportunity, which becomes more interesting as the uncertainty becomes greater (Cristofari et al. 2018). Most often, these benchmarks are not established in isolation at the farm but within a group of farmers (Chantre et al. 2015). Therefore, as many studies have shown (e.g. Bouttes et al. 2018), farmers can experience the AET as a welcome opportunity for learning, escaping from their routine, and facing a new challenge. Maintaining such an experimentation and learning dynamic is essential if they are to adapt to changing economic, social, and ecological conditions beyond the transition phase (Vogl et al. 2015).

Reconfiguring Exchange Networks and "Advisory" Devices: Is a Shift Towards a New Regime of Agricultural Knowledge Taking Place?

Agroecology is based on renewing knowledge on agricultural techniques by drawing more on farmers' knowledge (Warner 2007; Meynard 2017). Local experiments within different knowledge exchange communities are devices to create and spread agroecological knowledge. Analysing the features of these communities in depth allows us to understand what determines the capacity of these communities to promote the spreading of new knowledge and practices (Ingram 2018).

Beginning an AET implies that the farmer has taken a step back from the technical advice that goes hand-in-hand with purchasing inputs, and with respect to dominant technical benchmarks and standards. This distance allows farmers to build or join other networks of stakeholders (peer networks as well as farmer adviser/facilitator networks) that are in phase with the production methods being implemented (Chantre and Cardona 2014; Coquil et al. 2014). These peer exchange groups, which may or may not be hosted by an "adviser", facilitate creativity, reassurance faced with the uncertainty over the expected results, the building of a new framework of reference for action, and underlying values and reflexivity. In doing so, they facilitate the learning process (cf. Olsson et al. 2004), the management of risks and uncertainty, and the alignment of values and practices (Plumecocq et al. 2018). Therefore, learning and experimentation are largely promoted via a collective process in which production methods are proposed and discussed, ideas are integrated, practices are implemented at individual farms, and results are discussed in groups (Lamine 2011). These exchange devices are used in particular in the networks resulting from niches (*Groupements d'Agriculteurs Biologiques* (organic farmer groups), *Centres d'Initiatives pour Valoriser l'Agriculture et le Milieu rural* (centres for initiatives to develop agriculture and the rural space), etc.) and in instances of production under contract supported by strengthened advisory services and farmer field days on farms (Cholez et al. 2017).

These networks of stakeholders isolated from the dominant system (niches) constitute learning networks, i.e., privileged spaces for experimenting with and progressively spreading novel knowledge and practices to the dominant sociotechnical regime (Ingram 2015, 2018): "*By taking a different innovation and learning direction, niches, through the actions of their knowledge systems, challenge the dominance of the AKS [Agricultural Knowledge System], and seek to change it through diffusion of more radical ideas and practices*" (Ingram 2018: 3). This innovation dynamic is based on the AKS regime (Agricultural Knowledge System), in which knowledge production is primarily built around relations between "*all relevant knowledge producers and stakeholders, including the farmers*" (van Mierlo et al. 2017: 9). It differs from the KBBE (Knowledge-Based Bio-Economy) regime, in which knowledge production is primarily based on combining "*life science with techno-scientific innovations to develop the means for an efficient use of agricultural resources*" (Levidow et al. 2012; van Mierlo et al. 2017 cited in Elzen et al.

2017). These two postures define two types of knowledge production regime, through which niches conveying novel knowledge are likely to be integrated in different ways into the sociotechnical regime.

The case of selecting and evaluating varieties illustrates the existence of these two knowledge production regimes, and namely: (i) a dominant regime founded on the selection of lineages carried out by seed companies operating across very large geographic scales that tend to homogenise knowledge; and (ii) an innovation niche founded on selecting populations adapted to the local context through seed exchange between peers (country seeds network), primarily based on situated knowledge. While today, these two regimes tend to oppose one another, it is possible to imagine a process to hybridise these knowledge regimes in the context of a transition, in view of the values driving them (Fenzi and Bonneuil 2016).

The reconfiguration of knowledge regimes also calls into question the place that agricultural education could occupy in constructing this new knowledge, for example with respect to the education system's adaptive capacity to integrate knowledge established in these communities of practices or in agricultural innovation networks (Simonneaux et al. 2016). This adaptive capacity remains largely tied to actor networks that are structured between advisors and these niches (Ingram 2015).

Using Technological Innovations to Aid the AET of Farms

While the AET aims at developing practices to provide ecosystem services, this does not mean that it uses no technological innovations. Identifying needs for technologies necessary or favourable to the AET is a major research concern (Therond et al. 2017). For example, selecting plant species or varieties that provide ecosystem services (e.g. soil structuring or coverage), or genetic animal breeds suitable for developing plant diversity (e.g. through greater ability to move around while grazing) remain priority research pillars to facilitate the AET. This is also the case for self-guided autonomous hoeing or parcel weeding robots that do not use phytopharmaceutical products. Progress in robotics may promote the farming of certain diversification plants (e.g. weeding crops with little coverage) or may free up farmers' time during their routine activities, thus increasing their involvement in other tasks. There are many examples of advances in robotics that can assist the AET, and agriculture has become the second largest market for professional robotics services (Bellon-Maurel and Huyghe 2016). Yet, with a view to achieving the best possible remuneration for farmers and reducing agricultural labour, the utility of these technologies must be considered with respect to the global cost of adopting new technologies.

Another type of technological innovation concerns knowledge capitalisation. As noted above, the changes needed to implement an AET are complex, situated, and uncertain for farmers. Information and communication technologies can contribute to understanding this complexity and reducing the uncertainty associated with it by making use of automatic data collection and processing. This is becoming more significant as calculation capacities become stronger (Bellon-Maurel and Huyghe 2016). Combined with these massive new datasets, the development of serious games such as Rami Fourrager® (literally, *Forage rummy*) (Martin et al. 2011) allows farmers to use simulations to design and evaluate alternatives to their strategy and current practices. Another example is the GECO application developed in France (Soulignac et al. 2017), which allows farmers to formalise knowledge and make it available or discuss it within the broader community.

The contribution of technological innovations to the AET is therefore largely a function of their capacity to promote the construction of agroecological knowledge and practices. These technological innovations are also fundamentally dependent on farmers' capacities to adopt them or to integrate the results of these technologies into their decision-making systems.

Conclusion

Without aiming to be exhaustive, this chapter has highlighted a set of concerns and determinants around which various agroecological transition trajectories could develop. Note that this transition takes place over time intervals that vary, depending on the analysis scale used (the farm or the agri-food system as a whole), but it combines changes in food and agriculture to achieve the goals of sustainable food, according to the definition of the United Nations.

By focusing on a plurality of factors and stakeholders at work in these processes, we have presented the complexity of this transition, particularly in view of the interactions that develop between innovation niches and the dominant sociotechnical regime, as well as the capacity of stakeholders to develop and spread new practices. We have specifically emphasised the role of the values that can orient these different processes, as well as individuals' abilities to adapt to the change.

This complexity at work therefore makes it necessary to implement support methods on different scales, to help stakeholders to design and execute the necessary changes. We propose to conclude this chapter by mentioning a few concerns specific to the issue of transition governance.

First of all, the shift in values associated with the transition shows that it is still necessary to create space for debate. The idea is not to create a uniform and shared vision but rather to democratically resolve controversies, the diversity of which may potentially spawn innovations, in order to allow each person to develop his or her project and to best position it within the landscape under reconstruction. This process of public revealing contributes to legitimising these values, because even though they are progressing in favour of ecologisation, their legitimacy to support a large and coherent societal transformation movement remains uncertain (Borrás and Edler 2014). This is all the truer considering that some stakeholders doubt the very possibility of achieving change, as van Mierlo et al. (2017: 11) have emphasised "*a large divide exists between those who think that we should and can change our ways of producing and consuming food, and those who doubt the potential of alternative ways of farming to halt climate change and radically reduce environmental risks*". Additionally, the question of the legitimacy of the technological innovations chosen is coming up increasingly frequently, including in research communities (Schlaile et al. 2017). It therefore remains crucial to develop these arenas of discussion (i.e. transition arenas, Boulanger 2008) between different stakeholders to hone the new goals that a society, sector, or territory wishes to achieve. Constructing a common vision of the desired future, even if nobody can exactly foresee the system that the transition process will bring about, is an essential stage in the process. Moreover, even though starting these meetings may be difficult, their perpetuation over time will allow for the development of new and shared reflections that will address the legitimacy of the choices made. In this sense, the *Etats Généraux de l'Alimentation* launched by the French government in 2017 at national and regional level, as well as the participatory workshops of the TATA-BOX project across a delimited territory, constitute experiments in discussion arenas to facilitate the convergence of different visions. These "transition arenas" must receive the support of public resources (Smith et al. 2005), for converging actions will stem from converging visions.

It is also necessary to define different modes of governance based on the intentionality attributed to this transition (Smith et al. 2005). Given that it is embedded within a political intention, the governance mode can be more directive and coordinated than in the context of a transition resulting from more contingent processes. Therefore, the state must increase pressure on the dominant regime to promote the change via regulatory measures, providing new resources, or the recognition of distinctive markings. It is also necessary to increase the allocation of resources to networks of stakeholders that constitute niches, in order to allow them to evaluate their devices and continue to develop. It also falls upon niches to exert pressure on the institutional framework, to drive it to in turn exert pressure on the dominant regime so that the latter is more inclined to adopt the practices resulting from niches (cf. the "stretch and transform" process in Smith and Raven 2012). The ability of niches to interact with the institutional framework, in particular through the acknowledgement of elected officials, is recognised as an essential condition in the regime transformation process (Beers and Van Mierlo 2017).

Given that the spreading of innovation niches conveying radical innovations to the sociotechnical regime is essential to the transition process, it remains necessary to better understand the complexity of the relations that are woven between stakeholders in the dominant regime and niches. Actor network approaches may be able to help with this.

This point also suggests that the adaptive capacities of the dominant regime are just as important as those of niches or of the farm positioned as an essential link in experimenting with alternatives. Research is therefore needed to better understand the sources of this capacity of regime stakeholders to adapt to and integrate new knowledge and practices built around a new conception of cropping systems or of commercialisation and retailing.

Even though this chapter has focused on a conception of the agricultural system in relation to the food system, based on a sociotechnical transition approach, it nonetheless remains interesting also to focus on the socio-ecological systems literature. Transition governance must also be defined in a way that includes the management of the natural resources affected by this agroecological transition.

Acknowledgment This research was supported by the TATA-BOX project funded by the French Research Agency (ANR-13-AGRO-0006; 2014–2018) and by the PSDR 4 Programme (Project ATA-RI. 2016–2020) funded by INRA and the Region Occitanie.

References

Anses (2016) Actualisation des repères du PNNS: révision des repères de consommations alimentaires, Avis de l'Anses, Rapport d'expertise collective, December

Arthur WB (1989) Competing technologies, increasing returns, and lock-in by historical events. Econ J 99:116–131. https://doi.org/10.2307/2234208

Barbier M, Elzen B (eds) (2012) System innovations, knowledge regimes, and design practices towards transitions for sustainable agriculture, INRA Editi. Paris, France

Barbier C, Cerf M, Lusson JM (2015) Cours de vie d'agriculteurs allant vers l'économie en intrants: les plaisirs associés aux changements de pratiques. Activites. https://doi.org/10.4000/activites.1081

Beers PJ, Van Mierlo B (2017) Reflexivity, reflection and learning in the context of system innovation: prying loose entangled concepts. In: Elzen B, Augustyn AM, Barbier M, van Mierlo B (eds) AgroEcological transitions. Wageningen University & Research, Wageningen, pp 243–256

Bellon-Maurel V, Huyghe C (2016) L'innovation technologique dans l'agriculture. Géoéconomie 80:159–180. https://doi.org/10.3917/geoec.080.0159

Bize N (2017) Trajectoires de conversion à l'Agriculture Biologique en élevage bovin lait: Analyse rétrospective et actuelle. Rapport de stage

Bonaudo T, Bendahan AB, Sabatier R et al (2014) Agroecological principles for the redesign of integrated crop–livestock systems. Eur J Agron 57:43–51. https://doi.org/10.1016/j.eja.2013.09.010

Borrás S, Edler J (eds) (2014) The governance of socio-technical systems: explaining change. Edward Elgar Publishing, Cheltenham

Boulanger PM (2008) Une gouvernance du changement sociétal: le transition management. La Rev Nouv 11:61–73

Bouttes M, Darnhofer I, Martin G (2018) Converting to organic farming as a way to enhance adaptive capacity. Accepted with minor revision

Brauch HG, Spring ÚO, Grin J, Scheffran J (eds) (2016) Handbook on sustainability transition and sustainable peace. Springer International Publishing, Cham

Bremmer B, Bos B (2017) Creating niches by applying reflexive interactive design. In: Elzen B, Augustyn AM, Barbier M, van Mierlo B (eds) AgroEcological transitions. Wageningen University & Research, Wageningen

Brown G, Kraftl P, Pickerill J, Upton C (2012) Holding the future together: towards a theorisation of the spaces and times of transition. Environ Plan A 44:1607–1623. https://doi.org/10.1068/a44608

Brown JP, Goetz SJ, Ahearn MC, Liang C (2013) Linkages between community-focused agriculture, farm sales, and regional growth. Econ Dev Q 28:5–16. https://doi.org/10.1177/0891242413506610

Bui S, Cardona A, Lamine C, Cerf M (2016) Sustainability transitions: insights on processes of niche-regime interaction and regime reconfiguration in agri-food systems. J Rural Stud 48:92–103. https://doi.org/10.1016/j.jrurstud.2016.10.003

Buman T (2013) Opportunity now: integrate conservation with precision agriculture. J Soil Water Conserv 68:96A–98A

Campbell H (2009) The challenge of corporate environmentalism: social legitimacy, ecological feedbacks and the 'food from some where' regime? Agric Hum Values 26(4):309–319

Caron P, Biénabe E, Hainzelin E (2014) Making transition towards ecological intensification of agriculture a reality: the gaps in and the role of scientific knowledge. Curr Opin Environ Sustain 8:44–52. https://doi.org/10.1016/j.cosust.2014.08.004

Cayre P, Michaud A, Theau JP, Rigolot C (2018) The coexistence of multiple worldviews in livestock farming drives agroecological transition. A case study in French protected designation of origin (PDO) cheese mountain areas. Sustainability 10:1097. https://doi.org/10.3390/su10041097

Chang R, Zuo J, Zhao Z et al (2017) Approaches for transitions towards sustainable development: status quo and challenges. Sustain Dev 25:359–371. https://doi.org/10.1002/sd.1661

Chantre E, Cardona A (2014) Trajectories of French field crop farmers moving toward sustainable farming practices: change, learning, and links with the advisory services. Agroecol Sustain Food Syst 38:573–602. https://doi.org/10.1080/21683565.2013.876483

Chantre E, Cerf M, Le Bail M (2015) Transitional pathways towards input reduction on French field crop farms. Int J Agric Sustain 13:69–86. https://doi.org/10.1080/14735903.2014.945316

Charrier F, Hannachi M, Barbier M, Casabianca F (2017) Décider ensemble pour que l'impossible devienne possible. La co-construction d'un dispositif de gestion d'une maladie animale en Corse. In: 7th OPDE conference "des Outils pour Décider ensemble"; Montpellier, 26–27 of October 2017, 2017. p 21

Cholez C, Magrini MB, Galliano D (2017) Field crop production contracts. Incentives and coordination under technical uncertainty, in French cooperatives. Économie Rural 360(4):65–83. https://doi.org/10.4000/economierurale.5260

Coquil X, Béguin P, Dedieu B (2014) Transition to self-sufficient mixed crop–dairy farming systems. Renew Agric Food Syst 29:195–205. https://doi.org/10.1017/S1742170513000458

Coquil X, Dedieu B, Béguin P (2017) Professional transitions towards sustainable farming systems: the development of farmers' professional worlds. Work 57:325–337. https://doi.org/10.3233/WOR-172565

Cowan R, Gunby P (1996) Sprayed to death: path dependence, lock-in and pest control strategies. Econ J 106:521–542. https://doi.org/10.2307/2235561

Cristofari H, Girard N, Magda D (2018) How agroecological farmers develop their own practices: a framework to describe their learning processes. Agroecol Sustain Food Syst 42:777–795. https://doi.org/10.1080/21683565.2018.1448032

D'Alisa G, Demaria F, Kallis G (eds) (2014) Degrowth: a vocabulary for a new era. Routledge, New York/London

Darnhofer I, Bellon S, Dedieu B, Milestad R (2010) Adaptiveness to enhance the sustainability of farming systems. A review. Agron Sustain Dev 30:545–555. https://doi.org/10.1051/agro/2009053

de Boer J, Aiking H (2018) Prospects for pro-environmental protein consumption in Europe: cultural, culinary, economic and psychological factors. Appetite 121:29–40. https://doi.org/10.1016/j.appet.2017.10.042

Dessein J, Nevens F (2007) "I'm sad to be glad". An analysis of farmers' pride in Flanders. Sociol Ruralis 47:273–292. https://doi.org/10.1111/j.1467-9523.2007.00437.x

Diaz M, Darnhofer I, Darrot C, Beuret JE (2013) Green tides in Brittany: what can we learn about niche–regime interactions? Environ Innov Soc Trans 8:62–75. https://doi.org/10.1016/j.eist.2013.04.002

Duru M, Therond O, Fares M (2015a) Designing agroecological transitions; a review. Agron Sustain Dev 35:1237–1257. https://doi.org/10.1007/s13593-015-0318-x

Duru M, Therond O, Martin G et al (2015b) How to implement biodiversity-based agriculture to enhance ecosystem services: a review. Agron Sustain Dev 35:1259–1281. https://doi.org/10.1007/s13593-015-0306-1

Elzen B, Geels FW, Green K (2004) System innovation and the transition to sustainability: theory, evidence and policy. Edward Elgar Publishing, Cheltenham

Elzen B, Leeuwis C, van Mierlo BC (2008) Anchorage of innovations: assessing Dutch efforts to use the greenhouse effect as an energy source. Wageningen University and Research, Wageningen

Elzen B, Geels FW, Leeuwis C, van Mierlo B (2011) Normative contestation in transitions 'in the making': animal welfare concerns and system innovation in pig husbandry. Res Policy 40:263–275. https://doi.org/10.1016/j.respol.2010.09.018

Elzen B, Augustyn AM, Barbier M, van Mierlo B (eds) (2017) AgroEcological transitions. Wageningen University & Research, Wageningen

FAO (2012) Sustainable diets and biodiversity. Corporate authors: FAO, Rome and biodiversity international, Rome, Italy

Fenzi M, Bonneuil C (2016) From "genetic resources" to "ecosystems services": a century of science and global policies for crop diversity conservation. Cult Agric Food Environ 38:72–83. https://doi.org/10.1111/cuag.12072

Foxon T (2011) A coevolutionary framework for analysing a transition to a sustainable low carbon economy. Ecol Econ 70:2258–2267

Francis C, Lieblein G, Gliessman S et al (2003) Agroecology: the ecology of food systems. J Sustain Agric 22:99–118. https://doi.org/10.1300/J064v22n03_10

Geels FW (2004) From sectoral systems of innovation to socio-technical systems. Res Policy 33:897–920. https://doi.org/10.1016/j.respol.2004.01.015

Geels FW (2011) The multi-level perspective on sustainability transitions: responses to seven criticisms. Environ Innov Soc Trans 1:24–40. https://doi.org/10.1016/j.eist.2011.02.002

Geels FW (2012) A socio-technical analysis of low-carbon transitions: introducing the multi-level perspective into transport studies. J Transp Geogr 24:471–482. https://doi.org/10.1016/j.jtrangeo.2012.01.021

Gliessman SR (2015) Agroecology: the ecology of sustainable food systems, third edit. Taylor & Francis, Boca Raton

Gomant F (2017) Images des agriculteurs auprès du grand public. In: Demeter 2017, 23rd edn, pp 6–17

Hartmann C, Siegrist M (2017) Consumer perception and behaviour regarding sustainable protein consumption: a systematic review. Trends Food Sci Technol 61:11–25. https://doi.org/10.1016/j.tifs.2016.12.006

Hinrichs CC (2014) Transitions to sustainability: a change in thinking about food systems change? Agric Human Values 31:143–155. https://doi.org/10.1007/s10460-014-9479-5

Hoibian S (2010) Enquête sur les attitudes et comportements des Français en matière d'environnement (2010 edition), Study conducted by the ADEME, Collection des rapports du CRÉDOC, n° 270, October 2010,

Horlings LG, Marsden TK (2011) Towards the real green revolution? Exploring the conceptual dimension of a new ecological modernisation of agriculture that couls "feed the world". Glob Environ Chang 21:441–452. https://doi.org/10.1016/j.gloenvcha.2011.01.004

Ingram J (2008) Agronomist–farmer knowledge encounters: an analysis of knowledge exchange in the context of best management practices in England. Agric Human Values 25:405–418. https://doi.org/10.1007/s10460-008-9134-0

Ingram J (2015) Framing niche-regime linkage as adaptation: an analysis of learning and innovation networks for sustainable agriculture across Europe. J Rural Stud 40:59–75. https://doi.org/10.1016/j.jrurstud.2015.06.003

Ingram J (2018) Agricultural transition: niche and regime knowledge systems' boundary dynamics. Environ Innov Soc Trans 26:117–135. https://doi.org/10.1016/j.eist.2017.05.001

Klerkx L, van Mierlo B, Leeuwis C (2012) Evolution of systems approaches to agricultural innovation: concepts, analysis and interventions. In: Darnhofer I, Gibbon D, Dedieu B (eds) Farming systems research into the 21st century: the new dynamic. Springer, Dordrecht, pp 359–385

Kremen C, Iles A, Bacon C (2012) Diversified farming systems: an agroecological, systems-based alternative to modern industrial agriculture. Ecol Soc 17:44. https://doi.org/10.5751/ES-05103-170444

Lachman DA (2013) A survey and review of approaches to study transitions. Energy Policy 58:269–276. https://doi.org/10.1016/j.enpol.2013.03.013

Lamine C (2011) Anticiper ou temporiser. Injonctions environnementales et recompositions des identités professionnelles en céréaliculture. Sociol Trav 53:75–92

Lascialfari M, Magrini MB, Triboulet P (Forthcoming 2019) The drivers of product innovations in pulsed-based foods: insights from cases studies in France, Italy and USA, Revue Innovations

Levidow L, Birch K, Papaioannou T (2012) Divergent paradigms of European agro-food innovation: the knowledge-based bio-economy (KBBE) as an R&D agenda. Sci Technol Hum Values 38:94–125. https://doi.org/10.1177/0162243912438143

Lhoste V (2017) La diffusion du concept d'agroécologie au sein des standards de produits alimentaires en France. Rapport ODR

Loorbach D, Rotmans J (2010) The practice of transition management: examples and lessons from four distinct cases. Futures 42:237–246. https://doi.org/10.1016/j.futures.2009.11.009

Lubello P, Falque A, Temri L (eds) (2017) Systèmes agroalimentaires en transition, éditions QUAE

Lusson J, Coquil X, Frappat B, Falaise D (2014) 40 itinéraires vers des systèmes herbagers: comprendre les transitions pour mieux les accompagner. Fourrages 219:213–220

Magne MA, Thénard V, Mihout S (2016) Initial insights on the performances and management of dairy cattle herds combining two breeds with contrasting features. Animal 10:892–901. https://doi.org/10.1017/S1751731115002840

Magne MA, Ollion E, Cournut S, et al (2017) Some key research questions about the interest of animal diversity for the agroecological transition of livestock farming systems. In: First agroecology Europe forum fostering synergies between movement, science and practice, 25–27 October 2017, Lyon, France

Magrini MB, Duru M (2015) Diffusion d'une niche d'innovation dans les systèmes sociotechniques laitiers : coordination, standards et co-évolution. Une analyse de la démarche Bleu-Blanc-Coeur en France. Rev Innov 48:187–210

Magrini MB, Anton M, Cholez C et al (2016) Why are grain-legumes rarely present in cropping systems despite their environmental and nutritional benefits? Analyzing lock-in in the French agrifood system. Ecol Econ 126:152–162. https://doi.org/10.1016/j.ecolecon.2016.03.024

Magrini MB, Anton M, Cholez C et al (2017) Transition vers des systèmes agricole et agroalimentaire durables: quelle place et qualification pour les légumineuses à graines? Rev Française Socio-Économie 18:53–75. https://doi.org/10.3917/rfse.018.0053

Magrini MB, Befort N, Nieddu M (2018) Economic dynamics of technological trajectories and pathways of crop diversification in bio-economy. In: Lemaire G, Carvalho P, Kronberg S, Recous S (eds) Agro-ecosystem diversity: reconciling contemporary agriculture and environment quality. Elsevier Academic Press, Amsterdam, p 478

Marshall NA, Stokes CJ, Webb NP et al (2014) Social vulnerability to climate change in primary producers: a typology approach. Agric Ecosyst Environ 186:86–93. https://doi.org/10.1016/j.agee.2014.01.004

Martin G, Magne MA (2015) Agricultural diversity to increase adaptive capacity and reduce vulnerability of livestock systems against weather variability – a farm-scale simulation study. Agric Ecosyst Environ 199:301–311. https://doi.org/10.1016/j.agee.2014.10.006

Martin G, Felten B, Duru M (2011) Forage rummy: a game to support the participatory design of adapted livestock systems. Environ Model Softw 26:1442–1453. https://doi.org/10.1016/j.envsoft.2011.08.013

Mazé A (2017) Standard-setting activities and new institutional economics. J Inst Econ 13:599–621. https://doi.org/10.1017/S174413741600045X

Meybeck A, Gitz V (2017) L'expérience du programme FAO/PNUE pour des systèmes alimentaires durables. Rev AES 7:69–73

Meynard JM (2017) L'agroécologie, un nouveau rapport aux savoirs et à l'innovation. OCL 24:D303. https://doi.org/10.1051/ocl/2017021

Meynard J, Messéan A, Charlier A, et al (2015) La diversification des cultures: lever les obstacles agronomiques et économiques, Quae. Versailles

Meynard JM, Jeuffroy MH, Le Bail M et al (2017) Designing coupled innovations for the sustainability transition of agrifood systems. Agric Syst 157:330–339. https://doi.org/10.1016/j.agsy.2016.08.002

Moser SC, Ekstrom JA (2010) A framework to diagnose barriers to climate change adaptation. Proc Natl Acad Sci 107:22026–22031. https://doi.org/10.1073/pnas.1007887107

Nazari S, Pezeshki G, Sedighi H, Azadi H (2015) Vulnerability of wheat farmers: toward a conceptual framework. Ecol Indic 52:517–532. https://doi.org/10.1016/j.ecolind.2015.01.006

Nelson DR, Adger WN, Brown K (2007) Adaptation to environmental change: contributions of a resilience framework. Annu Rev Environ Resour 32:395–419. https://doi.org/10.1146/annurev.energy.32.051807.090348

Nozières MO, Baritaux V, Cournut S, et al (2014) Describing the evolutions, in a territory, of the interactions between livestock farming systems and downstream operators. Proposal for a methodological framework, based on the comparison of 4 territories and 2 types of production: milk and meat. In: 11th European IFSA Symposium "Farming systems facing global challenges: Capacities and strategies", 01 to 04 of April 2014, Berlin. p 12

OECD (2010) Eco-innovation in industry: enabling green growth. OECD Publishing, Paris

Ollion E, Brives H, Cloet E, Magne MA (2018) Suitable cows for grass-based systems: what stakeholders do in France? In: 28th EGF General Meeting on "Sustainable Meat and Milk Production from Grasslands", Ireland, 17th–21st June 2018

Olsson P, Folke C, Berkes F (2004) Adaptive comanagement for building resilience in socialâ€"ecological systems. Environ Manag 34:75–90

Padel S (2001) Conversion to organic farming: a typical example of the diffusion of an innovation? Sociol Ruralis 41:40–61. https://doi.org/10.1111/1467-9523.00169

Pérole G (2017) A Mouans-Sartoux, une restauration collective issue intégralement de l'agriculture biologique depuis 2012. Rev AE&S 7:119–121

Picard F, Tanguy C (2016) Innovations and techno-ecological transition. Wiley, Hoboken, pp 59–86

Plumecocq G, Debril T, Duru M et al (2018) The plurality of values in sustainable agriculture models: diverse lock-in and coevolution patterns. Ecol Soc 23. https://doi.org/10.5751/ES-09881-230121

Raven R, Kern F, Verhees B, Smith A (2016) Niche construction and empowerment through sociopolitical work. A meta-analysis of six low-carbon technology cases. Environ Innov Soc Trans 18:164–180. https://doi.org/10.1016/j.eist.2015.02.002

Reed MS, Allen K, Attlee A et al (2017) A place-based approach to payments for ecosystem services. Glob Environ Chang 43:92–106. https://doi.org/10.1016/j.gloenvcha.2016.12.009

Schlaile M, Urmetzer S, Blok V et al (2017) Innovation systems for transformations towards sustainability? Taking the normative dimension seriously. Sustainability 9:2253. https://doi.org/10.3390/su9122253

Schmitt E, Galli F, Menozzi D et al (2017) Comparing the sustainability of local and global food products in Europe. J Clean Prod 165:346–359. https://doi.org/10.1016/j.jclepro.2017.07.039

Simonneaux L, Simonneaux J, Cancian N (2016) QSV Agro-environnementales et changements de société: Transition éducative pour une transition de société via la transition agroécologique

Singh JS, Pandey VC, Singh DP (2011) Efficient soil microorganisms: a new dimension for sustainable agriculture and environmental development. Agric Ecosyst Environ 140:339–353. https://doi.org/10.1016/j.agee.2011.01.017

Smith A, Raven R (2012) What is protective space? Reconsidering niches in transitions to sustainability. Res Policy 41:1025–1036. https://doi.org/10.1016/j.respol.2011.12.012

Smith A, Stirling A, Berkhout F (2005) The governance of sustainable socio-technical transitions. Res Policy 34:1491–1510. https://doi.org/10.1016/j.respol.2005.07.005

Soufflet JF (1990) Compétitivité et stratégies agro-industrielles dans la filière viande bovine européenne en constitution. Économie Rural 197:42–48. https://doi.org/10.3406/ecoru.1990.4060

Soulignac V, Pinet F, Lambert E et al (2017) GECO, the French web-based application for knowledge management in agroecology. Elsevier Comput Electron Agric 29. https://doi.org/10.1016/j.compag.2017.10.028

Sovacool BK (2016) How long will it take? Conceptualizing the temporal dynamics of energy transitions. Energy Res Soc Sci 13:202–215. https://doi.org/10.1016/j.erss.2015.12.020

Therond O, Duru M, Roger-Estrade J, Richard G (2017) A new analytical framework of farming system and agriculture model diversities. A review. Agron Sustain Dev 37:21. https://doi.org/10.1007/s13593-017-0429-7

Triboulet P, Cuq B, Lullien-Pellerin V et al (2018) Les français et le blé dur. Ind des céréales 206:8–13

van Mierlo B, Augustyn A, Elzen B, Barbier M (2017) AgroEcological transitions: changes and breakthroughs in the making. In: Elzen B, Augustyn A, Barbier M, van Mierlo B (eds) Agroecological transitions: changes and breakthroughs in the making. Wageningen University & Research, Wageningen, pp 9–16

Vankeerberghen A, Stassart PM (2016) The transition to conservation agriculture: an insularization process towards sustainability. Int J Agric Sustain 14:392–407. https://doi.org/10.1080/14735903.2016.1141561

Vanloqueren G, Baret PV (2009) How agricultural research systems shape a technological regime that develops genetic engineering but locks out agroecological innovations. Res Policy 38:971–983. https://doi.org/10.1016/j.respol.2009.02.008

Vogl CR, Kummer S, Leitgeb F et al (2015) Keeping the actors in the organic system learning: the role of organic farmers' experiments. Sustain Agric Res 4:140–148. https://doi.org/10.5539/sar.v4n3p140

Warner K (2007) Agroecology in action: extending alternative agriculture through social networks. MIT Press, Cambridge

Weiner J (2017) Applying plant ecological knowledge to increase agricultural sustainability. J Ecol 105:865–870. https://doi.org/10.1111/1365-2745.12792

A Plurality of Viewpoints Regarding the Uncertainties of the Agroecological Transition

Danièle Magda, Nathalie Girard, Valérie Angeon, Célia Cholez, Nathalie Raulet-Croset, Régis Sabbadin, Nicolas Salliou, Cécile Barnaud, Claude Monteil, and Nathalie Peyrard

Abstract The concept of agroecological transition revives debates on how to deal with complexity and uncertainty. While the adaptive approach and its "adjust along the way" principle have been adopted as a relevant general framework to deal with partially irreducible uncertainty, the different approaches to the definition and management of uncertainty are rarely explicitated. In this chapter we highlight the diversity of these stances through brief presentations of research work that is related to agroecology and sustainable development, and anchored in various disciplines (modelling, management sciences, economics, ecology). This gives us a first glimpse of the variety of concepts used to describe uncertainty, characterising nature and the different approaches to manage it. It shows also that these definitions of uncertainties, clearly derived from particular disciplines or school of thought, can be applied together in a more or less complementary way. Finally, we discuss how

D. Magda (✉) · N. Girard · C. Cholez
AGIR, Université de Toulouse, INRA, Castanet-Tolosan, France
e-mail: daniele.magda@inra.fr; nathalie.girard@inra.fr; celia.cholez@inra.fr

V. Angeon
URZ, INRA, Petit-Bourg, Guadeloupe, France

Ecodéveloppement, INRA, Avignon, France
e-mail: valerie.angeon@inra.fr

N. Raulet-Croset
IAE de Paris, Université Paris 1 Panthéon Sorbonne, Paris, France
e-mail: raulet.croset.iae@univ-paris1.fr

R. Sabbadin · N. Peyrard
MIAT, Université de Toulouse, INRA, Castanet-Tolosan, France
e-mail: regis.sabbadin@inra.fr; nathalie.dubois-peyrard@inra.fr

N. Salliou · C. Barnaud · C. Monteil
DYNAFOR, Université de Toulouse, INRA, Castanet-Tolosan, France
e-mail: nsalliou@ethz.ch; cecile.barnaud@inra.fr; claude.monteil@ensat.fr

J.-E. Bergez et al. (eds.), *Agroecological Transitions: From Theory to Practice in Local Participatory Design*, https://doi.org/10.1007/978-3-030-01953-2_6

this explicitation of the diversity of approaches to uncertainty contributes to highlighting different ways of defining the agroecological transition itself – especially between determinist or more open-ended approaches–, and identifies interdisciplinary research issues.

Introduction

Uncertainty and complexity were at the heart of the first debates around sustainable development (Godard 2001; Hubert 2002). Today, the agroecological transition (AET) is once again reviving the full extent of the problem of dealing with the uncertainty tied to the complexity introduced by the joint management of the different dimensions of a change process. There is nothing new about analysing the uncertain, or uncertainty in the broad sense; it has even resulted in the development of fields of research advancing a particular point of view, for instance around the notion of a risk (Motet 2010) or even more recently, ignorance (Roberts 2013; Girel 2016)). Here, we focus on dealing with the uncertainty or, more specifically, the uncertainties, in management processes *s.l.* involved in the AET. The questions that have emerged around the methods of governance and management of the AET are a continuation of a long-established critical analysis of the bases of the management methods that prevailed prior to sustainable development (Voß et al. 2007). Previously based on the principles of anticipating, predicting, and predetermining goals and means, these management methods followed a "command-and-control" philosophy (Pahl-Wostl et al. 2010) which therefore sought to reduce uncertainty overall (Holling and Meffe 1996). Today, these methods are faced with the necessity of assuming the management of various types of uncertainty that are emerging on the global scale as the result of new sustainability paradigms, and specifically the AET. The uncertainty due to the unpredictable nature of the behaviour of complex managed systems is thus combined with uncertainties tied to the indeterminacy and ambiguity in play in both individual and collective decision processes.

Much research has sought to highlight, design, or implement in the field other "management philosophies" (Hatchuel and Weil 1992) for dealing with uncertainties without reducing the importance of sustainability. Forms of management and governance referred to as "adaptive" have thus become part of this debate (Voß and Bornemann 2011). The founding principle of the adaptive method is that the best strategy when faced with an irreducible uncertainty is to make the best of management experience to adjust along the way (Holling 1978). However, behind this extremely general framework, a wide array of proposals has developed around the way of adapting, and these proposals often have very different ways of dealing with uncertainties without these truly being elucidated. For example, many variants of the adaptive management method have emerged from different disciplines, without, however, providing an analysis of the particular different viewpoints adopted with regard to uncertainties. The significance of these different viewpoints is often relegated to a secondary level, with the focus being instead on the objects/points of entry through which the question is posed, or the levels at which it is addressed (concerning an object and its behaviour, on the scale of an action, an individual, or a group). Yet these different proposals or viewpoints, which are sometimes presented as being complementary, have stemmed from epistemologies/paradigms of uncertainty that are radically different or even difficult to reconcile. This lack of explication generates ambiguities from one researcher to the next, especially when they interact within multidisciplinary research initiatives, or in a support capacity. The TATA-BOX project met these criteria exactly, as a process in which a multidisciplinary research team supported local actors (cf. chapter "TATA-BOX at a Glance").

Little work has been done on the diversity of uncertainties and of the ways of dealing with them. Yet they are a structuring element in the analysis and support of the AET, and more broadly, of transformation processes engaging complex systems and multiple interacting dimensions.

This chapter sheds light on the diversity of viewpoints on uncertainty as regards the AET, based the work of researchers in different disciplines (modelling, management science, economics, ecology, etc.) (Girard and Magda 2016).[1] Each section relates different authors' explication of their relationship to uncertainty in their work. Depending on the author, they draw either on the concepts and approaches of their discipline, or on an approach developed around a given issue. The discussion section offers a synthesis and analyses these different viewpoints to identify elements that may inform reflection on the AET.

Understanding the Agroecological Transition as an Economic Situation of Radical Uncertainty

Dealing with uncertainty is central in economic analysis, which focuses on the rational behaviour of agents. An abundant literature in neoclassical economics addresses uncertainty probabilistic terms (cf. Postel (2008) for a literature review). In this school of thought, the world in which agents make decisions is known (or partially known) insofar as it can be characterised through a data set (whether objective or subjective data). Agents' decisions are predictable. The complete (i.e. maximum information available) and perfect (i.e. accurate) nature of the information is the cornerstone of this decision-making model, which describes a substantive rationality. In contrast with this approach formalising calculable uncertainty, others have focused on situations of radical uncertainty that do not offer a possibility of predicting economic behaviours (Keynes 1921, 1936; Simon 1964, 1978). The decisions to take demand "a wager on the future" due to the impossibility of presently

[1] This chapter is based on the presentations, conversations, and summary of a seminar organised as a part of the TATA-BOX project on 16 February 2016 entitled "Is it possible to adapt to uncertainties in the context of the agroecological transition and how can it be done?" ("*Peut-on et comment s'adapter aux incertitudes dans le cadre de la transition agroécologique?*").

possessing the information necessary for decision-making as defined by the neoclassical approach. In this case, agents' rationality is qualified as being limited or procedural (Postel 2008). It describes their ability to deliberate, that is, to construct and legitimise their choices.

The AET illustrates this situation of radical uncertainty. It urges people to produce and consume differently. This need to do things differently sets the terms of the change and its management in order to move beyond the conventional production and consumption model. As shown below, it implies differing decision and action logics in a context of greater uncertainty related to the way of redesigning the dynamics of human-nature relations and of legitimising the production and consumption models to promote. Dealing with uncertainty in production and consumption models in the AET aims at answering the following questions: How is uncertainty removed? In other words, what decision-making and action processes are clarified by these agroecological production and consumption models under construction? On what bases are these models legitimised?

Two transition pathways characterised by weak versus strong ecological engagement are proposed to implement new production and consumption systems (Horlings and Marsden 2011; Duru et al. 2014, 2015a). Depending on which of these transition pathways is preferred, their relationship to uncertainty differs. We posit that the construction of production-consumption models with weak ecological engagement is a part of an approach aimed at reducing uncertainty, promoted by a small number of actors whose rules for decision-making and action are based on the production and accumulation of scientific knowledge. By contrast, models with strong ecological engagement aim to explore uncertainty, involving a broader diversity of actors to network and a process of combining/recombining knowledge on multiple scales of time and space.

In its weak version, the AET shares the desire to control nature with the so-called "conventional" model, although through the development of technological artefacts that are more respectful of the environment. It integrates these ecological considerations into existing consumption and production models. Questions relating to the goals and definition of the production and consumption models to construct are therefore clearly identified from the beginning. They are aligned around the principle of promoting technical efficiency to improve production and yields.

These types of agroecological models are underpinned by a logic of reducing uncertainty. They therefore identify decision and action principles that are similar to those of the conventional model, and do not challenge the system of actors in the conventional model or their technico-economic values. The technico-economic efficiency and performance standards inherited from the conventional model control the organisation of the production and consumption of agroecological goods and services. Models with weak ecological engagement can thus emerge from within the economic and social order (ESO) governed according to the principles of industrial rationality (Thévenot 1989; Boltanski and Thevenot 1991). Within this ESO, the functions and roles of the different categories are specified. Large companies manufacture the technological solutions developed by specialised research institutes and used by farmers through predefined procedures. For each of these actors, uncertainty presents itself in a very limited form: production questions are identified at the start; and the objectives of the production and consumption models to build are known and are conveyed by a set of technical solutions pertaining to a process of producing and accumulating knowledge.

In its strong version, the AET aims at managing changes in order to effect an in-depth transformation of production and consumption models. It thus challenges the capacity and legitimacy of the incumbent system to fulfil society's aspirations. The models to design must however take into account the ecological aspirations that the actors must agree upon, both in their formulation and in the concrete mechanisms of achieving and evaluating them. Nothing allows to predict if agents will be successful in coordinating their goals and their actions – nor within what scale of time or space. Uncertainty is related to individual and collective capabilities to steer the change.

Given the impossibility of predicting the future, the AET places agents engaged in constructing production and consumption models with strong ecological engagement in a situation of radical uncertainty. They have to explore transition pathways by proposing response paths that are concrete in terms of technology, products, production systems, etc. The strong AET is therefore an axiomatic system for action that postulates that it is by exploring uncertainty through experimentation that legitimate production and consumption models can emerge.

This version of the AET breaks with the uncertainty reduction logic and the associated principles of industrial rationale. Agents challenging the incumbent model and positioning themselves in such a way as to promote the emergence and legitimisation of the new are numerous and do not act according to an established ESO. The strong AET renders the conventional model's methods to solve production problems and its evaluation method obsolete. In its strong version, the AET implies the need to break away from a logic of reducing uncertainty, based on the production and accumulation of knowledge by a small number of actors acting within an established ESO. As Crevoisier and Jeannerat (2009) show with regard to industrial activities, it requires a logic of networking a large diversity of actors and of combining/recombining knowledge on multiple scales of time and space.

Analysing New Contractual Forms as an Organisational Response to Behavioural and Technical Uncertainties in Agro-industrial Diversification Supply Chains

While multiple branches of economics are interested in situations of radical uncertainty (cf. *supra*), they may nonetheless grant different roles to it. For example, innovation economics sees uncertainty as an opportunity inherent to all processes of change (Pavitt 2005). New institutional economics, on the other hand, is based on the hypothesis that economic actors wish to reduce the uncertainty in which they operate (without, however, being able to ascribe probabilities to the occurrence of future events). Uncertainty thus explains the creation of institutions (North 2005). Within new institutional economics, the governance (or transaction cost theory) stream allows to analyse the organisational structures implemented by actors to frame their interactions, taking into account uncertainties that are both "behavioural" and "environmental" (Williamson 1996). Behavioural uncertainty relates to the fact that one of the parties to a transaction can potentially take advantage of the resulting situation of interdependency at the expense of the other; in other words, they can behave opportunistically. Environmental uncertainty relates to the events (or exogenous elements) that can potentially affect the transaction but which are not dependent on the parties to it. These can include unpredictable climactic aspects as well as variations in the cost of raw materials on the global market. Implementing contractual forms that are more coordinated than the market restricts the actions of stakeholders, by defining "*an agreement under which two parties make reciprocal commitments in terms of their behaviour – a bilateral coordination arrangement*" (Brousseau and Glachant 2008). At the same time, these contractual forms can encourage the specific investments necessary for value creation and innovation. Ultimately, the contract is a compromise between security and flexibility. Securing the investments of the parties appears to be necessary, considering that opportunistic behaviour is not eliminated; moreover, maintaining flexibility in interactions appears to be fundamental in a context of change. In a static approach based on the transaction cost theory, uncertainty is ultimately an attribute of the transaction that determines actors' organisational choices. However, as Yvrande-Billon and Saussier (2011) have pointed out, several attempts have been made to expand this framework in order to analyse how the organisational forms chosen also support learning on production techniques. Analysing contracts from this angle introduces a change in stance: the organisational form is thus understood as a way of having an impact on the state of technical knowledge, and hence of reducing the level of uncertainty around production techniques and consequently the transaction.

While to date, the vertical coordination of agri-food supply chains has mainly been studied in relation to the emergence of a quality economy on globalised markets, the AET revives the question of the coordination between actors in an uncertain context. In particular, the uncertainty surrounding production practices raises questions on the way that the chosen organisation methods contribute to creating and transmitting the technical knowledge for production. Reintroducing new species into crop systems is a prime example of this, since it simultaneously involves uncertainties related to the development of new commercialisation supply chains, and uncertainties related to the change in practices. In large-scale farming, a diversity of contracts structures exchanges between farmers, storage organisations, and transformation industries. In this diversity, we studied production contracts[2]

[2] Production contracts are arrangements that define the conditions for selling products but which also allow for anticipation and structuring, to varying degrees, of the production conditions of the crop. In this sense, they are different from classic sale contracts found in the sector, which only define the conditions for putting the seed on the market and for compensation. According to a

supporting the development of diversification supply chains and the way that they allow actors to coordinate with one another in a context of change and uncertainty.

The case study of a fava bean supply chain is illustrative of a change in crop systems moving towards a greater diversity of farmed crops, which is a key principle of agroecology (Altieri 1999). Moreover, as a pulse crop, the cultivation of fava beans has specific agro-environmental effects (related to the fixation of atmospheric nitrogen and decreasing greenhouse gases) (Jensen et al. 2010). This supply chain, initiated by a processor in western France, is emblematic of a form of governance combining vertical contracts and collective territorial governance within an association. The association, which groups together the manufacturer and several of its suppliers (cooperative or private storage organisations), appears to be complementary to the formal contracts signed between the manufacturer and each of its storage organisations. This form of governance ensures that actors have adequate flexibility to adapt to the uncertainty surrounding crops (unpredictable environmental and behavioural aspects), while providing them with guarantees (quantity and price guaranteed prior to sowing). It also supports a dynamic of creating and exchanging the technical knowledge necessary for production. By reducing behavioural uncertainty, signing production contracts encourages intangible investments coordinated among farmers (experimentation), storage organisations (training of technical and business actors and the acquisition of internal agronomic benchmarks) and the industrial firm (R&D), thus contributing to renewing the knowledge available on the crop. Furthermore, the governance of contracts is based on face-to-face interactions multiple times per year, facilitating the transmission and exchange of knowledge. During negotiations within the association, collectively defining contractual requirements regarding plant choices and production conditions supports the exchange of technical knowledge between storage organisations, in relation to the manufacturer's requirements concerning the technological qualities of the fava beans. Annually holding events bringing together farmers under contract also contributes to the exchange of experiences between farmers belonging to competing collection structures. Last of all, production under contract contributes to the acquisition of benchmarks relating to technical itineraries (by means of individual information sheets), which are analysed and then returned to the collective.

First of all, as the organisational forms at work imply a selection of stakeholders, it can in turn generate forms of exclusion. So, the status of the knowledge produced thanks to those organisational forms is neither totally private, nor public (which is characteristic of a club good (Buchanan 1965), so it limits the possibility of disseminating the knowledge to other territorial actors. This therefore raises the issue of the scope of the supply chains covered by the contracts, which is often that of niches, as in the case studied. Moreover, the agro-industrial nature of the supply chain reveals the underlying tension between the need for situated technical knowledge to diversify farmers' production systems, on the one hand, and the desire to standardise the products harvested in order to meet the requirements of industrial

survey we conducted on 20 cooperative leaders, in the large-scale cropping sector, in France, production contracts represent 0–40% of collection, according to organisations (Cholez et al. 2017).

transformation, on the other hand. The collective governance of contracts on a production basin basis nevertheless contributes to the emergence of compromises around this tension. Lastly, structuring the conditions of the appropriability of the knowledge exchanged between members of the association is seldom discussed between these actors, but appears to be crucial in ensuring a long-term collective dynamic.

Sensemaking in Management Situations Subject to Ambiguity and Uncertainty

In the case of an AET, if we want to increase knowledge on the ins and outs of a new farming practice or on the best forms of learning and experimentation, this constitutes a case of reducing the uncertainty of the situation. If, on the contrary, we want to trigger a change in viewpoints so that certain farmers focus on different issues or see them differently (different target?), this constitutes a case of reducing ambiguity, in order, for example, to allow for action involving more cooperation based on the broader sharing of the meaning ascribed to the situation and of the target.

In the case of situations commonly considered "uncertain", management science provides an in-depth reflection on the concept of a situation. The term "situation" is commonly used in business and management language, often in a metaphorical sense. One must "control the situation", "become more familiar with the situation", address a "situation of crisis", a "complex situation", and so on. It is nonetheless interesting to move beyond this metaphorical approach to look at the scope of the notion of a situation (Journé and Raulet-Croset 2008), in particular to understand how it can shine light on individual or collective action.

In management science, the notion of a *situation* was proposed and elaborated by Jacques Girin (1990) to account for a specific category of *situations*, internal and external to organisations, which can be the subject of management analysis. From this angle, Girin uses three elements to describe the situation: "*the participants, a space (the place or places where it takes place, the physical objects found there), and a time frame (a beginning, an end, a roll-out, and potentially a frequency)*" (Girin 1990: 59).[3] Introducing the purpose of the action, he proposes the situation we are dealing with to be considered a management situation when "*the participants are united and must accomplish, in a determined time, a collective action leading to a result submitted to an external evaluation*" (Girin 2011: 198). Actors internal to organisations, as well as other stakeholders such as suppliers or clients, can evaluate the engagement in a given situation.

Therefore, thinking in terms of situations enables one to identify their ingredients – the participants, the goal, the place or territory of action, the time frame, the evaluation – as well as their greater whole, the issue to which they relate, and the

[3] Our translation.

meaning given to them. Therefore, a single given situation can be understood as a "whole" in different ways, which opens to a plurality of interpretations of the same situation. With regard to "sensemaking", Karl Weick (1995) differentiates the case of uncertain situations and that of ambiguous situations. The two are often presented as being similar, even though they do not relate to the same reality and call for different types of actions. According to him, "*[i]n the case of ambiguity, people engage in sensemaking because they are confused by too many interpretations, whereas in the case of uncertainty, they do so because they are ignorant of any interpretations*" (Weick 1995: 91). He therefore contrasts situations that are difficult to manage because they are the subject of multiple interpretations (ambiguous situations), with situations that are difficult to manage due to a lack of information or knowledge to understand them (uncertain situations). Accordingly, in the case of uncertainty, more information must be sought to be able to better deal with the problematic situation. In the case of ambiguity, there is no use in seeking more information, because the ambiguity is the result of the multiplicity of interpretations: "*The problem in ambiguity is not that the real world is imperfectly understood and that more information will remedy that. The problem is that information may not resolve misunderstandings*" (Weick 1995: 92).

When faced with an ambiguous situation, the collective action can therefore consist in triggering changes in interpretations and, for an actor that is a driver of a situation, in triggering changes in the ingredients of the situation or in enriching their interpretation thanks to the interpretations of others. Drawing inspiration from pragmatist approaches, we can consider that it is a matter of examining the reason behind the *undetermined* nature of the situation (Journé and Raulet-Croset 2008). According to Dewey, the components of a situation often "*do not hold together*". Inquiry is therefore the process that allows one to move past this initial indeterminacy to a point of possessing enough structure to allow a coherent and meaningful unit to emerge. The situation is thus progressively defined through the interplay of connections between objects, events, and individuals, forming a "contextual whole" (Dewey 1993), and evolves in line with the actions of each person: "*what is designated by the word 'situation' is not a single object or event or set of objects or events. For we never experience nor form judgements about objects and events in isolation, but only in connection with a contextual whole. This latter is what is called a 'situation'. [...] In real life, these singular and isolated objects or events do not exist; an object or an event is always a part, a phase, or a particular aspect of an experienced surrounding world, that is, of a situation [...]*" (Dewey 1938: 66).

In agroecology, situations of managing life forms are by nature very complex, and much research seeks to better understand the interactions within the system by reducing uncertainty in adding new knowledge. However, these can also be the subject of multiple interpretations, because the issue associated with them, and namely the meaning given to the actions to "manage" the situation, often does not come up. What is commonly referred to as "uncertainty" therefore sometimes corresponds to "ambiguity" as Weick defined it. Different actors that are stakeholders in a situation of managing the living world can provide different interpretations of the same situation. Uncertainty and ambiguity can also be linked. In uncertain situations, additional knowledge can undoubtedly reduce uncertainty, but it can also enable a new interpretation of the situation. For example, understanding the influence of a farming practice in a territory or its effects on the environment does not necessarily make a problem considered from a technical angle obsolete, but it does enable other perspectives (territorial, environmental) of a problem. It is therefore up to the overseer/manager/person in charge of a situation to mobilise these different perspectives, either to enrich his or her own analysis of the facts or to construct a shared meaning, which despite being shared is liable to be a trade-off between multiple interpretations.

Modelling Uncertainties to Design Management Methods

After World War II, the mathematical modelling of decision-making emerged with Operations Research (Morse and Kimball 1951). Since then, it has experienced huge success in industrial production or services. Artificial Intelligence later extended its successes to decision-making problems involving the resolution of combinatorial problems that are more complex or that may require the implementation of learning methods (Sutton and Barto 1998).

In the domains of ecology and later agroecology, mathematical models to design management strategies emerged more recently (Wilson et al. 2006). This delay is mainly due to the significant uncertainty weighing down the dynamics of agroecosystems, as well as the interactions between biophysical practices and processes, which makes it complex to model them for management purposes. In the field of modelling, it is possible to distinguish two main sources of uncertainty in the input data for these models.

The first type of uncertainty, called "environmental", is a component of agroecological processes. In agronomics, crop models depend on "random" climate variables (temperatures, rainfall, etc.). Likewise, in ecology, changes in populations, communities, or meta-populations are uncertain because they are subject to uncontrollable aspects of the climate. This environmental uncertainty influences the effects of management methods in terms of yields, impacts on ecosystems, and so on. The second type of uncertainty is related to the quality of the observations, often referred to as partial observability. The modelling of the dynamics of agroecosystems under the effect of steering methods is made even more difficult by the fact that the evolution of these systems is observed with limited accuracy, or because not all of the elements of the system are observable. For example, the state of a crop's health of is often imperfectly observed, because the symptoms of a disease may be noticed late. Disregarding this latency can lead to poor disease management. The case of partial observation is clear in the example of a "seed bank", which is an element in the system that is currently not observed but which has a strong impact on the dynamics of the self-propagating plants in a crop. In this case as well, disregarding this aspect can lead to an abusive conclusion of eradicating the self-propagating species.

Stochastic modelling enables the representation of these first two sources of uncertainty, but requires knowledge of the laws of probability. These probabilities may be unknown and difficult to evaluate. They may also evolve over time. For example, the effects of climate change on the biophysical processes involved in crop models, or those of socio-economic changes on prices, are only measured in "real-time" and not upstream at the time of their modelling to design cropping methods. The uncertainty "on" models is therefore combined with the two types of uncertainty "within" models, as mentioned above.

Due to the sequential aspect of decision making (annually, monthly, etc.) and the uncertainty surrounding and within models, it is natural to address the topic of designing agroecology management strategies from the perspective of research on adaptive management methods, as they are adaptive to the current state of the system and adaptive to new knowledge that will be acquired in the process (cf. adaptive management, Williams 2011). Over the past 30 years, various mathematical approaches have been developed to design strategies to manage agro-ecosystems under conditions of uncertainty, oriented towards either at ecology or agronomics, or very recently, agroecology (Tixier et al. 2013), by integrating ecological networks and ecosystem services into agronomic models (Mulder et al. 2018). All of these approaches address the sequential aspect, and the most sophisticated of them address the adaptation of decision-making to new knowledge.

These approaches have often been based on the Markov Decision Processes (MDP) framework (Puterman 1994), which seeks to optimise sequential decisions under uncertain conditions. It permits the optimised design of steering methods, where at each time step decisions are made on the actions to carry out as a function of the current observed states of the system. It is therefore suitable for taking into account the first type of uncertainty (explicitly modelled) surrounding the future dynamics of the agroecological processes managed. It allows the construction of adaptive strategies for which the action to choose over the time interval t is only determined as of time $t-1$ as a function of the current state of the system, as opposed to defining action plans, which are determined in advance once and for all (cf., for example, Williams 2011). Following pioneering work on the use of MDP in farming and natural resource management (Kennedy 1986), the use of this framework was further developed in the domain of biodiversity conservation (Meir et al. 2004). It was then gradually spread by artificial intelligence researchers, in such a way as to take into consideration the different natures of uncertainty surrounding the input data of models, as mentioned above. For dealing with the uncertainty related to the partial observability of the state of the system, the extension of this framework to partially observable Markov decision processes (POMDP) was developed (it generated more complex mathematical problems) (Kaelbling et al. 1998). These POMDP were recently used in biodiversity conservation (Chadès et al. 2008). Reinforcement Learning (RL) approaches (Sutton and Barto 1998), often based on simulation, are suitable for solving problems related to the uncertainty surrounding the model of the system to manage. In this case, the implementation of management actions leads in turn to new observations of the system, which are useful for refining the model to better manage it. Reinforcement learning has been applied in the context of irrigation management (Crespo et al. 2011), for example.

The methodological tools to take uncertainty into consideration in modelling for agroecological design agroecological management methods are relatively mature. Several challenges nevertheless still have to be overcome. One of them is providing IT tools for modelling and designing management methods. Today, a few software toolboxes are available to modellers, dedicated to MDP, POMDP, and reinforcement learning for addressing spatialised management problems (cf. Chadès et al. 2014; Cros et al. 2017; Nicol et al. 2017 for examples of problems solved with these toolboxes). While the MDP framework is becoming increasingly known to agronomists and ecologists, the users of dedicated toolboxes still remain the modellers.

Finally, searching for management methods that respect a compromise between different ecosystem services generates problems, which continue to be difficult to resolve, in designing steering strategies, because they are the result of multi-criteria optimisation: the "values" of these different services are generally expressed in different, non-commensurable units (e.g. aesthetic value and gross margins). They cannot be aggregated into a single criterion to optimise, and in general, it is not possible to maximise all services simultaneously. To address adaptive decision-making issues on a multi-criteria basis, it is possible to use multi-criteria MDP approaches (e.g. Chatterjee et al. 2006), which are more difficult to solve than classic MDP.

Jointly Modelling Uncertainty and Ambiguity to Explore the Potential of an Agroecological Innovation

The French national action plan Ecophyto aimed at achieving a significant transition of French agriculture by cutting pesticide use by 50% within 10 years. Mid-term evaluation shows mixed results (Potier 2014) and the final assessment was that of a failure (Guichard et al. 2017) since pesticide use even increased. A potential alternative to pesticides involves using biodiversity to stimulate pest regulation services (Duru et al. 2015b). Biological pest control using natural enemies is nonetheless often related to ecological processes on larger scales than farm management (Pelosi et al. 2010), particularly at the landscape scale (Alignier et al. 2014). Many landscape ecology studies specifically demonstrate the beneficial effect of a landscape rich in semi-natural habitats (hedgerows, woods, meadows, etc.) on these biological pest control ecosystem services (Bianchi et al. 2006; Veres et al. 2013). Thus, agricultural actors could potentially co-design a landscape rich in these habitats to favour related ecosystem services and coordinate their actions, thus facilitating natural pest regulation rather than pesticide use (Schellhorn et al. 2015). However, there are significant hindrances surrounding such an innovation and they require the consideration of different types of uncertainty. Uncertainties are associated both with the variability of results (Barrett and Dannenberg 2013) and with differences in stakeholders' viewpoints (Mathevet et al. 2011). For example, landscape ecology findings are variable when considering the agricultural benefits of landscapes rich in semi-natural elements (Bianchi et al. 2006), and tend to be implied more than actually proven (Griffiths et al. 2008). Ecologists therefore investigate the factors explaining such variability (Tscharntke et al. 2016). We refer here to two fundamental types of uncertainty as defined by Walker et al. (2003) in relation to their work on decision support models: epistemological uncertainty and ontological uncertainty. Epistemological uncertainty is due to a lack of knowledge that can be corrected, for example, by research or data acquisition. On the other hand, ontological uncertainty encompasses the inherent variability of processes whose randomness cannot be reduced by acquiring more knowledge. This can be likened to of a dice toss, the result of which remains random beyond the knowledge that each side has a likelihood of one in six.

To these initial types of uncertainty is added another distinct form of uncertainty: *ambiguity*. Ambiguity refers to the simultaneous presence of equally-valid viewpoints about an issue (Brugnach et al. 2011). A viewpoint is the representation of a given situation by an actor (Weick 1995). In our case representations are about how an agroecosystem functions. This type of uncertainty has been explored in particular in the field of decision-making in natural resource co-management, in which stakeholders with different viewpoints are involved. Including ambiguity is particularly relevant when collective actions are at play because the convergence and divergence of interacting viewpoints can influence the successes or failures of these collective actions (Janis 1971). Therefore, when the intention is to explore co-management solutions, such as landscape-scale pest management, addressing the ambiguities of different stakeholders' viewpoint is critical.

The divergence or convergence status of different stakeholders' viewpoints can be reach by using Bayesian participatory modelling (Düspohl et al. 2012) adapted to the assessment of ambiguities (Salliou et al. 2017). Bayesian modelling explicitly takes ontological uncertainties into account because it is based on the elicitation of probabilities, thus integrating the variability of the phenomenon at stake. Ambiguity is taken into account by collecting probabilities specific to each actor, in order to parametrize a model structure common to all of them. Collecting and processing probabilities individually allows for a comparison of viewpoints. This is enabled by the fact that these individual probabilities are attached to a model structure (prior to parametrization) that has previously been co-constructed by actors in participatory collective workshops. This method is different from other participatory Bayesian modelling approaches that deal with ambiguities by integrating all viewpoints in a single parameterization (Henriksen et al. 2012). Keeping individual viewpoints apart enabled us to shed light on the convergence of actors regarding the low potential of using the landscape as a pest regulation tool. Beyond the ontological uncertainties described by each individual on ecological and social processes, the benefits of a landscape rich in semi-natural elements were always considered to be very limited.

This type of approach to uncertainties is particularly useful for an ex ante evaluation of the relevance of an agroecological innovation. Vuillot et al. (2016) have already pointed out the importance of *really* taking into account the representations of agricultural actors and farmers when creating public policies. Identifying innovation pathways fitting actors' interest in the agricultural world is essential to favour such innovation and potentially avoid significant failures like Ecophyto. This type of approach sheds light on the substantial gap often found between the intentions underlying public policies, and local representations.

Discussion and Prospects

These disciplinary and thematic clarifications are enough to demonstrate the diversity of notions of uncertainty used in research on the AET, whether in the case of strong or weak engagement, as noted by Angeon (cf. section "Understanding the agroecological transition as an economic situation of radical uncertainty"). This research clearly shows that the choice of one notion or another is derived from a specific stance with respect to uncertainty. Sometimes this stance is clearly anchored in a given discipline or school of thought, but the same notion can also be shared by different disciplines. Disciplines, via their concepts and methods, have constructed their own relationship to reality, complexity, and therefore uncertainty. However, our goal here was not to establish a typology of these notions by discipline, which would require more in-depth work both on the level of the epistemology of disciplines and on the ontology of each of the notions. Rather, we sought to shed light on the non-equivalency of these notions in their way of presenting and addressing uncertainty. By doing so, we can examine how these differences are related to different perspectives of the transition itself and its forms of support.

Different Stances in Dealing with Uncertainty

Through the elucidation of the definitions and ways of dealing with uncertainty used by the researchers whose work is discussed in this paper, it is possible to use the different notions to characterise three main types of relationship to uncertainty. The first relates to uncertainty that is considered environmental or exogenous to the system studied, but which has a varying degree of influence on this system. In this case, the relationship to uncertainty is distant: it is seen as something that is endured because the uncertainty is associated with external factors that are always beyond the actors' control and wishes. In this case, this uncertainty is a part of the context, which it is not possible to control, and not a part of the management situation described by Raulet-Croset (cf. section "Sensemaking in management situations subject to ambiguity and uncertainty"), to recall the distinction that Dewey makes between these two notions (Zask 2008). Ways of dealing with this environmental or contextual uncertainty vary and relate to different strategies for adapting to the unpredictable. Cholez (cf. section "Analysing new contractual forms as an organisational response to behavioural and technical uncertainties in agro-industrial diversification supply chains") presents environmental uncertainty (related to the climate or markets) as an element that cannot be quantified probabilistically, but that economic actors nonetheless take into account when they decide to coordinate with one another. It also constitutes a contextual element determining the type of coordination implemented in the new agroecological supply chains. Therefore, if uncertainty is present, forms with more coordination between actors will be sought. If uncertainty is low, the market will be the main factor in coordination. Other publications nonetheless attempt to describe this uncertainty in order to take it into account as a factor. For example, Sabbadin and Peyrard (cf. section "Modelling uncertainties to design management methods") describe how probability tools serve to describe this unpredictability in the form of random functions that it is possible to integrate into a modelled representation of the management system. In this case, environmental uncertainty is internalised.

Another relationship to uncertainty is built from knowledge on the objects and systems to be managed. This uncertainty emerges from a lack of information or knowledge on these objects, making their behaviour and their responses to actions unpredictable. Salliou et al. (cf. section "Jointly modelling uncertainty and ambiguity to explore the potential of an agroecological innovation") mention this uncertainty in ecological systems, whose organisation stems from a complex interplay between spatial and temporal interactions between a diversity of processes. They make use of the notion of ontological uncertainty to stress the fact that these behaviours will retain a certain amount of unpredictability, taking into account the incommensurability of the knowledge to produce in order to understand them. This uncertainty can be reinforced under the effect of environmental uncertainty in relation to factors directly affecting the dynamics of life forms. Reducing this uncertainty by producing knowledge on the mechanisms of the life forms at play nevertheless remains a goal. In this sense, Salliou et al. (cf. section "Jointly modelling uncertainty and ambiguity to explore the potential of an agroecological innovation") refer to the notion of epistemological uncertainty, in reference to Walker et al. (2003), who recall that uncertainty contains a component that may be at least partially reducible. Sabbadin maintains that this lack of knowledge is subjected to the limits of the observability of complex living systems, whose spatial and temporal organisation levels remain relatively intangible. The problem of measuring uncertainty in management is very directly related to the "quality" of the interplay between data as a prerequisite in a mathematical modelling process that aims at producing a robust (and not an accurate) representation of systems.

Lastly, a third category encompasses other research defining other types of uncertainty stemming from the actors themselves and the relationships that they maintain with objects, situations, and other actors. Therefore, differences in perspectives, goals, and knowledge but also in relationships to uncertainty itself are at play. This more subjective approach to uncertainty itself encompasses a diversity of proposals for addressing it, as the work presented in this chapter shows. Salliou et al. (cf. section "Jointly modelling uncertainty and ambiguity to explore the potential of an agroecological innovation") seek to elucidate the different viewpoints and representations of actors asked about the ability of the landscape to be able to regulate agricultural pests. They use the notion of ambiguity from one actor to the next, which they liken to the notion of variability in relation to their Bayesian modelling process. Concretely, this modelling method allows them to integrate different viewpoints as a function of variability. By doing so, this allows them to evaluate the divergences and convergences in viewpoints among actors and the evolution of these viewpoints when placed in a situation of knowledge sharing. One hypothesis is that divergences in viewpoints of a system are more or less obstructive to the implementation of a collective action. On the other hand, Angeon (cf. section "Understanding the agroecological transition as an economic situation of radical uncertainty") presents the question of the uncertainty of the AET as an exploration process involving a wide diversity of actors and a process of combining knowledge on different scales of time and space. Sabbadin and Peyrard (cf. section "Modelling uncertainties to design management methods") also mention the uncertainty that emerges from researchers' own representations surrounding the operation of agro-ecosystems, which propagate errors in the modelling process and therefore the output of models.

From a management science perspective, Raulet-Croset (cf. section "Sensemaking in management situations subject to ambiguity and uncertainty") also uses the notion of ambiguity to account for the differences in understanding among actors involved in a management project. These differences are simultaneously anchored in different representations, aspirations, goals, and knowledge bases. The notion of a "situation" conveys the idea that instead of reducing this ambiguity (even if convergences emerge on the collective level), the goal should be to understand how this ambiguity plays a role in constructing the meaning of the action of a given group at a given time. In this case, the ambiguity is therefore defined in terms of a "here and now" management situation, and by nature is differentiated from the ambiguity defined by the management object itself as well as for representations of the landscape collected during individual interviews.

Still with respect to this third category, other approaches have been developed in the field of economics to deal with the uncertainty tied to the interaction between actors jointly involved in AET projects. These approaches aim at establishing organisations that are institutionalised to varying degrees and aim both at reducing the behavioural uncertainty of individuals and at adapting to exogenous environmental uncertainty. They seek to organise or even regulate the relations and exchanges between actors to work towards a given goal, whether by networking actors, as per Angeon (cf. section "Understanding the agroecological transition as an economic situation of radical uncertainty"), or drawing up contracts between them, as described by Cholez (cf. section "Analysing new contractual forms as an organisational response to behavioural and technical uncertainties in agro-industrial diversification supply chains"). This author thus describes new forms of contracts implemented for the creation of agroecological crop supply chains and that give more place to the transmission and sharing of technical knowledge as a factor in reducing behavioural uncertainty and stabilising agreements.

It is clear from the different approaches detailed here that the diversity of notions is a result of differences in stances in understanding and dealing with uncertainty. They also demonstrate that differences can be identified within a single discipline or can be mobilised in dealing with a given subject matter. This is the case of the work of Salliou et al. (cf. section "Jointly modelling uncertainty and ambiguity to explore the potential of an agroecological innovation") on innovations on the landscape scale, in which a multidisciplinary research process involving elements simultaneously borrowed from ecology, social geography, and modelling attempts to mobilise two different points of entry and stances surrounding uncertainty, namely reducing the lack of knowledge on ecological systems and revealing ambiguities between actors. Inversely, different notions compete in reinforcing the same viewpoint of uncertainty. Nonetheless, this library of research is not capable of summarising the diversity of these stances on its own. In particular, considering uncertainty an opportunity and not a problem is not illustrated here. As mentioned by Cholez (cf. section "Analysing new contractual forms as an organisational response to behavioural and technical uncertainties in agro-industrial diversification supply chains"), other authors such as Pavitt (2005) or Gherardi (2008) acknowledge uncertainty or ambiguity as factors that are inherent to or even stimulating in innovation processes… In the context of transition studies, (Stirling 2014) addressed the topic of the diversity of innovation pathways for sustainable development by anchoring them in different relations to risk, ambiguity, uncertainty, and ignorance. For example, it is through this notion of ignorance – for him associated with the greater unknown within the field of possibilities – that he discusses adaptive learning as a source of systemic innovation and transformation. Research on the analysis and support of the AET has mainly sought to describe change processes. However, it is necessary to consider the obstacles and levers involved in these changes on different organisational levels (production systems, supply chains, the territory, etc.), as well as the trajectories and pathways of the transition, and in doing so, to consider methodologies for supporting actors in this transition.

Different Perspectives on the Agroecological Transition and Its Issues

This still incomplete description of the diversity of stances begs the question of their discussion in research on the AET. Due to its complexity, the AET implies management of uncertainty. The stance adopted with respect to uncertainty is rarely made explicit, and yet this choice of defining and dealing with uncertainty is directly related to the way of defining the AET and its issues. In this way, the fact of considering uncertainty a risk or an opportunity, of seeking to reduce it or to adapt to it, or of considering that progress will be the result of deepening knowledge on the objects to manage or of the capacity of actors to organise themselves and define that which is changing in their own situation, generates different perspectives on what the AET is, and on its levers.

Apart from qualifications of the "intensity" of the transition as being weak or strong (cf. section "Understanding the agroecological transition as an economic situation of radical uncertainty"), two perspectives of the transition process currently coexist and are the subject of debate in the scientific community. The first is described as deterministic and consists in achieving a relatively well-defined target, following the principles defined by agroecology. The precision with which this target is defined is often tied to the predetermination of the pathway and the process. The other perspective does not prejudge the state of the system to achieve, seeking instead to focus on the change process itself, that which it is capable of bringing on board as a dimension and triggering as a transformation along the way. These two perspectives are rooted in different representations or models of change that maintain relationships with uncertainty. In the first case, uncertainty is generally considered a risk of diverging from the goal. It therefore seeks to globally reduce this uncertainty whenever possible, to develop anticipation and forecasting, and to observe changing systems and capacities for reframing pathways. In the other case, the actors are the purveyors of the change. The uncertainty will be that which is felt, experienced, and managed by the actors themselves. Transformations in the system will be determined by the will and capacity of the different actors to collectively or individually organise. This perspective leaves more room for opportunity and surprise, even though the association between risk and uncertainty is still present. This initial interpretation should probably be nuanced inasmuch that these two perspectives – deterministic or indeterminate – both probably relate to a diversity of stances depending on the type of research undertaken.

This illustration raises the question however of the consequences of not elucidating these positions, despite the fact that multidisciplinary research is expected to address the issues of the AET. We believe that there is a risk tied to the ambiguity that may exist in mobilising different disciplines to address the different dimensions (social, ecological, technical, etc.) and objects involved in the transition, while omitting the lock-ins as well as the openings enabled by contrasting different perspectives on the relations between change and uncertainty. Controversies have emerged around these different perspectives, echoing tensions between life science, technological science, and social science approaches. Sometimes these approaches are simply incompatible, and at best can be made to coexist as different operational pathways for the AET. This raises a final question: to what extent is clarifying positions surrounding uncertainty capable of reinforcing this distancing or, on the contrary, of building a bridge to constructive dialogue in addressing issues pertaining to the transition?

References

Alignier A, Raymond L, Deconchat M et al (2014) The effect of semi-natural habitats on aphids and their natural enemies across spatial and temporal scales. Biol Control 77:76–82. https://doi.org/10.1016/j.biocontrol.2014.06.006

Altieri MA (1999) The ecological role of biodiversity in agroecosystems. Agric Ecosyst Environ 74:19–31. https://doi.org/10.1016/S0167-8809(99)00028-6

Barrett S, Dannenberg A (2013) Sensitivity of collective action to uncertainty about climate tipping points. Nat Clim Chang 4:36–39. https://doi.org/10.1038/nclimate2059

Bianchi FJJA, Booij CJH, Tscharntke T (2006) Sustainable pest regulation in agricultural landscapes: a review on landscape composition, biodiversity and natural pest control. Proc R Soc B Biol Sci 273:1715–1727. https://doi.org/10.1098/rspb.2006.3530

Boltanski L, Thevenot L (1991) De la justification. Les économies de la grandeur., NRF Essais

Brousseau E, Glachant JM (eds) (2008) New institutional economics: a guidebook. Cambridge University Press, Cambridge

Brugnach M, Dewulf A, Henriksen HJ, van der Keur P (2011) More is not always better: coping with ambiguity in natural resources management. J Environ Manag 92:78–84. https://doi.org/10.1016/j.jenvman.2010.08.029

Buchanan J (1965) An economic theory of club. Economica 32:1–14

Chadès I, McDonald-Madden E, McCarthy MA et al (2008) When to stop managing or surveying cryptic threatened species. Proc Natl Acad Sci 105:13936–13940

Chadès I, Chapron G, Cros MJ et al (2014) MDPtoolbox: a multi-platform toolbox to solve stochastic dynamic programming problems. Ecography (Cop) 37:916–920. https://doi.org/10.1111/ecog.00888

Chatterjee K, Majumdar R, Henzinger TA (2006) Markov decision processes with multiple objectives. In: Annual symposium on theoretical aspects of computer science. Springer, Berlin/Heidelberg, pp 325–336

Cholez C, Magrini MB, Galliano D (2017) Field crop production contracts. Incentives and coordination under technical uncertainty, in French cooperatives. Économie Rural 365:65–83. https://doi.org/10.4000/economierurale.5260

Crespo O, Bergez JE, Garcia F (2011) P2 hierarchical decomposition procedure: application to irrigation strategies design. Oper Res 11:19–39. https://doi.org/10.1007/s12351-009-0040-z

Crevoisier O, Jeannerat H (2009) Territorial knowledge dynamics: from the proximity paradigm to multi-location milieus. Eur Plan Stud 17:1223–1241. https://doi.org/10.1080/09654310902978231

Cros MJ, Aubertot JN, Peyrard N, Sabbadin R (2017) GMDPtoolbox: a matlab library for designing spatial management policies. Application to the long-term collective management of an airborne disease. PLos One 12(10):e0186014

Dewey (1938) Logic: the theory of inquiry. Henry Holt and Company, New York

Dewey J (1993) Logique. La théorie de l'enquête, (première édition 1938). PUF, Paris

Duru M, Fares M, Therond O (2014) Un cadre conceptuel pour penser maintenant (et organiser demain) la transition agroécologique de l'agriculture dans les territoires. Cah Agric 23:84–95

Duru M, Therond O, Fares M (2015a) Designing agroecological transitions: a review. Agron Sustain Dev 35:1237–1257. https://doi.org/10.1007/s13593-015-0318-x

Duru M, Therond O, Martin G et al (2015b) How to implement biodiversity-based agriculture to enhance ecosystem services: a review. Agron Sustain Dev 35:1259–1281. https://doi.org/10.1007/s13593-015-0306-1

Düspohl M, Frank S, Doell P (2012) A review of Bayesian Networks as a participatory modeling approach in support of sustainable environmental management. J Sustain Dev 5:1–18. https://doi.org/10.5539/jsd.v5n12p1

Gherardi S (2008) « aujourd'hui les plaques sont molles ! ». Savoir situé et ambiguïté dans une communauté de pratiques. Rev d'anthropologie des connaissances 2(1):3–35. https://doi.org/10.3917/rac.003.0003

Girard N, Magda D (2016) L432a: "Monograph of the relation between action, uncertainty and knowledge and of the modalities of natural resources qualification for each investigated collective actions" renommé "Analyse des principaux éléments de débats suite au séminaire organisés". Toulouse, p 156

Girel M (2016) Science et territoires de l'ignorance. Quae, Coll, Versailles

Girin J (1990) Problèmes du langage dans les organisations. In: Chanlat JF (ed) L'individu dans l'organisation. Les dimensions oubliées. Editions E. Québec, Québec, pp 37–77

Girin J (2011) Empirical analysis of management situations: elements of theory and method. Eur Manag Rev 8:197–212. https://doi.org/10.1111/j.1740-4762.2011.01022.x

Godard O (2001) Le développement durable et la recherche scientifique ou la difficile conciliation des logiques de l'action et de la connaissance. In: Jollivet M (dir) Le développement durable, de l'utopie au concept. De nouveaux chantiers pour la recherche. Elsevier (Coll. Environnement/NSS), Paris, pp 61–81

Griffiths GJK, Holland JM, Bailey A, Thomas MB (2008) Efficacy and economics of shelter habitats for conservation biological control. Biol Control 45:200–209. https://doi.org/10.1016/j.biocontrol.2007.09.002

Guichard L, Dedieu F, Jeuffroy MH et al (2017) Ecophyto, the French action plan to reduce pesticide use: a failure analyses and reasons for hoping. Cah Agric 26:14002. https://doi.org/10.1051/cagri/2017004

Hatchuel A, Weil B (1992) L'expert et le système. Gestion des savoirs et métamorphose des acteurs dans l'entreprise industrielle. In: Economica

Henriksen HJ, Zorrilla-Miras P, de la Hera A, Brugnach M (2012) Use of Bayesian belief networks for dealing with ambiguity in integrated groundwater management. Integr Environ Assess Manag 8:430–444. https://doi.org/10.1002/ieam.195

Holling CS (1978) Adaptive environmental assessment and management. Wiley, Chichester

Holling CS, Meffe GK (1996) Command and control and the pathology of natural resource management. Conserv Biol 10:328–337

Horlings LG, Marsden TK (2011) Towards the real green revolution? Exploring the conceptual dimension of a new ecological modernisation of agriculture that couls "feed the worl". Glob Environ Chang 21:441–452. https://doi.org/10.1016/j.gloenvcha.2011.01.004

Hubert B (2002) Sustainable development: think forward and Act now, Paris, France

Janis I (1971) Groupthink. Psychol Today 5:43–46, 74–76

Jensen ES, Peoples MB, Hauggaard-Nielsen H (2010) Faba bean in cropping systems. Field Crop Res 115:203–216. https://doi.org/10.1016/j.fcr.2009.10.008

Journé B, Raulet-Croset N (2008) Le concept de situation: contribution à l'analyse de l'activité managériale dans un contexte d'ambiguïté et d'incertitude. M@n@gement 11:27–55. https://doi.org/10.3917/mana.111.0027

Kaelbling LP, Littman ML, Cassandra AR (1998) Planning and acting in partially observable stochastic domains. Artif Intell 101:99–134. https://doi.org/10.1016/S0004-3702(98)00023-X

Kennedy J (1986) Dynamic programming. Springer, Dordrecht

Keynes JM (1921) A treatise on probability, reprinted in the collected Writings, vol 8

Keynes JM (1936) La Théorie générale de l'emploi, de l'intérêt et de la monnaie (trad. De Largentaye). Payot, Paris

Mathevet R, Etienne M, Lynam T, Calvet C (2011) Water management in the Camargue Biosphere Reserve: insights from comparative mental models analysis. Ecol Soc 16:43

Meir E, Andelman S, Possingham HP (2004) Does conservation planning matter in a dynamic and uncertain world? Ecol Lett 7:615–622. https://doi.org/10.1111/j.1461-0248.2004.00624.x

Morse MP, Kimball GE (1951) Methods of operations research. MIT Press, Cambridge, MA

Motet G (2010) Le concept de risque et son evolution. Ann des Mines – Responsab Environ 57:32–37. https://doi.org/10.3917/re.057.0032

Mulder C, Sechi V, Woodward G, Bohan DA (2018) Ecological networks in managed ecosystems: connecting structure to services. In: Moore JC, De Ruiter PCD, McCann KS, Wolters V (eds) Adaptive food webs: stability and transitions of real and model ecosystems. Cambridge University Press, New York, pp 214–227

Nicol S, Sabbadin R, Peyrard N, Chadès I (2017) Finding the best management policy to eradicate invasive species from spatial ecological networks with simultaneous actions. J Appl Ecol 54:1989–1999. https://doi.org/10.1111/1365-2664.12884

North DC (2005) Understanding the process of economic change. Princeton University Press, Princeton

Pahl-Wostl C, Holtz G, Kastens B, Knieper C (2010) Analyzing complex water governance regimes: the management and transition framework. Environ Sci Policy 13:571–581. https://doi.org/10.1016/j.envsci.2010.08.006

Pavitt K (2005) Innovation processes. In: Fagerberg J, Mowery DC (eds) The Oxford handbook of innovation. Oxford University Press, Oxford, pp 86–114

Pelosi C, Goulard M, Balent G (2010) The spatial scale mismatch between ecological processes and agricultural management: do difficulties come from underlying theoretical frameworks? Agric Ecosyst Environ 139:455–462. https://doi.org/10.1016/j.agee.2010.09.004

Postel N (2008) Incertitude, rationalité et institution. Rev économique 59:265. https://doi.org/10.3917/reco.592.0265

Potier D (2014) Pesticides et agro-écologie – Les champs du possible

Puterman ML (1994) Markov decision processes: discrete stochastic dynamic programming. Wiley, New York

Roberts J (2013) Organizational ignorance: towards a managerial perspective on the unknown. Manag Learn 44:215–236. https://doi.org/10.1177/1350507612443208

Salliou N, Barnaud C, Vialatte A, Monteil C (2017) A participatory Bayesian Belief Network approach to explore ambiguity among stakeholders about socio-ecological systems. Environ Model Softw 96:199–209. https://doi.org/10.1016/j.envsoft.2017.06.050

Schellhorn NA, Parry HR, Macfadyen S et al (2015) Connecting scales: achieving in-field pest control from areawide and landscape ecology studies: connecting scales. Insect Sci 22:35–51. https://doi.org/10.1111/1744-7917.12161

Simon HA (1964) Rationality. In: Gould J, Kolb WL (eds) Dictionnary of social science. Free Press, Glencoe, pp 573–574

Simon HA (1978) Rationality as a process and a product of thought. Am Econ Rev 68:1–16

Stirling A (2014) From sustainability to transformation: dynamics and diversity in reflexive governance of vulnerability. In: Hommels A, Mesman J, Bijker WE (eds) Vulnerability in technological cultures: new directions in research and governance. MIT Press, Cambridge, pp 1–61

Sutton RS, Barto AG (1998) Reinforcement learning: an introduction. MIT Press, Cambridge, MA

Thévenot L (1989) Équilibre Et Rationalité Dans Un Univers Complexe. Rev économique 40:147–198. https://doi.org/10.2307/3502113

Tixier P, Peyrard N, Aubertot JN et al (2013) Modelling interaction networks for enhanced ecosystem services in agroecosystems. In: Woodward GO, Bohan DA (eds) Ecological networks in an agricultural world. Academic, Amsterdam, pp 437–480

Tscharntke T, Karp DS, Chaplin-Kramer R et al (2016) When natural habitat fails to enhance biological pest control – five hypotheses. Biol Conserv 204:449–458. https://doi.org/10.1016/j.biocon.2016.10.001

Veres A, Petit S, Conord C, Lavigne C (2013) Does landscape composition affect pest abundance and their control by natural enemies? A review. Agric Ecosyst Environ 166:110–117. https://doi.org/10.1016/j.agee.2011.05.027

Voß JP, Bornemann B (2011) The politics of reflexive governance: challenges for designing adaptive management and transition management. Ecol Soc 16:9

Voß JP, Newig J, Kastens B et al (2007) Steering for sustainable development: a typology of problems and strategies with respect to ambivalence, uncertainty and distributed power. J Environ Policy Plan 9:193–212

Vuillot C, Coron N, Calatayud F et al (2016) Ways of farming and ways of thinking: do farmers' mental models of the landscape relate to their land management practices? Ecol Soc 21:35

Walker WE, Harremoës P, Rotmans J et al (2003) Defining uncertainty: a conceptual basis for uncertainty management in model-based decision support. Integr Assess 4:5–17
Weick KE (1995) Sensemaking in organizations. Sage, Thousand Oaks
Williams BK (2011) Adaptive management of natural resources—framework and issues. J Environ Manag 92:1346–1353. https://doi.org/10.1016/j.jenvman.2010.10.041
Williamson OE (1996) The mechanisms of governance. Oxford University Press, Cary
Wilson KA, McBride MF, Bode M, Possingham HP (2006) Prioritizing global conservation efforts. Nature 440:337–340. https://doi.org/10.1038/nature04366
Yvrande-Billon A, Saussier S (2011) Economie des coûts de transaction. Editions L, Paris
Zask J (2008) Situation or context? A reading of Dewey. Rev Int Philos 245:313–328

Towards an Integrated Framework for the Governance of a Territorialised Agroecological Transition

Pierre Triboulet, Jean-Pierre Del Corso, Michel Duru, Danielle Galliano, Amélie Gonçalves, Catherine Milou, and Gaël Plumecocq

Abstract This chapter aims to further our understanding of the governance mechanisms that might best support a territorialised agroecological transition (TAET). The challenge of governance is to coordinate the actions of a multitude of actors and to integrate different dimensions of agroecology. This challenge is portrayed as important in the sustainable agri-food systems literature, which seeks a convergence of governance approaches pertaining to either a Socio-Ecological Systems (SES) or a Socio-Technical Systems (STS)-oriented conception. Starting from a representation of the territory that combines these two approaches, we emphasize the importance of reflexive governance for collectively constructing a shared space of values and knowledge between actors. Case studies of eco-innovative food and energy projects in rural areas of Gers and Aveyron in France illustrate various governance mechanisms. Even if there are high expectations pertaining to the territory as a place for articulating public, market, and civil society actors around a shared vision of sustainable agri-food systems, there is still a long way to go before local governance of the transition becomes a reality, including from a long-term perspective.

P. Triboulet (✉) · D. Galliano · A. Gonçalves · G. Plumecocq
AGIR, Université de Toulouse, INRA, Castanet-Tolosan, France

LEREPS, Université de Toulouse, ENSFEA, Toulouse, France
e-mail: Pierre.Triboulet@inra.fr; Danielle.Galliano@inra.fr; Amelie.Goncalves@inra.fr; Gael.Plumecocq@inra.fr

J.-P. D. Corso
LEREPS, Université de Toulouse, ENSFEA, Toulouse, France
e-mail: Jean-Pierre.Del-corso@educagri.fr

M. Duru
AGIR, Université de Toulouse, INRA, Castanet-Tolosan, France
e-mail: Michel.Duru@inra.fr

C. Milou
LEREPS, Université de Toulouse, ENSFEA, Toulouse, France

Coopérative Qualisol, Castelsarrasin, France
e-mail: c.milou@qualisol.fr

J.-E. Bergez et al. (eds.), *Agroecological Transitions: From Theory to Practice in Local Participatory Design*, https://doi.org/10.1007/978-3-030-01953-2_7

Introduction

Governance can be defined as the set of mechanisms allowing a body of stakeholders to direct or steer a process in a desired direction. One of the pioneering aspects of governance is the emphasis that it places on the diversity of the actors involved or to be involved, in order to orient action. This diversity is presented as a safeguard against a single actor or type of actor seizing power, and as a way of integrating different viewpoints both on the process itself and on the way of steering it (hence, the introduction of a reflexive dimension in governance). In this sense, it contrasts with "governmentalities" mediated by representative democratic forms (Theys 2002). On the analytical level, the concept of governance requires that decision-making challenges be specified, which implies the identification of: (i) the boundaries of the system defining the social and political space within which these challenges are located; and (ii) the stakeholders in these challenges. On the operational level, governance relates to the mechanisms whereby these stakeholders drive the system to evolve toward the desired state.

This chapter aims at better understanding the governance mechanisms for supporting a territorialised agroecological transition (TAET). The challenge of governance is related to its capacity to coordinate the actions of a multitude of actors and to integrate different dimensions of agroecology, such as preserving biodiversity and agro-ecosystem resources, limiting pollution, developing product "quality", and so on. This challenge is portrayed as important in the sustainable agri-food systems literature, which seeks a convergence of governance approaches pertaining to either a Socio-Ecological Systems (SES)- or a Socio-Technical Systems (STS)-oriented conception (Ollivier et al. 2018). It is at the very core of the TATA-BOX project, which is based on a conceptual framework with three components – agriculture, natural resources, and the supply chain –, insofar as it stresses the importance of coordinating the dynamics of actors concerned by these three components of territories, in order to develop agroecological agriculture (Duru et al. 2015a). Within this framework, agriculture constitutes a meeting point between the productive, environmental, and agri-food dimensions of agroecology. However, the governance mechanisms of a TAET may vary, depending on the dimension prioritised or the way of qualifying different dimensions. The "agro-environmental" governance of interactions between agriculture and natural resources stems primarily from a socio-ecological system approach that emphasises medium- and long-term processes, as well as the role of government regulation. On the other hand, "agri-food" governance is more anchored in socio-technical system approaches, which put emphasis on the role of innovation as the driver of transitions.

The agroecological transition (AET) thus calls into question the political drive to integrate the productive and environmental dimensions of agriculture. The mechanisms for coupling these two dimensions constitute the nexus of the problem because they pertain to spatial and temporal scales, property rights, and modes of action that are not necessarily convergent (Hodge 2000). Hodge (2007) in particular raised the subject of this coupling by critically examining the production and value of an environmental good (landscape, biodiversity, carbon storage, flood prevention). An environmental good can be seen as a good coupled with an agricultural good; in other words, an increase in the production of the environmental good is expected, either as a result of the increased valuation of the agricultural good on agri-food or diversification markets, or because it belongs to a specific market related to ecosystem services, such as the one set up for carbon. Agricultural policies intended to modify agricultural practices are based on these underlying beliefs, with the goal of increasing the supply of ecosystem services treated as environmental goods (Hodge 2000). These policies may seek to penalise those that do not achieve the environmental standard, or to create incentives, via voluntary or systematic payments, to produce environmental goods. However, assessing the value of an environmental good is by no means simple. The ecological processes that underlie them have a significant degree of uncertainty; they are difficult to understand on the scale of a single actor, and even more on the scale of a group of actors; and they depend on different organisational levels. Knowledge requirements regarding environmental resources and associated management practices, and the mechanisms serving to orient practices in the desired direction have yet to be explored in more detail. It is therefore necessary to address the management of environmental resources within a flexible multi-actor framework offering the adaptive capacities necessary to manage complex systems (Folke et al. 2005). Hodge (2007) accordingly states that agri-environmental governance implies "*a mix of regulation, markets, government incentives and collective decisions, set within a context of social institutions and norms*".

Mount (2012) thus identifies the utility of a reflexive approach to governance, in which negotiation processes are a part of constructing the identity and legitimacy of the system implemented. Integrated agri-environmental governance in particular draws on this approach (Voß and Kemp 2006). Given that there does not exist "*'one' adequate problem framing, 'one' true prognosis of consequences, and 'one' best way to go that could be identified in an objective manner from a neutral, supervisory outlook*" (Voß and Bornemann 2011), this approach aims at integrating the diversity of strategies, viewpoints, and expectations. From this perspective, change can be perceived as the result of diverse efforts to shape it. Reflexive governance implies that actors regularly question their representations and expectations, and can integrate new expectations into the process, which implies overcoming power relations between actors and the potential influence of dominant actors (Voß and Bornemann 2011).

This conceptualisation of agri-environmental governance shines light on the need to integrate different fields and actors in order to address the problem raised. It may put more emphasis on either the socio-environmental or the socio-technical dimension, depending on the preferred angle of attack. Governance also questions the methods of mobilising actors to act on the problem at hand. For example, collaborative governance puts emphasis on a new mode of public action based on a public institution mobilising a diversity of actors in order to enable collective decision-making (Ansell and Gash 2008). More broadly, it focuses on management and decision-making mechanisms engaging a diversity of actors in collectively constructing a common goal that could not otherwise be accomplished (Emerson et al. 2012). By contrast, the governance of socio-technical transitions and of value chains tends to focus on private actors, even though the place of public actors is not absent from the analysis framework.

The last dimension of governance to take into account is its instrumental nature. It has a capacity to act on the means of coordinating different points of view with respect to both the process itself and the way of steering it. Governance implies paying attention to the mechanisms that will enable the emergence of shared viewpoints in the context of multiple perspectives. However, it also clearly raises questions around power, leadership, and the distribution of roles (Huxham 2003). The mechanisms of collaboration between actors constitute a key dimension of governance, because throughout a transition they will determine the capabilities required for collective action (Emerson et al. 2012). Lastly, the spatial scale also appears as a key element in the governance literature (Baron 2003). Certain works position themselves on very large scales initially, such as literature on global value chains or climate governance, whereas others highlight more limited scales, in which case the actors participating in the governance are expected to have concrete experience regarding the transformations in play in the space in question. There is general agreement that this type of work relates to the idea of territory, which corresponds to local or regional levels of action (Glaser and Glaeser 2014). The TATA-BOX project gives priority to this level, which we thus seek to explicate more specifically. That being said, environmental, economic, and social processes do not stop at the boundaries of a given territory, and it is important to take into account the effects of higher levels, not viewing the territory as a closed space.

We first present governance approaches from the point of view of socio-ecological systems, followed by that of socio-technical systems, later moving on to present the overlap between these two approaches and to propose an integrated analysis framework. We then explore different pillars to prioritise in the case of integrated environmental agri-food governance, which we illustrate based on case studies.

Different Approaches to Agri-environmental Governance

Socio-ecological Systems Governance

The subject of the governance of the AET is emerging as society develops an awareness that agricultural activities directly or indirectly draw on a wide variety of natural resources (soil, water, biodiversity, etc.). The result is a significant effect on these resources, in three different ways at the very least (Nesme and Withers 2016):

- By reducing resource availability via their exhaustion, in particular for non-renewable resources. Over-consumption can also disrupt the renewal of resources and reduce the capacities of future generations to meet their needs.

- By modifying the state of resources by changing the structure or functionality of the ecosystem. Agricultural activity results in varying degrees of modification of flows, generating pollution, contamination, and degradation.
- By changing the allocation of resources. For example, this could be forests transformed into pastures.

Awareness of the effects of agriculture on natural resources is gradually undermining the productivist agriculture model. This model is based on a linear conception of relations between the economy and the natural environment (Pearce and Turner 1990). By contrast, the agroecological model that the public authorities aim to promote takes into account the dynamic and complex interrelations between agriculture or natural areas that are anthropised to varying extents. Swinton et al. (2007) thus demonstrate that agriculture benefits from and produces ecosystem services and disservices. Whereas on the one hand, to work properly, it depends on the quality of ecosystem services (Zhang et al. 2007), on the other hand, as a multifunctional activity (Wilson 2008), agriculture provides ecosystem services (carbon sequestration, landscape aesthetics, preservation of biodiversity, etc.). It can also cause ecological nuisances (water pollution, reduction in biodiversity, aggravation of health risks, etc.). This production of services and disservices will have either a positive or a negative effect on the well-being of other actors.

Agroecology, in its most accomplished form in the production and use of ecosystem services, therefore constitutes an eminently social and relational challenge (Le Roux et al. 2008). The various actors of a territory must coordinate to define the mechanisms of collective governance of ecosystem services. This is crucial, considering that the majority of these services are common goods[1] for which the usage rights divided between actors are often poorly defined (Salles 2010).

Implementing governance of this type is however not self-evident. It requires a profound revision of ways of thinking and acting (cf. Fig. 1.) with, in particular: (i) a switch from a rationale of reducing the negative impacts of agriculture on the environment, to one of producing ecosystem services via biodiversity (Duru et al. 2015b), which implies shifting short-term strategies to long-term reasoning; (ii) and in parallel, a switch from a rationale of managing private goods (technical capital, chemical inputs, etc.) to one of managing common goods (water, biodiversity, etc.).

In this governance process, action is guided not only by individual interests, but also progressively by collective strategies. The market is therefore not the only form of coordinating relations between individuals, and more varied modes of organisation come to be included.

The challenge of the governance of agro-ecosystems in providing ecosystem services requires, in parallel and in addition to a change in the modes of organising collective action, a renewal of forms of public action. Therefore, the "Command and Control"-type public lever, based on a top-down logic, does not appear to be the most suitable one for supporting the social process of selecting the ecosystem services

[1] Rivalry and non-exclusion characterise a common good.

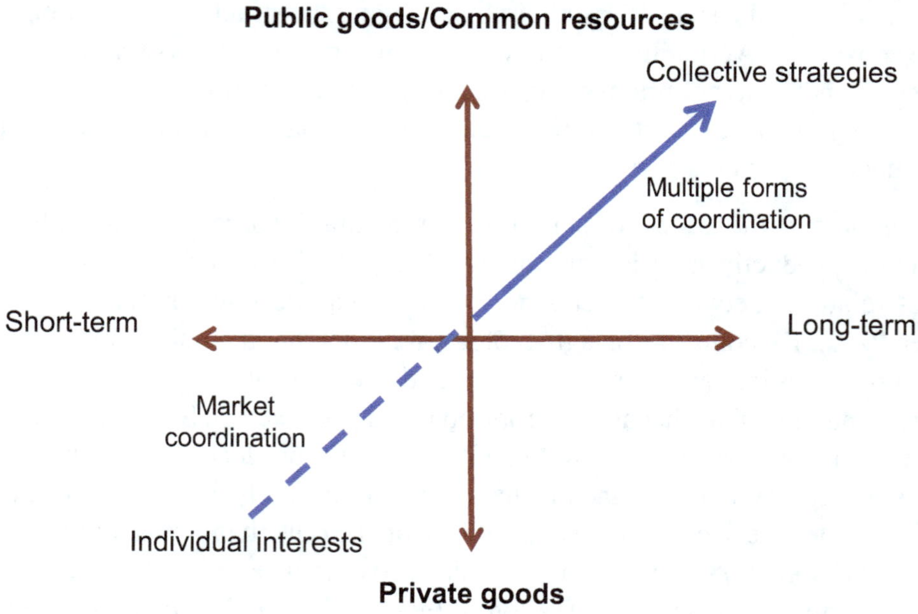

Fig. 1 The challenge of the governance of ecosystem services on the territorial scale

deemed to be worthy of attention on the territorial level. This process requires taking into account the features of the biophysical environments and socio-ecological relations specific to a given territory (Méndez et al. 2013). Moreover, it requires trade-offs between uses and values that are often multiple and competing. In this regard, Rodríguez et al. (2006) speak of "*ecosystem services trade-offs*" in the sense that the choice to preserve certain functionalities producing an ecosystem service will most often compete with that of other ecosystem services (Constans and Del Corso 2015).

Trade-offs between ecosystem services complicate the decision. Faced with this complexity, the role of public policies is above all to "orient" rather than to "steer" public action. Public bodies therefore make use of non-oriented regulation tools (Lascoumes and Simard 2011). These open up spaces facilitating coordination between territorial actors, on the basis of which they are able to choose the governance rules for producing ecosystem services. The choice of these rules is crucial, because it determines: (i) the scope of relevant actors (in particular the types of actors eligible to establish environmental goals); (ii) the mechanisms of the decision-making process; (iii) the way that ecosystem services are provided; and (iv) the way that costs and benefits are distributed between actors (Vatn et al. 2011; Vatn 2015). Consequently, these governance rules determine the level of engagement of actors in such a way as to encourage the preservation of environmental goods and services, and by doing so, give actors the chance to move beyond their personal interests.

The Governance of Socio-technical Systems

The introduction of sociotechnical systems prioritises an approach to change driven by technical innovations and the organisational innovations supporting them. The company is the actor that is the focus of the system, and the goal is to better understand the determinants that will incentivise firms to eco-innovate, in other words, to produce innovations that prevent or reduce negative impacts on the environment (Horbach 2008). In addition to firms' internal features, these determinants relate to factors of a regulatory, technological, and market nature (Galliano and Nadel 2016). The combination of these factors, and namely the push/pull (constraint/incentive) effect of regulation, the pull effect of demand, and the push effect of technology are presented in the literature as having a positive influence on firms' engagement in eco-innovation (Horbach 2008). The literature has focused on mechanisms favouring the transition from a socio-technical system towards increased sustainability. The focal point becomes the socio-technical regime, which takes into account the fact that firms and technologies are integrated within a broader set of institutions, actors, and values contributing to organising a socio-technical system (Rip and Kemp 1998). The issue of the change to more sustainable practices thus raises the question not only of the nature of this change, and namely how radical it is, but also of the nature of the actors driving this change, which can be located in different positions at the centre or on the periphery of the dominant socio-technical regime (cf. chapter "Agroecological Transition from Farms to Territorialised Agri-Food Systems: Issues and Drivers").

Smith et al. (2005) posit that the governance of the sustainable transition of socio-technical systems will be based on two goals. First, it must aim at articulating the pressure around selection that drives a socio-technical regime to change. This selection pressure can stem from a broader economic or political landscape, such as a change in modes of consumption, or on the contrary, from small niches driving radical innovation. Depending on the case, transition governance could aim at protecting and strengthening innovation niches to prevent them from disappearing, or on the contrary, at incentivising their integration into the dominant regime. It is by articulating the different selection pressures that effective decisions can be taken as to the choices to prioritise. Second, the governance of a sustainable transition must undertake to strengthen the adaptive capacity of the socio-technical regime. In other words, not all regimes have the same predispositions to embed themselves within open innovation logics allowing them to integrate multiple viewpoints, decompartmentalise actors, and open themselves up to new ideas and knowledge. The networks and devices contributing to strengthening the adaptive capacities of a sociotechnical system are becoming crucial stakes for transition governance.

Whether it aims at driving changes within the dominant regime or via innovation niches, the governance of sustainable transition raises the question of the dynamics and modes of interaction between actors in socio-technical systems in a context of globalisation. Literature on global value chains and private standards has contributed significantly to shedding light on the new modes of agri-food governance being implemented, and on their impact on a territorial scale.

Research on Global Value Chains (GVC) stems from the observation of a reconfiguration underway in the relations between actors in globalised supply chains. A GVC can be defined as "*the full range of activities, including coordination, that are required to bring a specific product from its conception to its end use and beyond*" (Gibbon and Ponte 2005: 77). This research has demonstrated the role of leading actors in the structuring and governance of these supply chains, and in particular in constructing normalisation and standardisation processes (Ponte and Gibbon 2005; Gereffi et al. 2005; Gibbon et al. 2008). It emphasises the complexity of this standardisation, which combines elements of public and private regulations, and which is integrating increasingly broader criteria (Bain et al. 2013). Gereffi and Lee (2009) thus distinguish three types of standards having a direct impact on the modes of coordination of actors: (i) health safety standards, which are mainly included in the decisions of supply chain actors via regulatory devices; (ii) product quality standards, which are mainly managed privately through quality coding devices (whether regulatory or not); and (iii) environmental and social standards, which require broader supervision of the production and transformation process to guarantee the quality of the standard in the eyes of consumers-citizens. The rapid increase in private standards – in particular environmental standards – requires more coordination between supply chain actors, and highlights the growing role of certification and accreditation bodies in guaranteeing that the stated quality standard is indeed complied with by the different parties involved in the production, processing, and sale of the product (Hatanaka and Busch 2008; Konefal and Hatanaka 2010). The development of this private governance of environmental standards on the international scale raises the question of the place of local actors and of the territorial dimension.

Therefore, critical approaches to food governance seen through the lens of environmental standards highlight the fact that this governance was developed by large leading actors in supply chains (big corporations, distributors), and large environmental NGOs in the service of a narrow vision of the environment, associated with increasingly formalised procedures (Busch 2014). What followed was the rise to power of a standardised certification regime that could contribute to excluding local actors (Hatanaka 2014). Concretely, even if certification facilitates access to markets for local farmers and organisations by optimising management practices, the associated costs can be prohibitive and the benefits are not necessarily proven, as many publications in southern countries demonstrate (Konefal and Hatanaka 2010). Moreover, these certification practices can also contribute to imposing standards that fail to take local social and environmental problems into account (Bush et al. 2013) and that are detrimental to the perpetuation/creation of modes of production that do not comply with these norms. A case in point is fruit and vegetable supply chains, in which visual quality and size criteria are large constraints on practices (Bressoud and Parès 2010).

Global supply chains are not, however, entirely devoid of local aspects, and they provide elements for reflecting on the implications of standardisation and local-global articulation inherent to seeking value for food products produced by agroecology (Gereffi et al. 2005). Loconto (2015) shows how standards imposed remotely must be aligned with the interests of local actors to meet sustainability goals: "*Local institutions and interests are stronger than 'rules' written into standards and the differences that we see in the practice of complying with standards is not so much about locally appropriating standards, but more about how governance at a distance is permitted because it is temporarily aligned with the interests, resources and obligations of the local actors*" (Loconto 2015). It is this alignment between local and distant actors that could allow pathways to be anchored, to enable a sustainable transition in time and space. The adaptive capacities of the governance methods of global chains are at play here, in their ability to adapt "*to local social and ecological contexts of production and consumption*" (Boström et al. 2015). This also calls into question the dominant regime's capacity to include a diversity of local initiatives that may possibly challenge it. The local thus appears to be a focal point for reflexive governance aimed at better articulating the diversity of actors and dimensions (environmental, social, nutrition-health) contributing to sustainable food (Marsden 2013).

An Integrated SES-STS Framework and Questions About Governance

As illustrated in the previous two sections, SES and STS governance approaches are based on a conception of systems that mobilise various actors and resources (Duru et al. 2015a). SES governance emphasises the collective management of natural resources relating to multiple coordination methods. It de facto mobilises the actors and resources within a territory. STS governance places the emphasis on the coordination of economic actors (between them, with consumers, with civil society) in their capacity to integrate the environmental dimension as a driving factor in the transition of systems. These instances of coordination between actors vary in their restriction to a given territory and often raise the question of the articulation between the local and the global.[2]

The approximation of the two approaches is born of a dual tension related to the activity of agricultural production: on one hand, taking the productive dimension into consideration as an important component of natural resource management requiring the inclusion of the issue of agricultural production and its valorisation in the support of the ecological transition; and on the other hand, increasing awareness of the negative consequences of globalised food systems, from both the environmental and the health point of view (De Schutter 2017). Moving towards healthy and sustainable food systems has thus become a part of the research and public

[2] Note however that economic theory still has trouble integrating the territory into its analysis (Zimmermann 2008).

policy agenda. This is why many voices have called for the expansion of the SES analysis framework to include a productive dimension. For example, McGinnis and Ostrom (2014) propose a change in the conceptual framework for socio-ecological systems that highlights the importance of "action situations". The goal is to examine how the social and ecological characteristics within a territory determine the actions of different actors and ultimately the achievement of different objectives and performances on the individual and collective scale. Marshall (2015) formalised the role of "transformation systems and products" for ecosystem goods by introducing a technical and technological system compartment (e.g. supply chains) at the interface between ecological and social systems, to address situations in which technologies are largely deterministic in the mode of exploiting natural resources (McGinnis and Ostrom 2014) and food systems(Vallejo-Rojas et al. 2016). Reciprocally, the literature on socio-technical systems has focused on the mechanisms to associate a wider diversity of actors and to integrate larger time scales in order to encourage sustainable transitions. Transition governance can therefore be conceived of within a two-dimensional space, the first dimension of which more specifically pertains to the SES component, and the other of which pertains to the STS component (cf. Fig. 2).

The governance of the agroecological transition (AET) must therefore aim at expanding the range of actors mobilised and at seeking better integration of components related to the market and public regulation (Hodge 2007). Research on agri-food systems demonstrates the existence of a diversity of agricultural systems that primarily use inputs that are biological and/or based on biodiversity and embedded within local versus globalised food systems. As the transition supports the idea of a change, it could therefore consist in endeavouring to identify the particularities of a mode of governance in favour of changes towards an ideal type of territorialised agri-food system. However, that would imply that the "ideal" agri-food system has been defined, and that the path to achieve it has been identified. Due to the complexity and overlap of the processes at play, transition paths are numerous and actors must be allowed to construct the paths that they wish to take themselves in order to promote the TAET (Duru et al. 2015a).

Adopting a reflexive perspective for the AET implies taking into account this diversity of ways of conceiving of agroecology and its embeddedness within agri-food systems, as well as the diversity of possible paths for this transition. The first key dimension of governance will therefore be to articulate a diversity of viewpoints or even oppositions between actors in order to allow actions to emerge in the context of the uncertainty and incompleteness of agroecology knowledge. The second dimension will be to encourage knowledge production and learning in such a way that the various stakeholders are drivers and pilots of changes, in particular farmers. The territory potentially becomes a key scale affirming the dynamics of collective actions to create added value and to strengthen the autonomy of actors (cf. chapter "An Integrated Approach to Livestock Farming Systems' Autonomy to Design and Manage Agroecological Transition at the Farm and Territorial Levels").

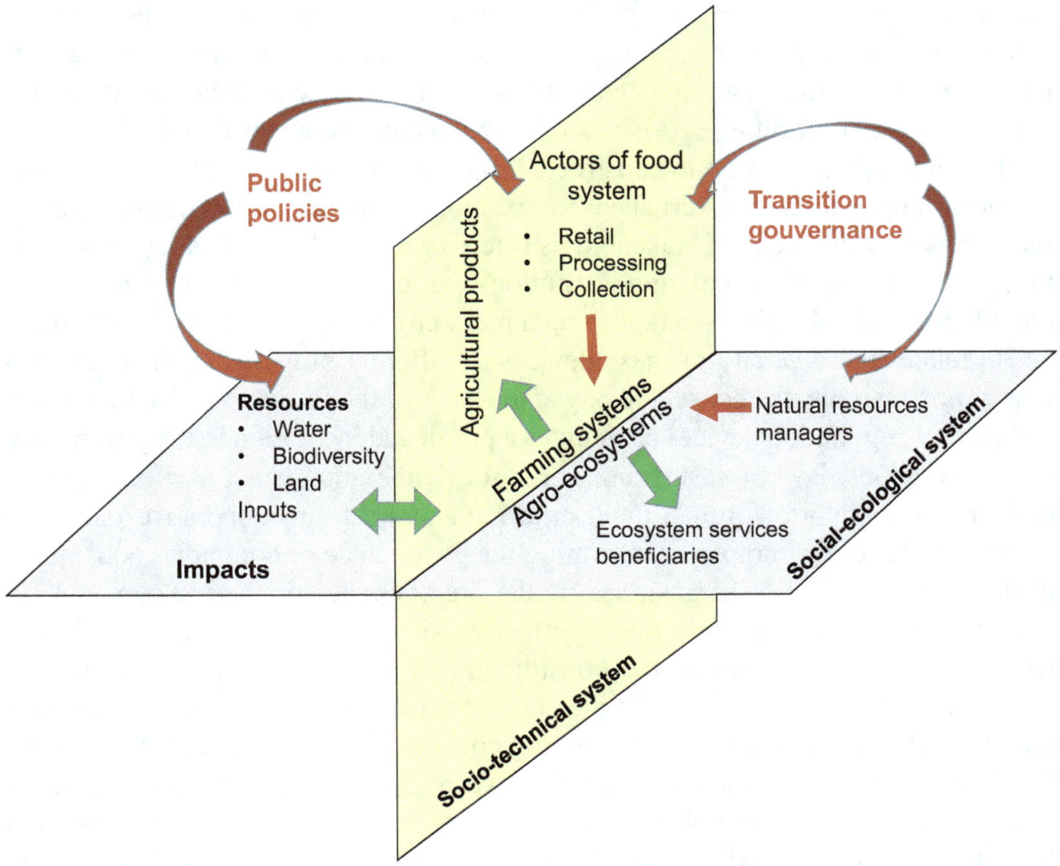

Fig. 2 Representation of a socio-ecological system (horizontal) and a socio-technical system (vertical) within a territory
Legend: farms and the agro-ecosystems that they manage are at the interface between the two systems. Green arrows indicate flows of products and services; burgundy arrows indicate decisions (rules, standards, subsidies). Transition governance mobilises actors in socio-ecological and socio-technical systems as well as in public policy

The Pillars to Prioritise for Integrated Environmental Agri-food Governance

Stemming from the territory (SES) and the agri-food system (STS), this consists in examining which governance mechanisms a diversity of actors, including public actors, mobilise to promote agroecology, as well as the place of public policies and the local/territorial levels in agri-food systems.

Reflexive Governance to Identify Value-Articulating Institutions

The governance of the AET of a territory implies that the various concerns conveyed by different actors are taken into consideration. These concerns are justified by different systems of values or interests, which are potentially conflicting. Making the values or interests underpinning different transition strategies compatible with one another appears to be both a prerequisite and a result of governance. This governance can be articulated around "conventions" (in the sense of collective cognitive devices that support differing visions of agroecology (Plumecocq et al. 2018)), or on the contrary, can pertain to constructed rule sets. These rules define the mechanisms for implementing governance (for example, more or less restrictive/voluntarist), as well as the scope of stakeholders (more or less inclusive). Vatn (2005) calls these rule sets "value-articulating institutions", in other words, devices that aim at articulating and collectively establishing a hierarchy of different transition paths.

He relates the diversity of these devices to different property regimes (cf. diagram). For example, the consent to pay at the basis of the formation of market value and market negotiation devices constitutes a privileged institution in the governance of private goods. By contrast, frameworks for deliberation, such as citizen juries, deliberative or hybrid forums, multi-criteria valuations, or deliberative monetary valuations, have the purpose of informing the governance of common goods. Each of these frameworks is characterised by the pre-eminence granted to communications processes as a precedent to collective decision-making. In this sense, frameworks for deliberation reveal the construction of a system of values shared by governance stakeholders.[3] The implementation of value-articulating institutions therefore ultimately relates to constructing collectively accepted governance solutions (Douai and Montalban 2012; Del Corso et al. 2017). Yet, upstream from deliberative processes, this search for legitimacy inherent to the governance process leads to two paradoxes: (i) what legitimate procedures support the decision on the form that value-articulating institutions must take on? and (ii) what procedures support the designation of the guarantor of the proper functioning of frameworks for deliberation? Let us consider these:

(i) In reality, if governance solutions draw their legitimacy from the deliberative nature of governance institutions, what is the source of legitimacy in the choice of the form of these institutions themselves? This choice is even more significant considering that the form of these institutions has an influence on mechanisms for expressing values, and to a certain extent on governance mechanisms. For example, expressing values in monetary terms seems to engage market coordination mechanisms.

(ii) It is important to point out that even if the ranking of the different roles of stakeholders in the governance of transitions may appear to be the result of deliberative processes, it can also reproduce relations of domination. The actor (potentially collective) that oversees the operation of value-articulating institutions themselves also possesses resources that it may use to its exclusive benefit. How is it possible to justify the legitimacy of the person fulfilling this role? While certain academic publications warn against the powers granted to experts

[3] However, beyond an agreement, it can also aim at establishing an entente, in the sense that each party understands the point of view of the other without necessarily being in agreement with them.

in these devices (Van Tilbeurgh 2015), other publications highlight the role of scientists in steering deliberative devices (Fung 2006).

The involvement of scientists in the implementation of governance structures opens up a debate around the aspects that may be neglected by actors directly involved, such as the representation of the interests of absent third parties (future generations, other countries, the biosphere, etc.). By doing so, this involvement can create the necessary conditions for what Amartya Sen (2009) calls "*open impartiality*", because it contributes to taking into consideration "*voices from far*". The legitimacy of these researchers' role is however consubstantial with their professional ethics. Chapter "Evaluation of the Operationalisation of the TATA-BOX Process" provides hints on how such legitimacy is constructed. Ideally, stakeholders should be the judge of the well-founded nature of scientists' ethics and impartiality when taking into account the interests of absent third parties. If they are not well-founded, it is not possible to guarantee the legitimacy of the governance solutions resulting from a deliberative process steered by scientists.

Symmetrically, in certain situations characterised by the existence of scientific controversies surrounding essential aspects of the problem relevant to the implementation of governance, and which require rapid decision-making (thus precluding the production of stable knowledge), deliberative processes involving a large set of stakeholders are claimed to enrich scientific results. These situations pertain to what Godard (1993) calls controversial universes. When actors' perception of governance stakes is directly influenced by scientific knowledge (as in the case of the use of glyphosate or GMOs), it may be necessary to promote the hybridisation of expert and place-based knowledge. Deliberative forums can thus offer spaces for multi-actor collaboration allowing the renewal of registers of knowledge and action to deal with uncertainty (Callon et al. 2001). However, certain actors in the field may not have an interest in engaging in such collaboration. To preserve their individual interests, they may be tempted to sow doubt regarding the legitimacy of scientists in producing useful knowledge for governance.

The Agri-food System as an Element Integrating Environmental, Social, and Economic Dimensions

There is an abundant literature on local food systems as an agricultural production format that is more diversified and more respectful of the environment (Mount 2012). These systems can be seen as innovation niches where actors experiment with new modes of governance. However, Mount (2012) argues in favour of moving beyond a caricatural approach to the governance of local agri-food systems based on the premises of reconnection between production and consumption, direct links between farmers and consumers, and shared goals and values among these two types of actors. The actors that mobilise to reconnect agriculture, food, and the environment can be involved on different spatial scales and within different configurations, which generate different *place-based reflexive governance* configurations

(Marsden 2013). Therefore, there is also the question of how different innovation niches coexist or are placed in tension with the dominant regime, and how the institutional landscape influences the various configurations observed.

One way of exploring TAET governance is by examining the range of actors ("private" actors, civil society, public actors, etc.) that will be stakeholders in the territorialised transition process. Vermeulen and Kok (2012) thus show that the different scenarios favouring sustainable products can be foreseen in terms of public-private regulation determining associated modes of governance. They examine these scenarios according to the role assigned to the public authorities. The first scenario pertains to classic public regulation based on restrictions and incentives. It is a form of **central regulation** that territorial actors can integrate into the governance, but over which they hold little sway. The second scenario of **interactive regulation** is based on the idea of collaboration between public and private actors aiming for private actors to be associated with or even drivers in the creation of measures promoting sustainable products, as is clearly seen in the development of private standards. The third scenario of **self-regulation** is based on initiatives by market and civil society actors. It is these actors that steer the processes, as public actors only play the role of supporting and facilitating these initiatives. Last of all, the authors identify a fourth scenario relating to the fact that public actors can be important economic actors as active consumers, for instance by recommending, through incentives or obligations, products for their cafeteria establishments.

Vermeulen and Kok (2012) examine how these four strategies are deployed in Holland around two supply chains: wood and coffee. By characterising the different phases of the development of environmental certifications for these two supply chains, they highlight the competition existing between various governance scenarios, and show that the place and role of public actors is essential, including to provide clear support for self-regulation. The development of organic agriculture is a good example of this coexistence of governance modes corresponding to different configurations of actors. The official labelling established by public policies is juxtaposed with collective private labelling resulting from coordination between actors that is at times territorialised. While public policies as well as markets and consumers support the dynamics of organic agriculture today, many questions remain unanswered as to related modes of governance promoting agroecological practices. In particular, this calls into question the relationship between public policies targeting production and the environment (via the CAP), and environmental certification that is increasingly tied to private market actors (Forney 2016). Many expectations concern the territory as a place for articulating public, market, and civil society actors around a shared vision of sustainable agri-food systems. Yet the local governance of these systems remains a major challenge, including over the long term. For more than 30 years, the Biovalley project in the Drôme region has been mobilising local institutions and a set of producers to develop an "organic" territory. Its success is related primarily to its capacity to mobilise a significant number of European funding mechanisms over the long term (Lamine 2015). However, many conflicts exist between project promoters and local institutions, as well as between actors included within the project and those excluded. Lamine (2015) thus notes that the inclusion of citizens and NGOs in the Biovalley project is likely to be a significant challenge in the near future in building a shared vision of the territory.

A few examples from the rural territories of Aveyron and neighbouring *départements* concretely demonstrate the various dimensions of governance considered in this chapter.

The analysis of the trajectory of eco-innovative agri-food projects and of methanation allows us to characterise different governance mechanisms use in steering these projects (Box 1). Stemming from local initiatives involving farmers, these

Box 1: The Central Role of Farmers in the Governance of Eco-innovative Food and Energy Projects

This study is on five eco-innovative projects in rural areas of Aveyron and Gers (Nuts 3 level) (Galliano et al. 2017). The environmental dimension is central in two of these projects, as they are collective methanation projects in Aveyron. The other three are agri-food projects with an environmental dimension (a local bread supply chain in Aveyron with agroecological practices, a sheep cheese supply chain in Aveyron partially from organic agriculture, and a large-scale organic crop supply chain in Gers).

While all of these projects involve an environmental component, they neither position themselves as nor necessarily lay claim to being AET actions. Concretely, it is primarily their economic and social impact on the territory that stakeholders highlight. These five initiatives convey the will of actors to engage in a collective process aimed at regaining leeway with respect to a global context that endangers the continuance of their activities or drives them towards strategies that they do not want to adopt. The same desire to create added value for farmers in the territory is present throughout these projects, in the context of the process steered by these farmers.

The central role of farmers is the common denominator of the governance of these initiatives. Yet the governance of the supply chains differs from that of methanation in terms of the scope of actors concerned, decision-making mechanisms, market integration, public-private relations, and time frames.

The three supply chains are characterised by **self-regulating governance** (Vermeulen and Kok 2012). They are projects stemming from the desire of economic actors within the same territory to establish a new offering combining agricultural production and transformation, and based on local know-how and the image of the territory. Therefore, the farmers and cooperatives involved are increasingly integrating the market into their strategy. This can either take place through the development of cooperative-run processing and distribution activities for the national or international market (cheese and grain supply chains), or through the creation of an inter-professional association grouping together all actors throughout the entire (bread) supply chain for local demand.

(continued)

Box 1 (continued)

All the stakeholders collectively define the goals of the three supply chains, the ways of achieving them, and the rules governing coordination. This takes place either through representatives or through the direct involvement of each individual (cheese supply chain). The rules of the collective's operation are quickly formalised (for instance through technical specifications) in order to guarantee the engagement of each person. This rapid formalisation reflects the time frame of these projects, which are implemented over a few months in order to rapidly provide outlets to the farmers in question. The speed of this implementation can also be explained by the convergence of the macro-economic context of the sectors in question with national and European public policies creating incentives or offering assistance for implementing such projects. Public actors in the territory (administrations or local governments) subsequently remain in the background, capable of facilitating initiatives on a one-time basis or getting involved by providing means, but not intervening in defining the project and its orientations. Given that the valorisation of the territory's resources (both tangible and intangible) is at the heart of these projects, significant support is also drawn from local professional networks (Chambers of Agriculture, professional training organisations, etc.) that do not play a direct role in governance but generally prove to be important in helping to define possible strategic options.

As for the two methanation projects, they are characterised by **interactive governance** that closely intertwines the involvement of public and private actors. These projects were initiated following European policies for promoting renewable energy transposed to the national level. Local elected officials were the spearheads for these projects alongside farmers, who also became leaders in the initiative. These initiatives were therefore initially structured via a very limited hard core of public and private actors that subsequently expanded to a large workgroup with many farmers. While the jurisdictions of local governments and municipal groups did not allow them to develop ad hoc policies to benefit methanation, they made a significant contribution by providing resources (financing, advising, logistics) and by contributing to defining the strategy via the participation of elected officials in workgroups. Throughout the lifespan of projects, national and European policies remained deterministic factors in their evolution, over which local actors sought to have an influence by mobilising regional elected officials or state agents. This regulatory context over which actors had little sway contributes to explaining the time frame of projects. These are initiatives that required several years to truly structure themselves. The first stage, consisting of defining major orientations, was characterised by coordination between actors that was largely informal. By contrast, the second phase, that of project implementation, was supported by a much higher degree of formalisation (establishment of firms,

(continued)

Box 1 (continued)

signing of contracts), in order to guarantee the perpetual engagement of each person (including farmers).

Even though these methanation projects do contain a territorial anchoring component, this appears to be less significant than for the three supply chain projects, due to the lower degree of mobilisation of local actors and resources. This is mainly explained by the fact that these projects, despite drawing support from local agricultural resources, used a technology that was new for the territories in question, and mainly intended to produce energy consumed outside of these territories. This also explains how external actors (public banking, methaniser manufacturers, etc.) gradually came to form a part of the project, informally at first and then by becoming shareholders and thus explicitly participating in governance. The farmers nevertheless remained heavily involved in governance. As key members of workgroups, they later became majority shareholders in the firms driving the project. As a whole the farmers concerned were consulted in important decisions/orientations that they nonetheless did not necessarily make, as their representatives were responsible for the more everyday aspect of governance.

projects are rooted in formal organisations that constitute the medium for knowledge exchange and value sharing among actors. Agri-food projects are the result of economic actors in a single territory wishing to construct a new offering combining agricultural production and transformation, as well as developing the image of a territory pushing for sustainable agriculture. The governance is self-regulating (Vermeulen and Kok 2012) and aims at associating the different links in the supply chain in order to agree on product quality goals and on the means and knowledge to develop and share, in order to achieve them. This governance is often based on the key role of a few individuals with the capacity to enrol and mobilise a diversity of actors and knowledge. Even though consumers are not stakeholders in these projects, a key element of their success is tied to the success of products on local or more distant markets, indicative of the capacity of alignment of the local interests of agricultural actors with demands that are often remote. Methanation projects are based on more interactive governance because they mobilise local public actors to a greater extent. However, even though there exist clearly-displayed public support policies and these projects consist in developing resources locally, they have trouble establishing themselves due to complex and constantly changing regulations (which generate uncertainty), and to the greater diversity of actors involved, which makes it difficult to establish a shared goal. The gap between environmental standards reflecting national methanation legislation and local interests, values, and resources clearly appears here to be an obstacle to these environmental projects.

The development of a dried legumes supply chain by a cooperative illustrates private governance aiming at establishing a system of shared values (Box 2). While

Box 2: Governance as Constructing a System of Shared Values: The Example of a Dried Legumes Supply Chain Implemented by an Agricultural Cooperative

The concrete case of an agricultural cooperative's development of a legume supply chain, which underlies an initiative embedded within the tenets of the AET, can offer a specific example of the challenges of governance founded on the creation of a system of shared values. The success of such an initiative essentially relies on multiple categories of actors, operators, farmers, and consumers.

1. **An agroecological transition driven by a cooperative facing a dual challenge**

Established in the Tarn et Garonne and Gers *départements (Nuts 3 level)*, the Qualisol agricultural cooperative wagered on the development of a dried legume supply chain. In light of the agronomic, food, and environmental benefits expected fromv the development of these legume crops, this project can be understood as a territorialised AET initiative. It nevertheless collided with multiple obstacles and uncertainty factors.

For the cooperative, the challenge proved to be twofold: first, it had to successfully take up a satisfactory position in a market that was still unstable and in which other competing operators were present; and second, it had to be capable of getting its farmer members to grow legume crops over the long term in order to lay the foundations of the supply chain.

With regard to the first challenge, the cooperative had to be successful in setting itself apart, to capture the attention of potential customers and to offer products capable of convincing consumers at the end of the chain. However, despite the growing body of knowledge on the nutritional advantages of legumes,[4] this type of food remains relatively unknown. Consumers also mention obstacles to consumption with regard to digestibility, the practicality of using them, and so on. Promoting the consumption of legumes was thus tied to improving the information on these products and modernising their image and the ways of using them.

Concerning the second challenge, garbanzo bean or lentil crops can offer relative financial security through the signing of contracts. This is however undermined by the uncertainty of the success of these crops, which require the acquisition of new technical know-how. This security is also limited by farmers' lack of understanding of their benefits on a rotational multi-year scale.

(continued)

[4] This information was updated as a part of the nutritional recommendations of the *Haut conseil de la santé publique* (the high council for public health) under the *Programme National Nutrition Santé* 2017–2021 (national health and nutrition plan 2017–2021).

Box 2 (continued)

Lastly, it can suffer due to an annual profitability considered to be unsatisfactory when compared with other better-controlled crops. Effectively, the agronomic benefits attributed to legumes, despite generally being known to farmers, are barely considered little in assessments of the direct value of these crops.

2. **Collective learning to reduce the uncertainty of producers and consumers**

The cooperative used multiple levers to overcome this dual challenge.

First of all, it supported itself by establishing its own brands in both the organic supply chain and the conventional supply chain, in order to hold extended control over the downstream part of the chain. The organic supply chain, developed first, and which had a higher security margin (better price stability, rotational approach more customary among farmers in the organic system, better informing of consumers, and popularity of the organic label), allowed the cooperative to acquire know-how on these products while limiting risk-taking. It was thus able to make use of this learning to later develop its supply chain in the conventional market. For the latter, the cooperative also wished to establish a positive image setting it apart in the eyes of customers and consumers through the "*Haute valeur environnementale*" (HVE, high environmental value) certification approved by a set of technical specifications, compliance with which is supervised by an independent certification body. Therefore, within both supply chains, the cooperative highlighted the origin of products, which by being associated with HVE or organic specifications, offered clients and consumers security in terms of production transparency.

The cooperative also aims at diversifying dried legumes crops (multiple species of beans, chickpeas, lentils) in such a way as not only to strengthen its appeal among clients through a broad product range, but also to play on consumers' curiosity and interest. The cooperative moreover purchased shares in a processing company to offer dried legumes in a form ready to use in salads, cooked dishes, and dough/pastry.

This process seeking to secure the desired production on the market through differentiation is reflected in its economic valuation, which is capable of making investments profitable and convincing farmers to produce these crops. It is also backed by a second lever directed at the latter: that of collective action. This is manifest in the *Groupes d'Intérêts Economiques et Environnementaux* (GIEE, economic and environmental interest groups) framework. This collective framework offers a form of security by throwing questions, failures encountered, and solutions tested into a communal pot in order to jointly identify factors in success. In its structure, the GIEE also encompasses other categories of partners (commune communities, associations, federations, economic organisations, etc.). Because of this, it represents

(continued)

Box 2 (continued)

an excellent method for exchange between multiple categories of actors that are stakeholders in the success of a territorialised supply chain.

Therefore, through its multi-actor and inclusive approach, the cooperative was successful in establishing an original mode of governance capable of articulating, between these multiple actors, the multiplicity of environmental, agronomic, and food values transmitted by the actors present. The concept of quality is at the heart of this articulation. Concretely, it is around a shared definition of quality that the convergence of producer and consumer preferences is able to take place. In fact, an agreement on quality as a value-articulating institution represents a crucial governance concern in and of itself. This agreement is such as to trigger a broadening of modes of thought and types of action. Because of this, it can appear to be a factor in reducing uncertainty for different categories of actors, thus securing their actions as well as orienting these actions in a direction that appears desirable to them.

the desire to be embedded within a TAET initiative is the driver of the cooperative's engagement, it is faced with the dual challenge of finding booming markets to sell these dried legumes and incentivising farmers to grow these crops, on which they lack knowledge and resources. For the first challenge, the cooperative developed various strategies for creating the value of these products around its own brand, and by constructing a specialised network allowing it to establish credible and lucrative quality markings. For the second challenge, it drew on the support of collective action by creating a multi-partner GIEE (*Groupe d'Intérêt Economique et Environnemental* (economic and environmental interest group) allowing it to create the conditions for exchanging knowledge on the practices to implement. In both cases, the quality agreement appears to be the value-articulating institution allowing the different stakeholders' preferences to emerge.

The last example stems from research on territorial protein autonomy based on interaction between grain and livestock farmers in the Aveyron Valley (Box 3). Several multi-partner participatory workshops served to establish scenarios for the reduction of irrigated maize crops, to be replaced by alfalfa for dairy farmers (Moraine et al. 2016). While actors, and in particular farmers, agreed on the benefits of such interaction, it was the concrete implementation of the governance necessary to set them up that constituted an obstacle in this case. No actor was identified as having the capabilities necessary to define the value-articulating institution supporting these interactions. Integrating a public actor such as the *Agence de l'eau* (water agency) could therefore constitute a solution for interactive governance between a public actor, an agricultural cooperative, and farmers.

These examples testify to the social challenges associated with the governance of the rural the TAET. Behind these initiatives are a diversity of actors, including farmers, seeking increased economic value for the local resources of their territory,

Box 3: Towards Governance of Exchanges Between Grain and Livestock Farmers: The Example of Multi-cropping-livestock Farming in Aveyron (Moraine et al. 2016)

The specialisation of regions and farms has compounded the environmental impacts of agriculture due to the mass use of inputs and an increase in the vulnerability of farms in a context of high inter-year climate variability. To way to meet these challenges is to diversify farm productions, which is associated with exchanges between specialised farms within small territories. This orientation requires not only a revision of the mode of managing resources within a territory to reduce their consumption and/or impacts, but also a deeper or shallower reorganisation of supply chains in order to adapt to this diversification.

By representing the agriculture in a given territory as a socio-ecological system, it is possible to identify environmental challenges, the actors concerned by the resource (consumption and impacts), and levers for action. Representing the dominant supply chains in a territory as a socio-technical system allows us to evaluate the degree of reorganisation necessary to achieve environmental goals as well as feasibility in economic terms and with regard to the organisation of work. In other words, these two frameworks of analysis cross-compare governance in socio-ecological and in socio-technical systems, to identify the shared aspects or incompatibilities between them.

This cross-comparison was applied in the Aveyron Valley, where the upstream is characterised by the concentration of surface areas as temporary and permanent grasslands, while the downstream is dominated by large-scale cropping areas for maize, grains, and sunflower. This juxtaposition within this basin of a zone specialised in livestock farming and one dominated by large-scale cropping is representative of many grassland/hillside situations in France. Each zone has a certain amount of diversity of production systems, but these systems tend to specialise.

A participatory diagnosis of concerns related to cropping-livestock farming integration in the Aveyron basin was carried out. This diagnosis was based on an initial workshop bringing together a wide variety of participants (farmers, advisers at chambers of agriculture, cooperative technicians, the *Agence de l'Eau* (water agency), and representatives of an environmental non-profit association). Iterative work at the workshop allowed participants to establish multiple scenarios, the common aspects of which were: (i) reducing the surface area of irrigated maize by replacing it with alfalfa sold to livestock farmers, thus allowing for water savings and the maintenance of the economic performance of large-scale cropping farms; (ii) strengthening the place of alfalfa on dairy farms, and thus increasing the protein autonomy of the territory.

(continued)

Box 3 (continued)

One challenge is designing the contractual mechanisms of interaction between crop and livestock farmers (prices of materials exchanged, price and volume guarantees, logistics, financing, etc.) and checking with the water agency that these changes in technical systems are consistent with the resource governance plan. This contractualisation could therefore be tripartite (farmers, collection agency, and water agency). Organisational innovations should consequently be devised to deploy and perpetuate such arrangements, taking into account that the farmers present do not have experience or benchmarks for this. It is therefore recommended to implement reflexive governance.

This example is relatively simple insofar as the technical change is limited to production and logistics, but does not result in changes in transformation, distribution, and food choices.

alongside environmental benefits. While the market dimension is strongly present in the projects used as an example in this chapter, particularly to develop the value of the agroecological production approach in the eyes of consumers, the latter are not closely associated with their governance. And although this illustrates a difference with respect to urban agri-food systems, in which consumers are often drivers (Sonnino 2017), the desire to open up to a diversity of supply chain and territorial actors, including regional governments, is testimony to changes in modes of governance of rural agri-food systems, with a shift from agri-industrial systems governance to territorialised food systems governance (Lamine et al. 2012).

Conclusion

This chapter is based on the argument that there is not only one archetype of governance for the TAET. On the territorial scale, a variety of initiatives exist, contributing to the agroecology of practices and embedded within various agri-food systems. This relates to a representation of the territory that combines a horizontal dimension pertaining to socio-ecological systems, and a vertical dimension pertaining to socio-technical systems. The challenge is therefore to identify the different governance mechanisms that will favour the AET process. The literature agrees on the importance of reflexive governance in collectively constructing a shared space of values and knowledge that set in motion increasingly agroecological practices. It also highlights the fact that environmental governance requires the association of a diversity of private and public actors, as well as the integration of a combination of regulations, markets, and collective action. In this context, the territory appears to be a place of tension, articulating the construction and reappropriation dynamics of local actors and the local redeployment dynamics of more global actors. One marker of these tensions is the entry of sustainable agri-food systems, from local niche systems up to globalised systems.

In the second section we first focused on the ways of making actors' preferences converge around a shared goal. In particular, we highlighted the importance of value-articulating institutions in establishing collectively accepted governance solutions. The undertaking to identify and legitimise value-articulating institutions is at the core of governance stakes, and in particular raises the question of the place of scientific knowledge. After that, we went over Vermeulen and Kok (2012) typology of modes of governance. This typology allows us to specify the role of public actors with regard to the governance problem posed, from a distant role by regulation (central governance), to participation in governance (interactive governance), the support of economic actors and civil society (self-regulated governance), and even an active role as a market actor (for instance through school cafeterias). A few case studies of eco-innovative food and energy projects in the rural territories of Aveyron and neighbouring *départements* allowed us to illustrate the mechanisms of interactive and self-regulated governance. For agri-food projects, the quality of products and associated standardisation processes constitute the value-articulating institution that orients the practices of all actors throughout the value chain. For projects for exchanges between grain and livestock farmers or methanation projects, difficulties were experience in setting up coordination around a value-articulating institution. Lastly, the success of projects appears to be related to the capacity of leading actors to integrate a diversity of actions and to mobilise stakeholders as a whole towards a common path.

Our conclusions are in line with those highlighted in the literature. Even if there are high expectations pertaining to the territory as a place for articulating public, market, and civil society actors around a shared vision of sustainable agri-food systems, there is still a long way to go before local governance of the transition becomes a reality, including from a long-term perspective (Lamine 2015). This relates to local actors' capacity for defining a goal shared by the different stakeholders of the territory and for providing themselves with the means to achieve this goal. Moreover, it also relates to their capacity for integrating expectations that are external to the territory in question, whether nearby towns or embedded within globalised agri-food systems. Regarding the latter point, Boström et al. (2015) notes that the major challenge of governance in moving towards increased sustainability: "*A broader social science view on supply chains is necessary if we are to understand how unsustainable practices (continue to) prevail and how more sustainable ones could be facilitated. Yet we are only beginning to understand the enormous governance challenges facing state and non-state actors, networks, organizations and individuals to – in a constructive and responsible manner – handle the economic, social and ecological complexities associated with global supply chains*".

References

Ansell C, Gash A (2008) Collaborative governance in theory and practice. J Public Adm Res Theory 18:543–571. https://doi.org/10.1093/jopart/mum032

Bain C, Ransom E, Higgins V (2013) Private agri-food standards: contestation, hybridity and the politics of standards. Int J Agric Food 20:1–10

Baron C (2003) La gouvernance: débats autour d'un concept polysémique. Droit et société 2:329–349

Boström M, Jönsson AM, Lockie S et al (2015) Sustainable and responsible supply chain governance: challenges and opportunities. J Clean Prod 107:1–7. https://doi.org/10.1016/j.jclepro.2014.11.050

Bressoud F, Parès L (2010) Quelles références pour une production de légumes de territoire ? In: Muchnik J, de Sainte Marie C (eds) Le temps des SYAL, Techniques. Cairn.info, Paris, pp 211–228

Busch L (2014) Governance in the age of global markets: challenges, limits, and consequences. Agric Hum Values 31:513–523. https://doi.org/10.1007/s10460-014-9510-x

Bush SR, Belton B, Hall D et al (2013) Global food supply. Certify sustainable aquaculture? Science 341:1067–1068. https://doi.org/10.1126/science.1237314

Callon M, Lascoumes P, Barthe Y (2001) Agir dans un monde incertain. Essai sur la démocratie technique. Seuil, Paris

Constans M, Del Corso JP (2015) L'évolution des paysages viticoles de Banyuls: les politiques publiques face à des enjeux environnementaux paradoxaux. In: Béringuier P, Blot F, Desailly B, Saqalli M (eds) Environment. France, Paris, pp 457–477

De Schutter O (2017) The political economy of food systems reform. Eur Rev Agric Econ 44:705–731. https://doi.org/10.1093/erae/jbx009

Del Corso J-P, Nguyen TDPG, Kephaliacos C (2017) Acceptance of a payment for ecosystem services scheme: the decisive influence of collective action. Environ Values 26:177–202. https://doi.org/10.3197/096327117X14847335385517

Douai A, Montalban M (2012) Institutions and the environment: the case for a political socio-economy of environmental conflicts. Camb J Econ 36:1199–1220. https://doi.org/10.1093/cje/bes046

Duru M, Therond O, Fares M (2015a) Designing agroecological transitions: a review. Agron Sustain Dev 35:1237–1257. https://doi.org/10.1007/s13593-015-0318-x

Duru M, Therond O, Martin G et al (2015b) How to implement biodiversity-based agriculture to enhance ecosystem services: a review. Agron Sustain Dev 35:1259–1281. https://doi.org/10.1007/s13593-015-0306-1

Emerson K, Nabatchi T, Balogh S (2012) An integrative framework for collaborative governance. J Public Adm Res Theory 22:1–29. https://doi.org/10.1093/jopart/mur011

Folke C, Hahn T, Olsson P, Norberg J (2005) Adaptive governance of social-ecological systems. Annu Rev Environ Resour 30:441–473. https://doi.org/10.1146/annurev.energy.30.050504.144511

Forney J (2016) Blind spots in agri-environmental governance: some reflections and suggestions from Switzerland. Rev Agric Food Environn Stud 97:1–13

Fung A (2006) Varieties of participation in complex governance. Publ Adm Rev 66:66–75. https://doi.org/10.1111/j.1540-6210.2006.00667.x

Galliano D, Nadel S (2016) Les processus sectoriels de l'innovation environnementale: les spécificités des firmes agroalimentaires françaises. Économie Rural 356:47–67. https://doi.org/10.4000/economierurale.5055

Galliano D, Gonçalves A, Triboulet P (2017) Eco-innovations in rural territories: organizational dynamics and resource mobilization in low density areas. J Innov Econ Manag 24:35–62. https://doi.org/10.3917/jie.pr1.0014

Gereffi G, Lee J (2009) A global value chain approach to food safety and quality standards. Working paper. Duke University, Durham

Gereffi G, Humphrey J, Sturgeon T (2005) The governance of global value chains. Rev Int Polit Econ 12:78–104. https://doi.org/10.1080/09692290500049805

Gibbon P, Ponte S (2005) Trading down: Africa, value chains, and the global economy. Temple University Press, Philadelphia

Gibbon P, Bair J, Ponte S (2008) Governing global value chains: an introduction. Econ Soc 37:315–338. https://doi.org/10.1080/03085140802172656

Glaser M, Glaeser B (2014) Towards a framework for cross-scale and multi-level analysis of coastal and marine social-ecological systems dynamics. Reg Environ Chang 14:2039–2052. https://doi.org/10.1007/s10113-014-0637-5

Godard O (1993) Stratégies industrielles et conventions d'environnement: de l'univers stabilisé aux univers controversés. INSEE Méthodes 39–40:145–174

Hatanaka M (2014) Standardized food governance? Reflections on the potential and limitations of chemical-free shrimp. Food Policy 45:138–145. https://doi.org/10.1016/j.foodpol.2013.04.013

Hatanaka M, Busch L (2008) Third-party certification in the global agrifood system: an objective or socially mediated governance mechanism? Sociol Ruralis 48:73–91. https://doi.org/10.1111/j.1467-9523.2008.00453.x

Hodge I (2000) Agri-environmental relationships and the choice of policy mechanism. World Econ 23:257–273. https://doi.org/10.1111/1467-9701.00271

Hodge I (2007) The governance of rural land in a liberalised world. J Agric Econ 58:409–432. https://doi.org/10.1111/j.1477-9552.2007.00124.x

Horbach J (2008) Determinants of environmental innovation-new evidence from German panel data sources. Res Policy 37:163–173. https://doi.org/10.1016/j.respol.2007.08.006

Huxham C (2003) Theorizing collaboration practice. Publ Manag Rev 5:401–423. https://doi.org/10.1080/1471903032000146964

Konefal J, Hatanaka M (2010) The Michigan State University School of Agrifood governance and technoscience: democracy, justice, and sustainablity in an age of scientism, marketism, and statism. J Rural Soc Sci 25:1–17

Lamine C (2015) Sustainability and resilience in agrifood systems: reconnecting agriculture, food and the environment. Sociol Ruralis 55:41–61. https://doi.org/10.1111/soru.12061

Lamine C, Renting H, Rossi A et al (2012) Agri-food systems and territorial development: innovations, new dynamics and changing governance mechanisms. In: Darnhofer I, Gibbon D, Dedieu B (eds) Farming systems research into the 21st century: the new dynamic. Springer, Dordrecht, pp 229–256

Lascoumes P, Simard L (2011) L'action publique au prisme de ses instruments. Rev française Sci Polit 61:5–21. https://doi.org/10.3917/rfsp.611.0005

Le Roux X, Barbault R, Baudry J et al (2008) Agriculture et biodiversité, ESCo. INRA France, Paris

Loconto A (2015) Assembling governance: the role of standards in the Tanzanian tea industry. J Clean Prod 107:64–73. https://doi.org/10.1016/j.jclepro.2014.05.090

Marsden T (2013) From post-productionism to reflexive governance: contested transitions in securing more sustainable food futures. J Rural Stud 29:123–134. https://doi.org/10.1016/j.jrurstud.2011.10.001

Marshall GR (2015) A social-ecological systems framework for food systems research: accommodating transformation systems and their products. Int J Commons 9:1–28. https://doi.org/10.18352/ijc.587

McGinnis M, Ostrom E (2014) Social-ecological system framework: initial changes and continuing challenges. Ecol Soc 19:30. https://doi.org/10.5751/ES-06387-190230

Méndez VE, Bacon CM, Cohen R (2013) Agroecology as a transdisciplinary, participatory, and action-oriented approach. Agroecol Sustain Food Syst 37:3–18. https://doi.org/10.1080/10440046.2012.736926

Moraine M, Grimaldi J, Murgue C et al (2016) Co-design and assessment of cropping systems for developing crop-livestock integration at the territory level. Agric Syst 147:87–97. https://doi.org/10.1016/j.agsy.2016.06.002

Mount P (2012) Growing local food: scale and local food systems governance. Agric Hum Values 29:107–121. https://doi.org/10.1007/s10460-011-9331-0

Nesme T, Withers PJA (2016) Sustainable strategies towards a phosphorus circular economy. Nutr Cycl Agroecosyst 104:259–264. https://doi.org/10.1007/s10705-016-9774-1

Ollivier G, Magda D, Mazé A et al (2018) Agroecological transitions: what can sustainability transition frameworks teach us? An ontological and empirical analysis. Ecol Soc 23:18. https://doi.org/10.5751/ES-09952-230205

Pearce DW, Turner RK (1990) Economics of natural resources and the environment. Harvester Wheatsheaf, London

Plumecocq G, Debril T, Duru M et al (2018) Caractérisation socio-économique des formes d'agriculture durable. Accepté Économie Rural:99–120. https://doi.org/10.4000/economierurale.5430

Ponte S, Gibbon P (2005) Quality standards, conventions and the governance of global value chains. Econ Soc 34:1–31. https://doi.org/10.1080/0308514042000329315

Rip A, Kemp R (1998) Technological change. In: Rayner S, Malone E (eds) Human choices and climate change, vol 2. Battelle Press, Columbus, pp 327–399

Rodríguez JP, Beard TD, Bennett EM et al (2006) Trade-offs across space, time, and ecosystem services. Ecol Soc. https://doi.org/10.2307/26267786

Salles J-M (2010) Évaluer la biodiversité et les services écosystémiques: pourquoi, comment et avec quels résultats? Natures Sci Sociétés 18:414–423. https://doi.org/10.1051/nss/2011005

Sen A (2009) The idea of justice. Belknap Press of Harvard University Press, Cambridge, MA

Smith A, Stirling A, Berkhout F (2005) The governance of sustainable socio-technical transitions. Res Policy 34:1491–1510. https://doi.org/10.1016/j.respol.2005.07.005

Sonnino R (2017) The cultural dynamics of urban food governance. City Cult Soc. https://doi.org/10.1016/J.CCS.2017.11.001 (in press)

Swinton SM, Lupi F, Robertson GP, Hamilton SK (2007) Ecosystem services and agriculture: cultivating agricultural ecosystems for diverse benefits. Ecol Econ 64:245–252. https://doi.org/10.1016/j.ecolecon.2007.09.020

Theys J (2002) La Gouvernance, entre innovation et impuissance. Développement durable Territ 2:28. https://doi.org/10.4000/developpementdurable.1523

Vallejo-Rojas V, Ravera F, Rivera-Ferre MG (2016) Developing an integrated framework to assess agri-food systems and its application in the Ecuadorian Andes. Reg Environ Chang 16:2171–2185. https://doi.org/10.1007/s10113-015-0887-x

Van Tilbeurgh V (2015) La négociation dans les dispositifs environnementaux: De la construction d'asymétries à l'imposition de préférences. ESO Travaux & Documents 38

Vatn A (2005) Institutions and the environment. Edward Elgar Publishing, Cheltenham

Vatn A (2015) Markets in environmental governance. From theory to practice. Ecol Econ 117:225–233. https://doi.org/10.1016/j.ecolecon.2014.07.017

Vatn A, Barton DN, Lindhjem H et al (2011) Can markets protect biodiversity? An evaluation of different financial mechanism, UMB Noragric Report 60. Aas

Vermeulen WJV, Kok MTJ (2012) Government interventions in sustainable supply chain governance: experience in Dutch front-running cases. Ecol Econ 83:183–196. https://doi.org/10.1016/j.ecolecon.2012.04.006

Voß JP, Bornemann B (2011) The politics of reflexive governance: challenges for designing adaptive management and transition management. Ecol Soc 16:9

Voß J, Kemp R (2006) Sustainability and reflexive governance: introduction. In: Voß J-P, Bauknecht D, Kemp R (eds) Reflexive governance for sustainable development. Edward Elgar, Cheltenham, pp 3–28

Wilson GA (2008) From 'weak' to 'strong' multifunctionality: conceptualising farm-level multifunctional transitional pathways. J Rural Stud 24:367–383. https://doi.org/10.1016/j.jrurstud.2007.12.010

Zhang W, Ricketts TH, Kremen C et al (2007) Ecosystem services and dis-services to agriculture. Ecol Econ 64:253–260. https://doi.org/10.1016/j.ecolecon.2007.02.024

Zimmermann JB (2008) Le territoire dans l'analyse économique. Rev française Gest 184:105–118

The Key Role of Actors in the Agroecological Transition of Farmers: A Case-Study in the Tarn-Aveyron Basin

Julie Ryschawy, Jean-Pierre Sarthou, Ariane Chabert, and Olivier Therond

Abstract For farmers, the transition towards agroecology implies redesigning both their production system and their commercialisation system. To engage in this type of transition, they need to develop new knowledge on practices adapted to local conditions, which will involve new actors in their network. This chapter explores the role of actors' networks in the agroecological transition of farmers, with a particular focus on farming practices and modes of commercialisation. We held semi-structured interviews to understand: (i) individual farmers' trajectories of change, considering practices at the farm and food system levels; (ii) the role of farmers' networks in their involvement in the agroecological transition; and (iii) the role of their networks on a broader scale. In the Tarn-Aveyron basin, we interviewed ten dairy farmers and 50 actors interacting with them in connection with their farming practices. We focus on two dairy farmers' trajectories: one who took a path towards agroecology, and the other who did not. We then show that the role of actors' network is crucial in facilitating or impeding the agroecological transition. We highlight the importance of considering actors' networks as a whole, including in the commercial sector, as having a key role in farmers' shift towards agroecological transition.

Introduction

Many actors in various spheres are increasingly coming to recognise agroecology as a relevant solution for the environmental and social problems posed by conventional agriculture (Wezel et al. 2009; Altieri et al. 2017). Agroecology is a relatively new

J. Ryschawy (✉) · J.-P. Sarthou · A. Chabert
AGIR, Université de Toulouse, INRA, Castanet-Tolosan, France
e-mail: julie.ryschawy@inra.fr; jean-pierre.sarthou@inra.fr; ariane.chabert@ensat.fr

O. Therond
LAE, Université de Lorraine, Inra, Colmar, France
e-mail: olivier.therond@inra.fr

J.-E. Bergez et al. (eds.), *Agroecological Transitions: From Theory to Practice in Local Participatory Design*, https://doi.org/10.1007/978-3-030-01953-2_8

concept, and its rise hides significant diverging viewpoints both around the definition of agroecology and how to encourage the agroecological transition (AET) of farming. In France, this rise of agroecology is conveyed by two largely prescriptive discourses. On one hand are the people that highlight all of the virtues of agroecology during the Anthropocene, and namely the inevitability of a solution that must be imposed in view of the problems posed by the conventional model (Vanloqueren and Baret 2009; Wezel et al. 2009). Taking ecological and social issues into account, this consists of recalling and explaining the rationality of the proposal for change, and even calls for increased responsibility with respect to it (Le Foll 2012; Duru et al. 2015). While the majority of actors recommend the combined improvement of economic and environmental performance, some are driving for a deeper transformation in systems, and in particular socio-economic systems (Ryschawy et al. 2015; Sanderson Bellamy and Ioris 2017). On the other hand are actors against agroecology, who seek to point out all of the problems raised by this paradigm shift and to recall the robustness and potential of conventional model solutions, in particular in facing the problem of pollution, via technological advancements such as precision agriculture (Sanderson Bellamy and Ioris 2017). Taking economic issues into account, they seek to disqualify proposals for change, which are judged to be ideas that are far-fetched or from "a few gurus" suspected of returning to the past and relabelling ancestral practices. Opponents of agroecology mention the efforts already made with respect to the complex situations of farms, and highlight technological proposals that are more compatible with "tomorrow's agriculture" (Bonny 2017). While it is now recognised that the industrial farming model that developed starting in the 1950s enabled agriculture to progress and modernise within a logic of confinement – a laboratory study in which its operation was not verified under real conditions (Aggeri and Hatchuel 2003) –, an increasing amount of research indicates that the organisation of agricultural advising and supply chains as well as the standards associated with them are locking out the transition of agriculture towards other models, in particular agroecology (Vanloqueren and Baret 2009).

These two prescriptive model definitions and the positions of the actors concerned hide a broad diversity of situations within a territory (Therond et al. 2017). In this paper, we hypothesise that the position of numerous actors, and of farmers in particular, remains a hybrid between these two perspectives, in which their involvement in the AET instead takes place through the combination of different exchanges with an evolving social network. To test this hypothesis, we developed a device for analysing the relations between the dynamics at farms and the nature and role of the social networks of the farmers concerned. In this study, we particularly focused on the way that farmers jointly reconfigure their networks and their practices in order to compromise with the uncertainty inherent to the ecologisation of their production system (Girard 2014). To do so, we assume that farmers' knowledge evolves, with the ecologisation of farming specifically implying hybridising empirical/situated knowledge and scientific/generic knowledge (Chevassus-au-Louis 2007; Duru et al. 2015). Therefore, our work follows research on the analysis of knowledge systems and innovation systems in farming (Klerkx et al. 2010) by focusing on the circulation of information between actors via actors' networks.

From this viewpoint, our analysis leads us to a better understanding of the nature of the social interrelations determining the transition (or not) of livestock farmers to agroecology, in terms of both agricultural practices and commercialisation practices. We combined two types of approach. Systemic agronomy allowed us to analyse farmers' trajectories over the long term (Coquil et al. 2013), while the sociology of organised action allowed us to better understand the social interrelations or organisational configurations within which farmers circulated during their trajectory (Crozier and Friedberg 1980). In terms of method, we held semi-structured interviews with a diversity of livestock farmers from the Tarn-Aveyron basin, as well as with the main actors with whom they interacted to design their practices. In this chapter, we present the methodology used and illustrate it by means of the example of two model trajectories of change in the livestock farming practices of farmers in the Tarn-Aveyron basin, and namely a farmer with little engagement in the AET, compared to one who is highly engaged in it. This example shows us how these two trajectories are related to two different types of exchange network. We then zero in our analysis and cross-compare changes in farming practices and changing commercialisation practices. Lastly, we zoom out to a more generic level to draw more general conclusions on the interrelations between actors on the level of the Tarn-Aveyron basin as a territory. To conclude, we discuss these results in light of additional research carried out under the TATA-BOX project.

Methodological Approach Developed

Sampling and Data Collection

In this study our goal was to understand a diversity of positions and roles of actors in the territory of the Tarn-Aveyron basin with respect to the AET. Furthermore, in order not to exclude key elements *a priori*, we adopted a broad interpretation of agroecology. We drew inspiration from the MAAF's[1] political definition of agroecology as a "*set of effective practices on the economic and environmental level*" (Le Foll 2012) that address ecological issues relating to the food system (Francis et al. 2003). This led us to take into account the perception of consumers and citizens, commercialisation systems for food products and, more generally, the social dimension of agroecology (Wezel et al. 2009; Sanderson Bellamy and Ioris 2017). Acknowledging this broad definition of agroecology, and based on information from local partners of the TATA-BOX project (the chamber of agriculture, cooperatives, etc.), we identified a diversity of livestock farmers in terms of both agricultural practices on the one hand, and commercialisation practices on the other (notion of the ecology of the food system).

To analyse the relations between dynamics of practices on farms, and social network dynamics, we created an interview device applied over two consecutive years.

[1] Ministry of Agriculture and Agri-Food and Forest.

This device was deployed in three main stages: (1) identification of a diversity of livestock farmers in terms of agroecological practices (farming and commercialisation practices); (2) identification of the actors belonging to their respective social networks via telephone preliminary interviews; and (3) semi-oriented interviews with all of these actors, which we recorded in order to subsequently replay and analyse using an inductive method (Glaser and Strauss 1967).

In the first year we interviewed five farmers, selected according to the gradient of agroecological farming practices on their farm, and in the following year five farmers, according to a gradient of commercialisation practices (Fig. 1). Through this sampling method, we avoided the classic trap of STSoc (science, technology, and society) approaches, which consists in always focusing on the people who are innovative in their field. Instead, we opened up the field of analysis to a broad diversity of actor positions. During interviews with livestock farmers, we chose to start with the question of the farmers' practices, without assuming what their position with respect to agroecology was, or what constitutes it. We then asked them to tell us the story of their farm to understand the paths taken throughout their trajectory (Coquil et al. 2013; Ryschawy et al. 2013), subsequently asking for clarifications regarding their practices, changes in their values and social networks, and the role of the members of these networks.

To establish the list of social actors to interview, we used a telephone preliminary interview to ask the ten farmers to identify the main actors in collaboration with whom they designed their practices. As a result, we were able to interview 50 actors from the social networks of the ten livestock farmers selected (23 the first year and 27 the second year). These actors were equally likely to be either classic advisory actors or actors from agricultural supply chains, or else neighbours, spouses, or

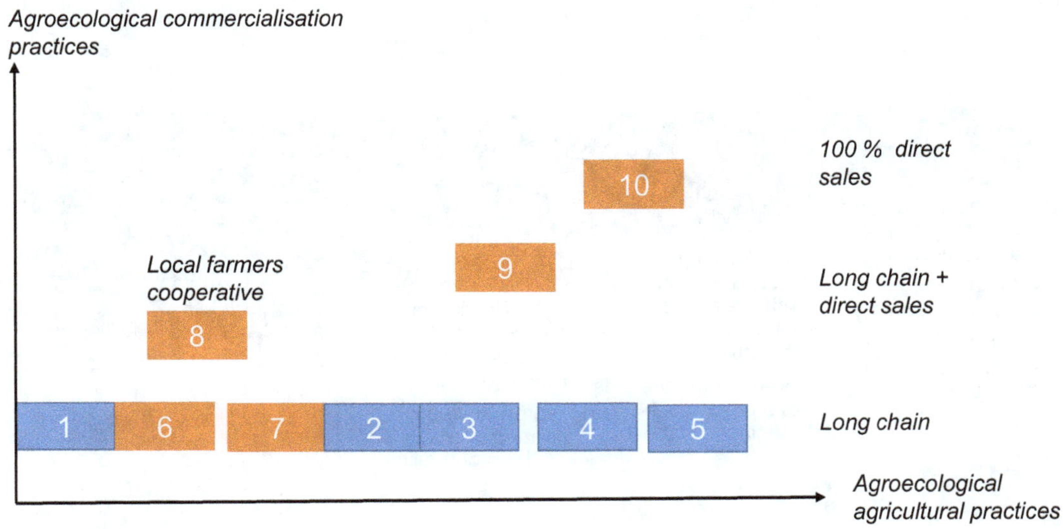

Fig. 1 Positions of the farmers interviewed. In blue are the farmers interviewed during the 1st year (gradient of farming practices); in orange are the farmers interviewed during the 2nd year (gradient of commercialisation practices, based on the number of intermediaries and social proximity between producers and consumers)

other territorial actors (tourism, etc.). We asked these actors to detail their role, their relations with the farmer who had mentioned them, and their relations with other actors in the territory (Box 1). The topic of agroecology was addressed only at the end of the interview, so as not to orientate their statements towards farming issues and their role in this dynamic.

Box 1 An Active Learning Process Involving Master's Students in the Second Half of Their Degree

This research was carried out as part of the "Territorial Engineering" module of the last year of the agronomic engineering degree at the INP-ENSAT specialising in AGREST: agroecology of the production system in the territory. In the form of a one-month PBL (problem-based learning) experience, the module allowed students to analyse the strategies and interrelations of actors underpinning the AETs at work in a territory. The educational utility of this work is to turn students into actors of their training by giving them a problem to resolve and supporting them in this process instead of handing them a theory to apply (Raucent et al. 2016). In this PBL, the theoretical elements involved the theoretical and methodological frameworks and the methods available, but not the topic of the AET and the position of the related actors, which are subjects of the analysis (Therond et al. 2010). To carry out their project, the students were supervised by two territorial agronomy researchers (Olivier Therond, INRA UMR AGIR and Julie Ryschawy, INPT ENSAT) and one researcher in the sociology of organised action (Thomas Debril, INRA UMR AGIR). The viewpoints of these two disciplines were combined to integrally cover the agronomic and socio-economic dimensions of the AET of a territory (cf. Fig. 2).

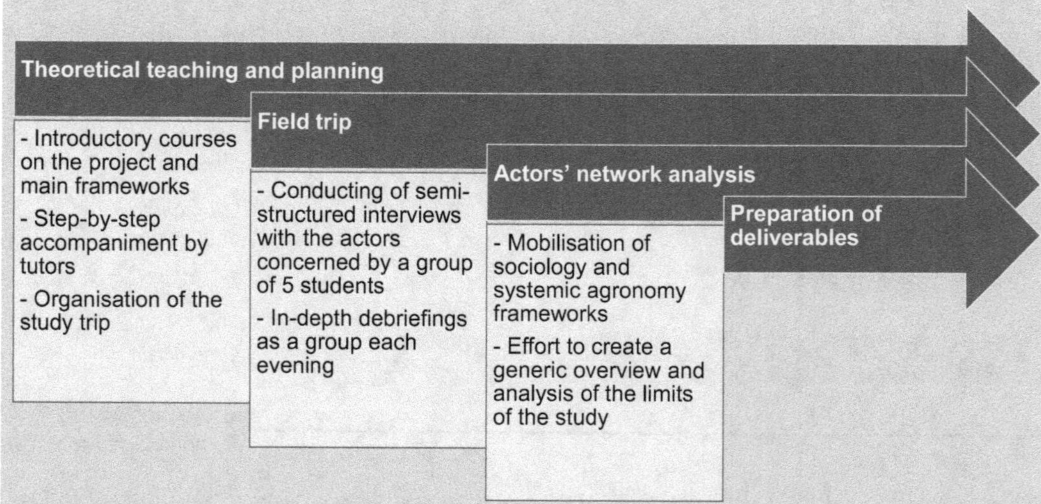

Fig. 2 PBL approach for Master students' active learning

Analysing Farmers' Past Trajectory of Change to Understand Their Transition

Our retrospective interviews allowed us to understand the long-term strategies of farmers and to identify the various "coherence phases" in which the farm had been engaged over time. These were the phases in which the farmer's practices were consistent with his or her values and objectives (Coquil et al. 2013). For each phase, we considered the farmer's objectives and way of thinking. Specific quotes were recorded and linked to technical-economic data on the farm and on the farmer's practices and commercialization system. This methodology has been adapted from the Sociology of Organized Action (Crozier and Friedberg 1980). It allows us to identify how the evolution of the farmer's values leads to changes of varying depth in the production system, resulting in a new coherence phase. We were particularly attentive to elements of the socio-economic context (e.g. new supply chains), the agronomic context (e.g. climate, soil erosion), and the influence of the farmer's social network leading to a transition in the livestock farming system. In particular, meeting with an actor or the dissemination of a piece of information can be key elements in the trajectory of the farmer in question and the path that he or she follows (Ryschawy et al. 2013).

Analysing the Role of the Actors' Networks in the Transition Towards Agroecological Practices

We used the framework of the sociology of organised action to analyse the interrelations between actors surrounding each livestock farmer and how each actor influences the farmer in his or her values, perspective of the livestock farming system, and/or choice of practices. To understand the role of the actors' networks, we first analysed the relationships that each farmer had with the actors interviewed, in terms of level of interaction (from once a year to daily), type of interaction (top-down expertise or knowledge exchange) and type of relationships (affinity or conflict). We then built on the same approach to analyse the relationships between actors themselves, and draw conclusions on the broader local network through a stakeholder analysis. Here we considered the involvement of the local actors, for or against agroecology, and the level of each actor's importance in local farmers' decisions and transitions in their practices.

Results: Actors' Networks as Obstacles or Levers to the Agroecological Transition

To present our results on the role of actors' networks in farmers' AET, we first present the farmers' trajectories and their influence on the evolution of these actors' networks, considering two extreme case studies: a farm that is not at all agroecological, and one that is highly agroecological. We then consider the broader analysis on the Tarn-Aveyron Basin, highlighting that some actors are basically "central" and unavoidable for most farmers. In particular, the actors involved in the commercialisation of inputs and products play a key role. We then focus on the other actors that are "peripheral", that is, who are not involved in the network of all the livestock farmers studied, but who have a major influence on their transition or not towards agroecology. These actors may be part of the agricultural sector, including researchers and farmers, but are not necessarily so.

Trajectories of Change and Individual Reconfiguration of the Network

Of the ten farmers interviewed, in this paper we have chosen to present the in-depth analysis of one farmer not engaged in the AET (called Mr. CONV) and his network, and then to compare this analysis to that of the trajectory and network of a farmer heavily engaged in the AET (called Mr. AE, for agroecology).

Configuration I – The Case of Mr. CONV: Agroecology Seen Through the Conventional Lens

Increasing the Coherence of a Model Integrated Throughout the Trajectory
The analysis of the trajectory of change of Mr. CONV (Farmer 1 in Fig. 1), a farmer not engaged in the AET (Fig. 3), shows how his embeddedness within the incumbent sociotechnical regime drives him to continuously and increasingly reinforce a highly segmented innovation logic.

Following the retirement of his parents, who had previously been his business partners, Mr. CONV continued farming with Prim'Holstein dairy cows, managing on his own a herd of 45 lactating animals on 72 ha, with a quota of approximately 300,000 litres of milk. According to certain farming technicians, "*his system [was] stable and produce[d] good-quality milk*". At the time, Mr. CONV's goal was to operate based on a logic of maximising milk production, typical of the "Colbertist" integrated system (Chevassus-au-Louis 2007; Girard 2014). He farmed maize and straw cereals (wheat and barley) to complete the fodder ration (which advisers also considered to be of good quality). Things really took off when he started to organise the arrival on the farm of his son, who had previously been involved only occasionally, during his studies. At the time, the innovation logic retained for the farm's future was to increase the volume of milk: this was followed by an increase in the herd to 60 cows. Therefore, his main goal would be to optimise his production tool to achieve the least expensive milk production possible and therefore to be competitive on large markets. Mr. CONV was therefore operating on the basis of a "technicist" logic advanced by the dominant sociotechnical regime (Plumecocq et al. 2018).

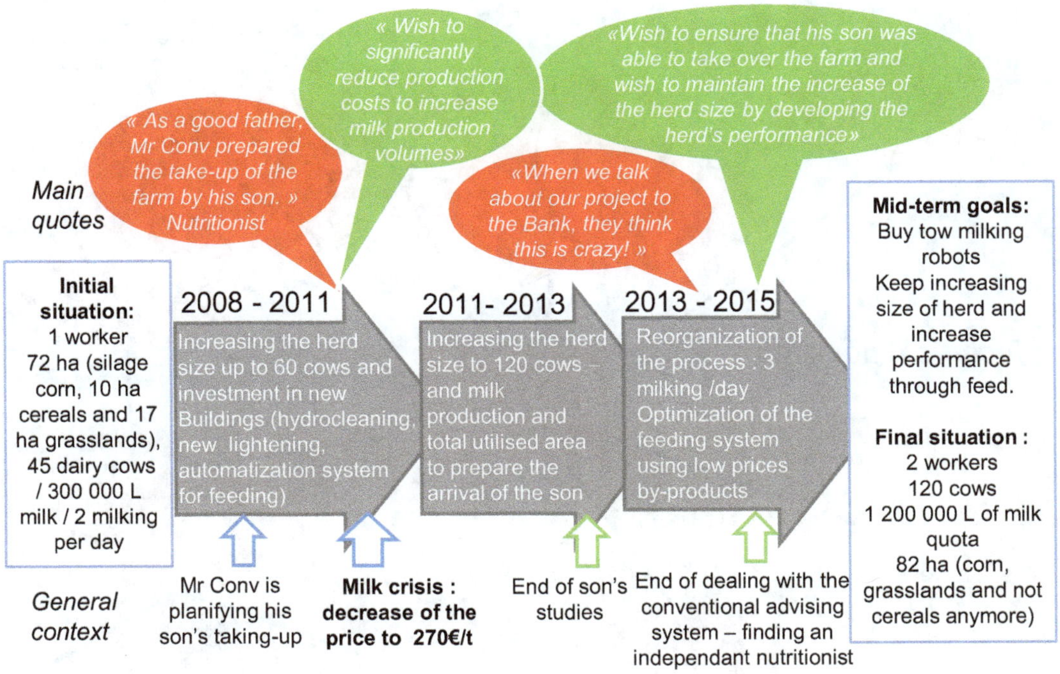

Fig. 3 Mr. AE trajectory diagram along time identifying the main coherence phases in his system practices and values and main quotes illustrating his way of thinking. Main important fact influencing his choices are represented below the arrow – adapted from Coquil et al. (2013)

In 2013 his son graduated and started working with his father full-time. Starting in 2015, the construction of a high-tech building and the installation of milking robots – both of which entailed huge costs – were hindered by accidents and mistakes by the workers. These choices led to a delicate financial situation which could have caused the system to go bankrupt or triggered a change of logic. This marked a new stage characterised by a significant reorganisation of the work, as they implemented a third daily milking round in order to produce more milk per day and to pay off their investments. This reorganisation maintained the objective of optimising milk production, as Mr. CONV emphasises: "*The investment is done, so now to pay off the expenses!*". This was a strategic decision that at the time allowed him to produce 10–15% more milk, that is, 400,000 litres of milk.

Mr. CONV prefers to purchase proteins and cereals rather than producing them on his lands, and does not seek nutritional autonomy that would limit his production levels. In parallel, the management of farmed areas, and in particular maize and cereal ensilage, are entrusted to an independent contractor: "*Today, the less we work on the land, the better things go*". While the volume of milk has effectively increased and expenses have somewhat decreased, the farm is still subject to heavy debt payments. This handicap makes the bank reluctant to provide the new loans necessary for establishing his son and implementing his plans. According to him, "*problem number one today is the banks*". The success of this undertaking, that many consider to be highly ambitious, depends on this lock-in. Even so, Mr. CONV and his son appear sure of their goals and are working towards them; they do not want to hear about agroecology, and consider "*that today, the priority is to provide for everyone*".

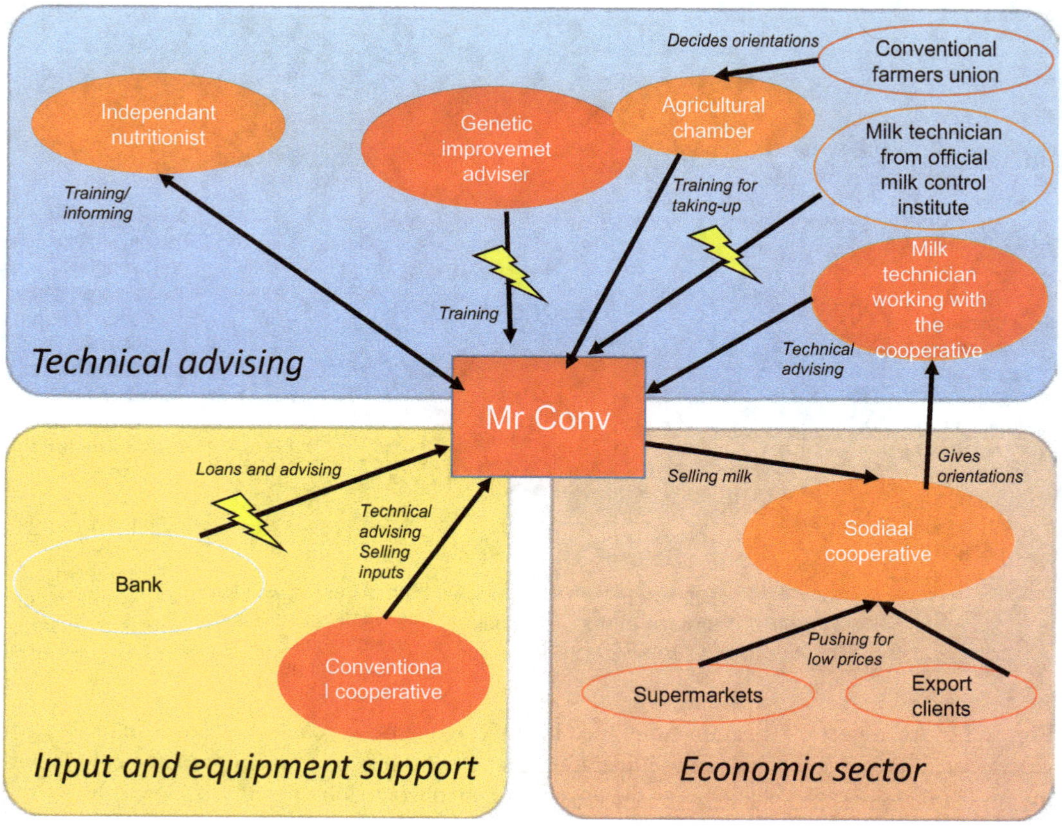

Fig. 4 Mr. CONV actors' network. Main actors are represented here by type, in plain colour are the actors that were met by the student. In green if supporting agroecology, in orange if intermediate and red if not. The storm signs are indicating a conflict and an ending of the discussions

Analysis of the Network of Mr. CONV and His Son: A Top-Down Network to Enable Real Technical Optimisation of the Dairy Workshop

These changes in Mr. CONV's trajectory resulted in an evolution in his network of actors, which progressively became more coherent with his innovation logic and his segmented-by-workshop perspective of his system (Fig. 4). By increasing milk production, Mr. CONV is following the same logic as one of his advisers, financed by his cooperative to help optimise production: "*It's the volume of milk that enables you to pay the bills*". This logic echoes that of the Sodiaal cooperative, to which he delivers his milk, because it transforms most of the milk that it collects into dairy products that it sells at purchasing hubs. These hubs are particularly sensitive to prices and largely determine the choices of local livestock farmers, although they do not necessarily prioritise local markets. Based on this "volume logic", Mr. CONV applies a strategy that is highly segmented by workshop, and to do so, surrounds himself with dairy production advisers, more or less automatically applying their logic as recommended by an "expert" council. He describes their advice above all as "technical". In addition, the genetic selection expert emphasises that "*the breed of dairy cows is designed to produce inexpensive milk in large quantities*". All the advisers agree on the fact that agroecology does not appear relevant for overcoming challenges in the global food supply, stressing that "*[i]n the majority of cases, it's more worthwhile to buy a bit of soy cake so as to produce more milk with the cows*". This logic is also apparent in the purchasing of high-quality feed and the establishment of very specific rations to optimise production levels per cow and to limit waste. In July 2014, Mr. CONV delved even deeper into this logic: he stopped using the technical support of the dairy management agency, which he considered not to be sufficiently effective, and sought out the services of an independent nutritionist outside of the typical conventional network. Even more technical than the others, the nutritionist noted a key element in the trust that Mr. CONV put in him: "*I can establish rations practically to the gram of digestible protein*", thus helping him to determine optimised and therefore less expensive rations based on co-products.

Mr. CONV or the Limits of "Conventional" Advisory Services of the Dominant Sociotechnical Model

Mr. CONV's dynamic is supported and strengthened by the other actors. His network is driving him to continue even further down the path of the technicist paradigm and to completely exclude any AET. Comments by the adviser from the cooperative ("*if somebody tells you about feed self-sufficiency, they're out of touch with reality! It's just a big fad.*") and the independent nutritionist ("*if agroecology means planting three trees in the middle of a cereal field, it's a joke*") attest to this. This prescriptive approach creates value assessment devices that rank practices and clearly depict "technical" progress as being the optimisation of productivity associated with technological innovation (Plumecocq et al. 2018). For example, Mr. CONV emphasises that his independent nutritionist "is *a part of the networks where it feels like people want to make progress*". These cognitive and normative frameworks carry weight in knowledge and influence the individual practices of livestock farmers, simultaneously playing a role in the reproduction of the norms of the dominant sociotechnical regime (van der Ploeg et al. 2009; Klerkx et al. 2010). Yet Mr. CONV no longer truly trusts the "experts" of the conventional model, because they do not go far enough. He stresses moreover that advisers "*are incapable of leading groups*" and that "*[i]f all of the dairy management agencies in France were highly competent, our job wouldn't exist*". As Chiffoleau (2009) points out, the analysis of Mr. CONV's network allows us to see who the real "experts" are. He nevertheless seeks out other sources, in this case the independent nutritionist with whom he has developed a horizontal relationship, which tends to be more usual in a agroecological model in the sense of Altieri et al. (2017).

Configuration II: The Case of Mr. AE: Agroecological Intensification as a Form of Hybridisation

Mr. AE is strongly engaged in the AET in terms of farming practices (Farmer 4 in Fig. 1). The Fig. 5 allows us to see how he is changing the actors in his network, along with his transition towards agroecology. He will limit involvement with actors that prevent him from transitioning towards agroecology, and bring in new actors that will facilitate the transition.

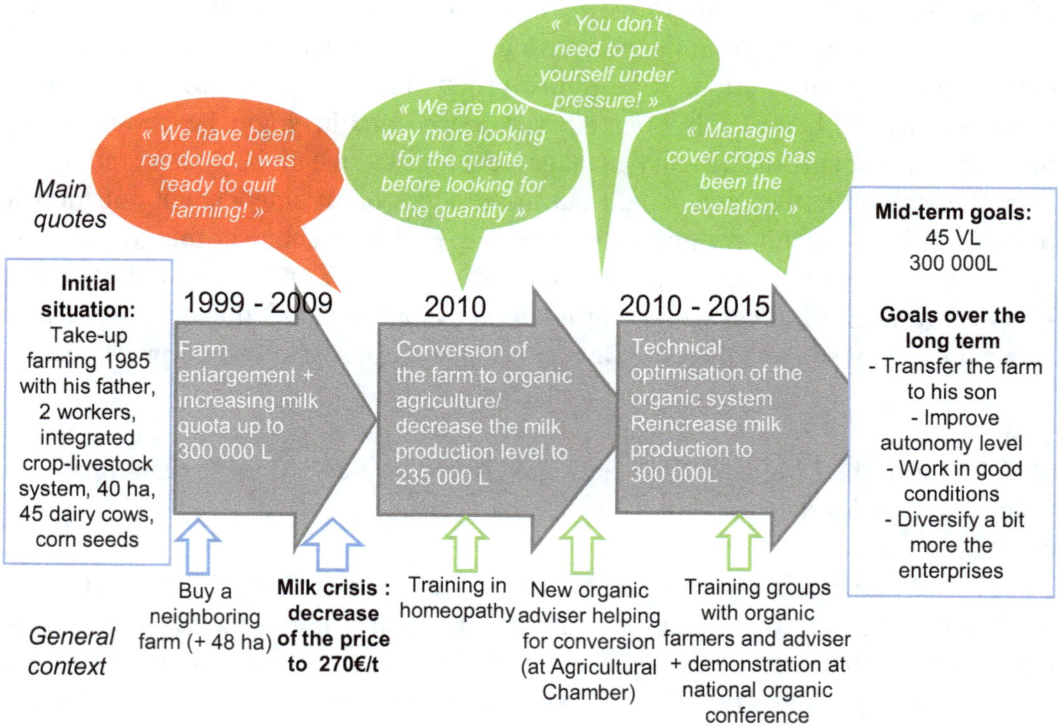

Fig. 5 Mr. AE trajectory diagram along time identifying the main coherence phases in his system practices and values and main quotes illustrating his way of thinking. Main important facts influencing his choices are represented below the arrow. – adapted from Coquil et al. (2013)

The Coherence of an Agroecological Model: A Trajectory Towards Technical and Decision-Making Autonomy

Mr. AE is a member of the *Groupe agricole d'exploitation en commun* (GAEC, joint agricultural group) established by his father in 1985. The 40-ha farm of his UAA produced 300,000 litres of milk with 45 Prim'Holstein dairy cows. The system was managed conventionally, with a large proportion of irrigated maize, among other crops. In 1999, Mr. AE purchased a neighbouring farm and thus increased his UAA to 88 ha. In 2005, following his father's retirement, he tried to team up with someone in the GAEC with whom he had no family connection, but this person left the farm after 6 months. Under pressure from the MSA (the agricultural social mutual society), Mr. AE chose to switch to the legal status of an EARL. In 2009, he gave it "one last shot" after losing €18,000 when the price of milk dropped to €270/t and cereals to €100/t, despite the heavy workload its production entailed. As he explains today, "*I was considering stopping everything because I'd run into a dead end*", and started to sell some of his livestock. After consulting with the organic agriculture adviser at the Chamber of Agriculture, he chose to embark on organic production, and explained the new logic underpinning his change of approach: "*We're going to be much more focused on quality before looking for quantity*".

On 1 November 2010, he started the conversion of his livestock and crop farming. The first delivery of organic milk was on 1 January 2013, but his crops were still sold conventionally at the time. As he progressed through the stages of his conversion to organic, Mr. AE took various training programs (homeopathy in 2005, management of plant coverage and intercropping in 2010, bioindicator plants in 2013, large-scale organic cropping in 2014, artificial insemination planned for 2015) and joined the organic dairy farmers' association of Aveyron. All of this reflects the significant change in his reasoning: "*When you're totally goal-oriented and see nothing else, you're not going to go take a class*". Since the conversion and thanks to this training, he is gradually trying to improve his system by implementing new tests, such as a recent combination of maize/soy to ensile, to ensure his protein autonomy over the long term: "*It takes five years to get the same yields as conventional farms. You can't put pressure on yourself. The limit is ourselves; there's a technical change taking place. You can't become good in every respect overnight*".

At one stage he attempted to cross a Brown Swiss bull with his Prim'Holstein cows to improve his milk production (protein and fat content) before switching to insemination, which he carries out himself. His intention is to be able to pass down the farm (hopefully to his son) and to have good working conditions, by diversifying the crops a little more. His way of going about things fits quite well with agroecology: organic production combined with simplified cropping techniques (SCT), permanent ground coverage (plant coverage and intermediate crops, seeding under coverage), and good rotation management, all the while seeking increasingly advanced autonomy. Ploughing is however still necessary to turn over coverage without using glyphosate. The major change in his innovation logic is the fact of having switched to a systematic perspective encompassing both production workshops – plant and dairy – in conjunction with one another. For example, he considers livestock farming as a means of ensuring an outlet for his plant production, even when there is a problem with them: "*whether it works or not, if it doesn't work, we ensile it!*". Feed self-sufficiency has become a key objective, because it is a way of reducing production costs by shielding oneself from market prices, which are very high and variable from one year to another: "*in organic, it stays regular, and that's really nice*". Limiting investments limits the financial risks that he had faced in the past: "*I won't say that we make a lot more money, but we spend a lot less, so when things go wrong it's a lot less serious*".

Analysis of Mr. AE's Network: Horizontalisation of Practices But no Changes in Terms of Commercialisation

Mr. AE's trajectory is marked by increasing embeddedness in peer learning networks, which has allowed him to acquire the knowledge to implement agroecological innovations (Fig. 5). The conversion to organic has allowed him to change his perspective, and to produce less while stabilising his income. In terms of production, the top-down recipe of mechanisation is not the only approach; there is also exchange between peers. Regarding his conservation agriculture network, Mr. AE explains that "*[i]t's a great technique; those guys are really passionate*". However, this new network is not his only source of innovation. Mr. AE also innovates in close collaboration with his adviser at the Chamber of Agriculture, who set up the group of livestock farmers with whom he is working to achieve more feed self-sufficiency and to prepare his organic transition. This adviser, despite being an employee of a structure belonging to the dominant sociotechnical regime, has a perspective that is based firmly on knowledge exchange between peers: "*Our philosophy is to make farming autonomous*" or "*Among organic or SCT farmers, we find the same state of mind, that they want to try out other things*". For example, Mr. AE is working towards autonomy and is substantially reducing his number of suppliers: "*In terms of suppliers, they've been reduced significantly. In reality, we're a lot more autonomous*". Mr. AE has thus switched to a systemic perspective, a long-term innovation logic, and a rather negative view of the conventional farming network of actors of the dominant sociotechnical regime: "*It's true that switching to organic makes you realise that there's something wrong with the system. You let yourself be had and you didn't even realise it*". Despite everything, Mr. AE remains highly critical of the new "agroecological" practices: "*Simplified cropping techniques, yeah, but the materials haven't been simplified…*".

In terms of commercialisation, Mr. AE has not modified his system and actors' network (Fig. 6): he took advantage of the opportunity for a new organic milk market offered by his cooperative, Sodiaal (the same one as Mr. CONV). The milk produced will be transformed into powder at a factory in Montauban and shipped to China.

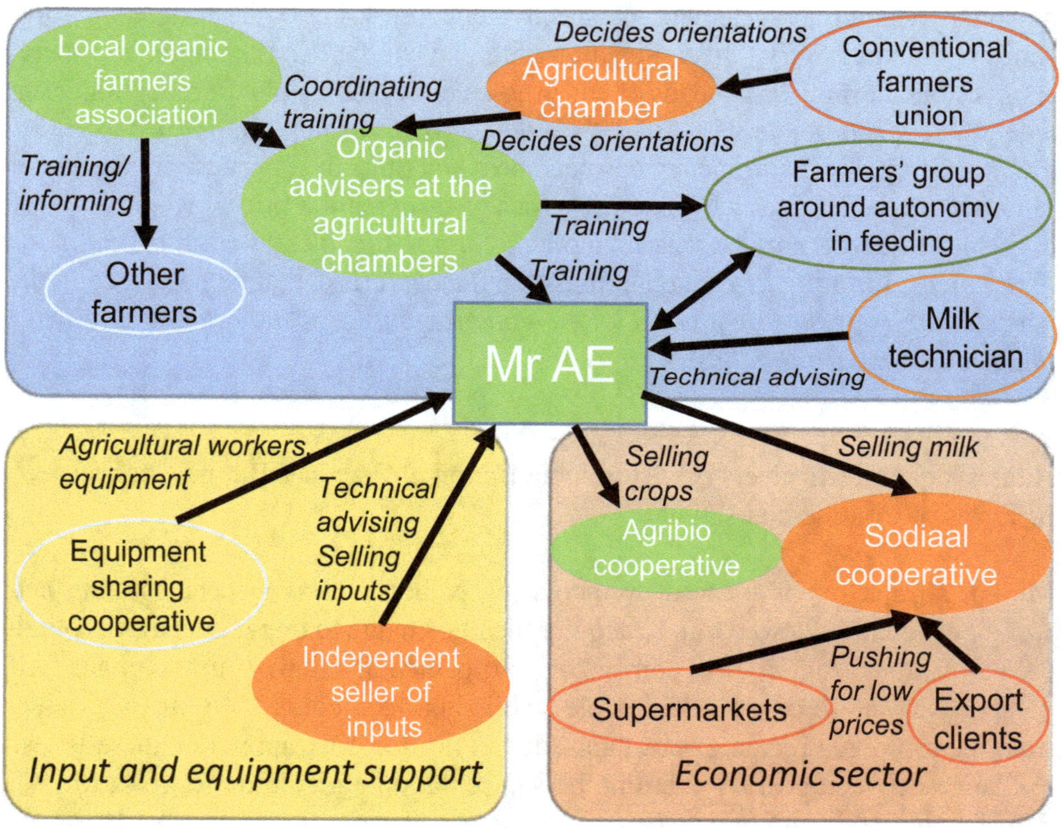

Fig. 6 Mr. AE actors' network. Main actors are represented here by type, in plain colour are the actors that were met by the student. In green if supporting agroecology, in orange if intermediate and red if not

Mr. AE: A Hybrid Farmer Who Moderates Two Caricatures of Agroecology

During his transition, Mr. AE needed to acquire new knowledge, especially on grass management. As also emphasised by Farmer 6, "*it's not obvious; you have to relearn how to manage grass*". To do so, Mr. AE adopted a knowledge-sharing and peer-knowledge exchange logic. The idea was to produce the "horizontal" knowledge recognised as being necessary for the development of agroecology, as the adviser from the Chamber of Agriculture pointed out: "*We try to address farmers' worries*"; "*For it to work, you have to really listen to farmers*"; "*You have to keep an open ear and make the best of opportunities*". The fact that this adviser belongs to a prominent structure in the dominant sociotechnical system does not ultimately prevent him from becoming a part of this horizontal dynamic of knowledge production and innovations. Ultimately, Mr. AE's trajectory allows us to nuance a caricatural representation that often sets agroecology and modernity against each other. Mr. AE remains a "technical" farmer in the meaning ascribed by the dominant system, all the while adopting the principles and practices of agroecology. This is supported by Bonny's argument (2017) that technology should not systematically be seen as an opposite of agroecology. By using the example of this farmer, we demonstrate that technological progress can be a tool that contributes to agroecology, in its definition as combinations of practices that are useful in promoting productivity and respect for the environment. In this sense, agroecology can be seen as a "modern" concept in which nature is used to contribute to the needs of human beings with two clearly separate categories, in the sense of Latour (2006).

The agroecology implemented on Mr. AE's farm appears to largely follow a productivist logic in terms of markets. Specifically, Mr. AE's commercialisation practices relate to the opening of a new market for organic powdered milk in China by the SODIAAL cooperative. The support of farmers in their organic conversion by the Chamber of Agriculture of Aveyron in order to supply this market plays a role in intensifying production that is commercialised by conventional actors. However, Mr. AE mentions "*that [he] would prefer to sell on short supply chains, but the excessively low demand forces [him] to stay on long chains*". It has turned out to be simpler to retain this historical farm model with collection by the local cooperative. Therefore, contrary to many ideas and as Therond et al. (2017) have emphasised, agroecological farming practices do not necessarily go hand-in-hand with short supply chains, and reciprocally, as we will show through the five farmers studied, are located along a gradient of commercialisation practices.

Agroecological Practices and Food Systems: Zooming in on the Case of Commercialisation Practices

The second year of our study allowed us to explore the supposedly classic link between agroecological farming practices and agroecological commercialisation practices. We found that both the actors supporting farmers in the commercialisation process, and consumers, allowed farmers to move away from the highly divisive notion of the "technical" and to open a much broader field of action for change. With farmers 8, 9, and 10, consumers or actors in direct sale have become new intermediaries, as emphasised by one farmer, a friend of farmer 9, who has his own cutting plant and is a member of a producer store ("*you have to bring quality products to consumers. There's no point in producing just to produce*") or livestock farmer 10, who sells directly, at her farm or at markets ("*contact with the consumer is a good way to learn*"). These examples clearly demonstrate a broadening of the perspective of the system and of the role of agricultural production in relation to the requests and expectations of consumers. In this way, these interactions between farmers and consumers can lead farmers to move beyond a production system-centred approach, to instead adopt a more all-encompassing consideration of food system issues (Francis et al. 2003; Plumecocq et al. 2018).

Concerning actors of the incumbent sociotechnical regime, Bonneuil and Joly (2013) discuss the neoliberal knowledge production regime. This converges with the ideas of Vanloqueren and Barret (2009), who argue that science in the way that it is currently conducted – in other words, strongly marked by hypothetical and deductive elements, technical standards, and optimisation goals – is locking out the AET. Given that the work of advising and development actors is also underpinned by this logic, it follows that other "niche" actors (according to Schot and Geels (2007)) would be necessary to make the dominant regime evolve.

We therefore found that conversely to Mr. AE's strategy, another strategy for producers was to establish a small cooperative (30 livestock farmers) on the local scale (three *Départements*; FADN NUTS III, http://ec.europa.eu/agriculture/rica/) to once again take charge of milk commercialisation and price setting modalities, in quest of greater stability and fairer financial compensation. As highlighted by livestock farmer 8, a member of the cooperative, "*[w]e trust consumers to choose the right product*".

This involves the creation of a niche market by making use of the image of a local product sold by livestock farmers themselves or at supermarkets where they carry out demonstrations to explain their method. These demonstrations foster trust, as indicated by a manager: "*Going even further than that would mean getting intimate with people. If the calves have received medals, the cows are good, and the farm is clean, I trust that*". The quality of the product offered is also an essential point in this strategy and is backed by a set of technical specifications shared by all the producers. Because the specifications are not very restrictive, certain livestock farmers give priority to practices that are similar to the organic specifications (grazing, no antibiotic administration, autonomous feeding at the farm, etc.), but the majority remain very close to conventional livestock farming, with one noteworthy exception, as mentioned by the president of the cooperative: "*There's something else that could be put on the packages as well [other than cows] and that nobody includes, even those that could include it: that it's 'GMO-free'*". Contrary to what could be expected from an agroecological method at a human-scale cooperative, there is no exchange between producers around agricultural practices, as emphasised by farmer 8: "*no, it's true that we don't visit each other's farms*".

Another strategy observed consists of transforming milk oneself and commercialising production via short supply chain networks. The logic in this case is no longer to offer an inexpensive product to consumers, but to target consumer satisfaction through a local quality product and social interaction (Therond et al. 2017; Plumecocq et al. 2018). Farmer 10 is an example of 100% direct sale commercialisation: "*when a customer tells me that they liked it, that it was good, that's how they pay me*", in other words, their recognition and satisfaction are more important than money. Value is extracted from milk volumes and quality by transforming the milk into cheese or yoghurt. As a sales adviser at the Aveyron Chamber of Agriculture pointed out, "*overall, they find each other. It makes it more work for them, but they're aware that they're not forced to use long supply chains. And they're often happy people, joyful people, entrepreneurs, creators*". Reducing the number of intermediaries assures a higher price set by the producers themselves. Yet this process implies know-how and the resulting additional time working, that can be included in the final price of products. Processing is not necessarily an easy stage to carry out, but the Chamber of Agriculture is organised to support this type of strategy through advisory and training services, and according to it, this support goes far beyond technical aspects: "*Behind it, I involve people, a pathway, problems, solutions to the problems… ultimately, I put a whole story behind it all*". We observed that in this type of direct commercialisation strategy, some of the milk is often not transformed, and remains sold on long commercialisation supply chains, which enables a compromise providing security as opposed to absolute dissociation from large-scale dairy corporations (Therond et al. 2017). Ultimately, the risk of this type of approach resides in exclusion from the local agricultural network, as a colleague of farmer 4 pointed out ("*we were quickly marginalised as soon as we set off in that direction*"), even though the members of the network of producers sharing these direct sale tendencies do support one another ("*there's a lot of mutual help, fortunately, otherwise we wouldn't make it*").

The Influence of Actor Interrelations on the Agroecological Transition in the Tarn-Aveyron Basin

The main conclusions presented above, based on our cross-analysis of the ten livestock farmers retained and their networks, allowed us to construct a stakeholder analysis of the role of actors in the AET (Fig. 7). This approach allows us to consider the actors interviewed as regards their involvement in favour of agroecology and their influence on local farmers. In Fig. 7, the actors in favour of agroecology are highlighted in green, whereas those against it are highlighted in red.

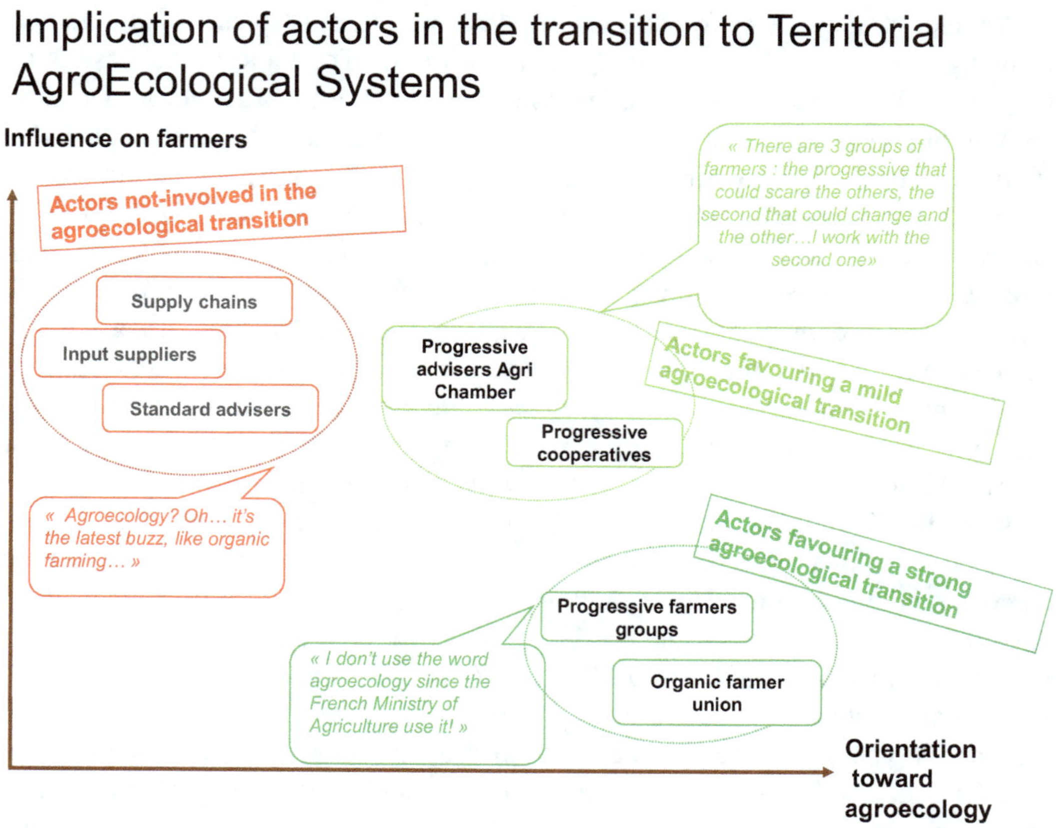

Fig. 7 Stakeholder analysis of the role of actors in the AE transition

Central Actors Are Difficult to Avoid and Not Always in Favour of the AET

Another representation of the network allowed us to consider the level and type of interactions and the types of relationships between all actors (Fig. 8). Applying this framework to the five farmers on a gradient of agricultural practices, we noted that some actors were "central", that is, difficult to avoid as they were in contact with all the farmers interviewed. The actors in the pink circle are considered to be "central" actors with whom all farmers interact for purchasing inputs or commercialisation. The actors in the yellow circle are "peripheral" actors who are specific to each farmer, depending on his or her personal stance with regard to agroecology. This helped us to understand the contrasting perspectives on agroecology and the actions linked to them. For instance, Mr. CONV (Farmer 1) is interacting only with actors in red, as "peripheral actors", whereas Mr. AE (Farmer 4) is interacting with more green actors who are in favour of agroecology, as "peripheral" actors, even though he is still connecting to red central actors through his commercialisation practices. We found that "central" actors play a key role in farmers' decisions, even if they are not necessarily in favour of agroecology. We illustrate this specifically for each central actor highlighted.

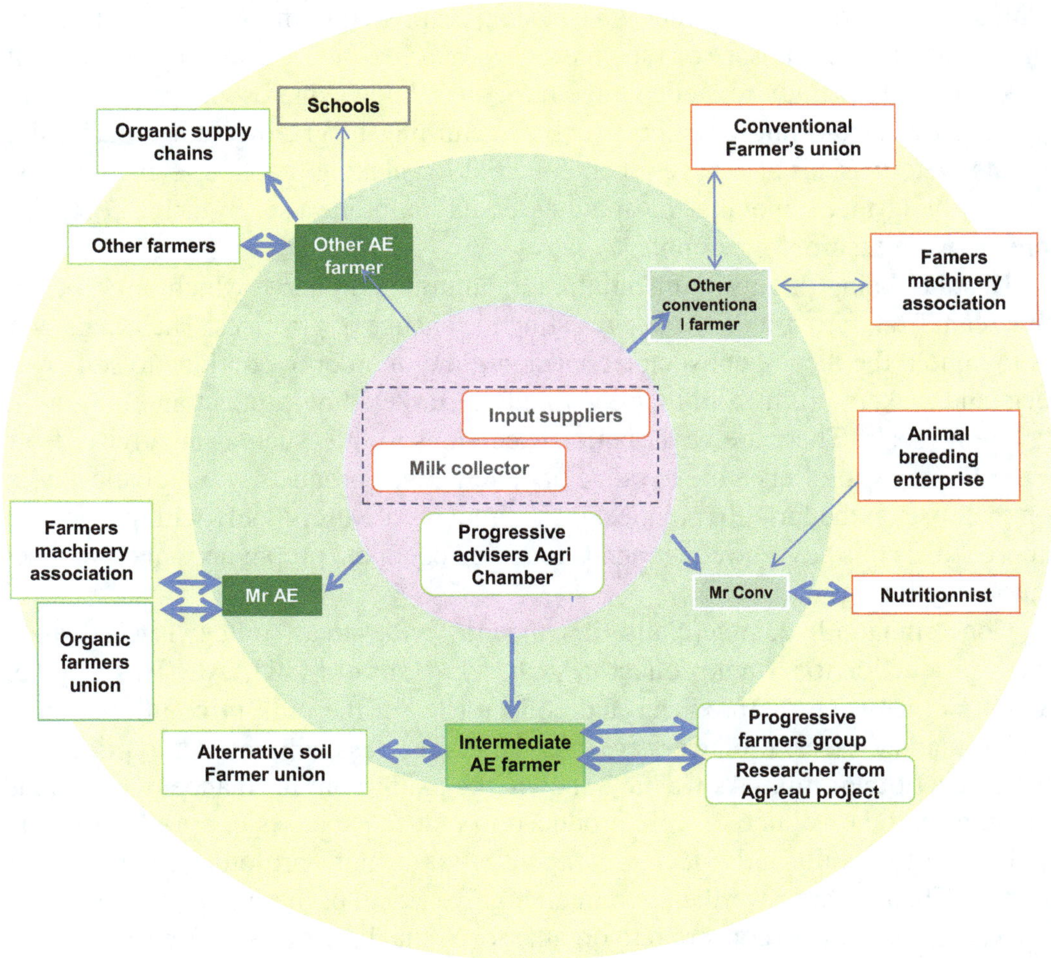

Fig. 8 Local actors' networks of the farmers interviewed, along a gradient of agricultural practices. The farmers studied are represented in the green circle. The light green areas represent a low level of involvement in agroecology and the dark green areas a strong one. The actors in favour of agroecology are indicated in green, whereas those against it are indicated in red. The actors in the pink circle are considered to be "central" actors, with whom all farmers interact when it comes to purchasing inputs or commercialisation. The actors in the yellow circle are "peripheral" actors, who are specific to each farmer, depending on his or her personal stance towards agroecology. This figure is not intended to be exhaustive; it presents a summary of the results of the five interviews carried out with the dairy farmers and the 27 interviews carried out with the main actors in their professional network within the Tarn-Aveyron territory. Only the major relationships which the farmers claimed had played a role in the adoption of new practices are represented here; other relationships may exist but are not considered central in farmers' decisions. The network analysis revealed the points of divergence between the different types of farmers, which are partially detailed below

This broader analysis based on the interviews with the ten farmers and their network shows that all of the farmers studied have varying degrees of contact with the Chamber of Agriculture, agricultural suppliers, and farmers' organisations. These three types of actors have a strong influence on the operation of these farms, because they are linked to a large number of farmers, who are relatively attentive to the advice given or who easily become involved in these structures. Therefore, the

Chamber of Agriculture currently has a significant weight in the AET, particularly in the case of the advisers of the *mission agriculture biologique* (organic farming task force). These advisers encourage horizontal knowledge exchange groups and agroecological farming practices, such as returning to grazing rather than using inputs. Veterinarians also have much potential for making practices evolve or even transforming them, moving towards decreased antibiotic use, by using food as a preventive measure, for example.

Despite being highly influential, agricultural suppliers, which are located upstream from farms, appear to be resisting the change of practices. Their main goal is to uphold the incumbent sociotechnical regime in order to continue to sell products and make a profit from them, with advice targeted by product and not on the "system" scale. There are nevertheless exceptions to this, such as an adviser from Euro Phyto, who offers a broad range of "alternative" products with holistic advice on their use on the farm. In practice, these products are useful only within a holistic approach, as this adviser recommends reducing chemical prophylaxis and promises autonomy for farms.

Concerning milk commercialisation, the most widespread strategy is to sell all of one's production to a single collector, such as Lactalis or SODIAAL. This involves a contract between the producer and collector to set the milk price in relation to global market prices. Producers are thus left defenceless with respect to prices and the future of their milk. As we have shown, the main room for manoeuvre is found in increasing the volume of milk produced to reduce expenses per production unit and/or try to significantly decrease expenses via a more profound change in the system. The farmers nevertheless remain highly critical of these large corporations, such as farmer 7, who converted to organic for Sodiaal: "*because there's the farmer and then there's the vultures. You can't have a conversation with the people at Lactalis. It's a multinational; it's a really particular mindset*"; or a livestock farmer who commercialises only on short supply chains: "*Sodiaal is only a cooperative in name*"; "*they're not interested in little niches*".

Actors Called "Peripheral" Yet Essential in Changing Practices

Figure 8 shows that "peripheral" actors may favour agroecology even if they are not in contact with a large majority of farmers locally. The farmers engaged in an AET seek out alternative advising actors, such as CIVAM (rural environment and farming development initiative centres), as well as exchange between peers via farmers' associations, as one of the farmers noted: "*when you stay in your bubble, you always think you're the best, and when you step out of it, you tell yourself, 'oh, that's not working,' and that opens up your mind a bit*". Exchange between peers also takes place informally by observing trial and error at neighbours' farms, which is essential for convincing people: "*I think that my neighbours are going to watch me, to see if it works, and then if it works, they'll change*" (cf. Box 2). These new exchanges are essential in limiting the isolation phenomenon, as a conventional farming technician pointed out: "*you feel isolated when you do direct seeding farming. It's not*

Box 2 The SERACC Network (*Services de Régulation en Agricultures de Conservation et Conventionnelle*) as an Example of Knowledge Exchange Between Scientists and Farmers

From 2013 to 2016, INRA Toulouse ran a PhD research project based on scientific and empirical knowledge exchange. In contact with farmers' associations that had undertaken empirical experiments and knowledge transfer on conservation agriculture, we identified specific needs from scientific research that could complement farmers' knowledge. Ecosystem functions and subsequent services that could benefit farmers were strongly acknowledged by producers engaged in conservation agriculture, but they lacked the tools and ecological expertise to assess the impact of their practices on them. Such agroecological experiments are moreover ill-suited to classical experimental platforms (often with short-term oligo-factorial experimental design) that allow for an exhaustive scientific comprehension of some of the processes involved, but which are far from farmers' expectations related to multifactorial and local features. It was thus decided that the project would be designed for farmers and with farmers. Fifty-four farmers engaged in the project, forming a network later called the SERACC network, each dedicated one of their own fields (1–1.5 ha) to the study for two growing seasons. Thirty-five of them were members of associations (21 from *Sol et Eau en Ségala*, 4 from *Association Occitane de Conservation des sols*, 2 from *Groupement des Agriculteurs Bio du Gers*, 5 from *Agro d'Oc* and 3 from *Groupement des Agriculteurs de la Gascogne Toulousaine*) with a gradual involvement in conservation and/or organic agricultures, while the remaining 19 were neighbouring farmers with more classical practices with regard to tillage and the non-use of cover crops or diversified rotation, for instance. To benefit from this wide diversity of systems, most of the decisions concerning cropping practices on the experimental field were left to the farmer, yet were closely monitored, and only a few restrictions were requested for the purposes of the experiment (crop cultivars and seeds' origin were identical for all farms and non-organic farmers had to leave an untreated area in their fields). Such design benefited from farmers' experience and knowledge, as well as ecological equilibria that can only be achieved in systems implemented in the long run. For farmers, this design allowed them: (i) to be actively involved in a research program; (ii) to have direct feedback from science with data explicitly related to their farm and practices; and (iii) to have access to comparative data from local farms with contrasting practices.

the most popular trend". However, it is important to note that even though these farmers seek new sources of knowledge and experience, they are never completely dissociated from central actors, in particular those in commercialisation. In other words, the networks of these farmers are hybrid: they are based on actors in the dominant sociotechnical system as well as those of innovation niches.

As we saw by illustrating the trajectories of farmers 7–10, the system can be unlocked by listening to actors other than those in the agricultural sector in the strict sense. They can be the actors supporting these agricultural actors in order to make their commercialisation practices evolve, but also actors in tourism (restaurateurs, cutlers, etc.), regional nature reserves or numerous environmental associations, or consumers outside of the agricultural environment (Box 3 –Beudou et al. 2017).

Box 3 Cultural and Territorial Vitality Services Play a Key Role in Livestock Agroecological Transition in France

In France, researchers and public policy makers are calling for the agroecological transition of livestock farming. This transition is facing technical, economic, social, and cultural obstacles. Whereas technical obstacles are studied extensively, other categories are receiving very little attention despite their potential role in this respect. This article analyses the livestock cultural and territorial vitality (dis)services (or negative impacts) perceived by local actors on two distinct French territories and understand how these services could act as levers for the AET of livestock. To do so, we interviewed 45 local actors from the livestock sector and local rural development in two French territories: Aubrac (24) and Pays de Rennes (21). We considered mainly farmers, advisors and supply chain actors, but also granted specific importance to local actors not in the agricultural sector (tourism, environment, gastronomy). We conducted inductive content analyses to draw on interviewees' perceptions and to link the cultural and territorial vitality services identified, to the AET of livestock.

Our work revealed 20 cultural and territorial vitality services, including the nurturing of social bonds and the creation of rural jobs, that can be organized into 11 categories (seven categories of cultural services and four categories of territorial vitality services). Among the 11 cultural services, cultural landscapes linked to livestock and gastronomy heritage were the most cited. Among the nine territorial vitality services, the contribution to social bonds on the territories was the most cited. Here, we show for the first time that the prioritisation of cultural and territorial vitality services differed between the territories studied. Emblematic cattle breeds, food know-how, and quality products were more important in Aubrac, whereas territorial vitality services such as on-farm jobs and social bonds linked to livestock were more cited in the Pays de Rennes. This methodological approach allowed us to highlight and prioritise the different cultural and vitality services that need to be supported by public policy and translated into action. Furthermore, the main findings of this study allowed us to highlight the importance of taking into account the point of view of actors that are not from the agricultural sector and that act in favour of or against the AET.

Conclusion

This study has highlighted the importance of studying actors' networks if we are to gain an in-depth understanding of the levers or lock-in underlying the decisions of farmers to change (or not) towards agroecology. We have suggested that agricultural practices toward agroecology are not necessarily linked to agroecological practices in terms of commercialisation. There is a need to consider the entire actors' network, including the agricultural sector and the other sectors, as playing a key role in farmers' AET. The actors with whom the farmers in our study were discussing their practices were not the same for agricultural practices and commercialisation – which could contribute to explaining such findings. We have also highlighted the fact that actors can act in favour or against AET with no regard for their influence on farmers through a stakeholder analysis. Enterprises commercializing inputs were, in particular, shown to develop barriers to agroecological practices, as they were opposed to autonomy in inputs.

Our analysis has shown that farmers were mostly hybrids on a gradient towards agroecology, who might rely to a greater or lesser extent on technology. This is linked to a hybridisation in the types of advice/exchange they get and the types of actors they include in their network. There is heavy emphasis on "central" actors, including all the farmers' networks studied, even if they were developing relationships with other specific "peripheral" actors, to develop specific practices.

The method we developed could be applied as a first step to understand the local context before implementing participative conception process. With whom should one work? What are their knowledge and motivations? What are the conflicts, power games or, on the contrary, affinities when it comes to working together? Who are the real experts to be considered? Are they official? Who are the actors excluded from the network? In line with Chiffoleau (2009), we consider that network analysis is a basis to develop participative work with local actors, and to highlight power games. Such results are also useful for policy makers, as they show that networks are more hybrids and evolving than supposed, and have a large impact on the AET. In line with Klerkx et al. (2010), we think this type of study highlights the need for policies that take the adaptiveness of innovation networks into consideration more.

Annex 1 Description of the Farmers Interviewed in the Study

Farmer considered	Agricultural practices	Commercialisation practices
Farmer 1	Specialised dairy cattle farmer, Holstein herd with high production goals, not engaged in agroecology	Long chain
Farmer 2	Livestock farmer with meat and dairy cattle, but with little integration between crops and livestock farming (purchase of feed, large quantities of mineral inputs)	Long chain
Farmer 3	Dairy goat farmer with little autonomy in terms of inputs, but who is trying to graze animals despite dependency on concentrates	Long chain
Farmer 4	Dairy cattle farmer with protein autonomy and organic farming	Long chain
Farmer 5	Dairy cattle farmer, with a beef and pork workshop, highly engaged in agroecology (agro-forestry, conservation agriculture, feed self-sufficiency, organic, member of local farmer networks)	Long chain for milk
Farmer 6	Conventional dairy cattle farmer (no agroecological practices)	Long chain
Farmer 7	Dairy cattle farmer in the process of converting his farm to input-based organic agriculture	Long chain – Potential market for exporting organic powdered milk to China, opened up by the Sodiaal cooperative.
Farmer 8	Conventional dairy cattle farmer, with few or no agroecological practices	Commercialises his production via a cooperative grouping together 30 producers across a territory covering three departments.
Farmer 9	Dairy cattle farmer with agroecological practices (grazing, food autonomy...)	Coexistence of two types of product outlets (long and short supply chains).
Farmer 10	Conventional dairy cattle farmer, but with agroecological practices	Commercialises entirely on short supply chains (sale at the farm, market, produce stores), carrying out transformation himself.

References

Aggeri F, Hatchuel A (2003) Ordres socio-économiques et polarisation de la recherche dans l'agriculture: pour une critique des rapports science/société. Sociol Trav 45:113–133. https://doi.org/10.1016/S0038-0296(02)01308-0

Altieri MA, Nicholls CI, Montalba R (2017) Technological approaches to sustainable agriculture at a crossroads: an agroecological perspective. Sustainability 9:1–13. https://doi.org/10.3390/su9030349

Beudou J, Martin G, Ryschawy J (2017) Cultural and territorial vitality services play a key role in livestock agroecological transition in France. Agron Sustain Dev 37. https://doi.org/10.1007/s13593-017-0436-8

Bonneuil C, Joly PB (2013) Sciences, techniques et société. La Découverte, Paris

Bonny S (2017) High-tech agriculture or agroecology for tomorrow's agriculture? Harvard Coll Rev Environ Soc 4:28–34

Chevassus-au-Louis B (2007) L'analyse des risques. L'expert, le décideur et le citoyen. éditions Quae

Chiffoleau Y (2009) La sociologie des réseaux au service d'une recherche engagée : Retour sur un travail d'équipe en viticulture languedocienne. In: Beguin P (Directeur), Cerf M (eds) Dynamique des savoirs, dynamique des changements, Collection. Toulouse, France, pp 111–127

Coquil X, Lusson JM, Beguin P, Dedieu B (2013) Itinéraires vers des systèmes autonomes et économes en intrants: motivations, transition, apprentissages. In: 20eme Rencontres Recherches Ruminants. Paris, France, pp 1–4

Crozier M, Friedberg E (1980) Actors and systems: the politics of collective action. University of Chicago Press (first published in 1977)., Chicago

Duru M, Therond O, Fares M (2015) Designing agroecological transitions; a review. Agron Sustain Dev 35:1237–1257. https://doi.org/10.1007/s13593-015-0318-x

Francis CA, Lieblein G, Gliessman SR et al (2003) Agroecology: the ecology of food systems. J Sustain Agric 22:99–118. https://doi.org/10.1300/J064v22n03_10

Girard N (2014) Quels sont les nouveaux enjeux de gestion des connaissances ? » L'exemple de la transition écologique des systèmes agricoles. Rev Int Psychosociologie Gest des Comport Organ XIX:51–78. doi: https://doi.org/10.3917/rips.049.0049

Glaser BG, Strauss AL (1967) The discovery of grounded theory: strategies for qualitative research. Aldine de Gruyter, Hawthorne

Klerkx L, Aarts N, Leeuwis C (2010) Adaptive management in agricultural innovation systems: the interactions between innovation networks and their environment. Agric Syst 103:390–400. https://doi.org/10.1016/j.agsy.2010.03.012

Latour B (2006) Nous n'avons jamais été modernes, Essai d'anthropologie symétrique, La Découve. Paris, France

Le Foll S (2012) La première graine, Calmann-Lé. Paris, France

Plumecocq G, Debril T, Duru M et al (2018) The plurality of values in sustainable agriculture models: diverse lock-in and coevolution patterns. Ecol Soc 23. https://doi.org/10.5751/ES-09881-230121

Raucent B, Milgrom E, Romano C (2016) Guide pratique pour une pédagogie active – Les APP… Apprentissages par Problèmes et par Projets

Ryschawy J, Choisis N, Choisis JP, Gibon A (2013) Paths to last in mixed crop-livestock farming: lessons from an assessment of farm trajectories of change. Animal 7:673–681

Ryschawy J, Debril T, Sarthou JP, Therond O (2015) Agriculture, jeux d'acteurs et transition écologique. Première approche dans le bassin Tarn-Aveyron. Fourrages:143–148

Sanderson Bellamy A, Ioris A (2017) Addressing the knowledge gaps in agroecology and identifying guiding principles for transforming conventional Agri-food systems. Sustainability 9:330. https://doi.org/10.3390/su9030330

Schot J, Geels FW (2007) Niches in evolutionary theories of technical change. J Evol Econ 17:605–622. https://doi.org/10.1007/s00191-007-0057-5

Therond O, Paillard D, Bergez JE, et al (2010) From farm, landscape and territory analysis to scenario exercise: an educational programme on participatory integrated analysis. In: 9th European IFSA symposium, building sustainable rural futures: the added value of systems approaches in times of change and uncertainty, 4–7 july 2010, Vienna, Austria, pp 2206–2216

Therond O, Duru M, Roger-Estrade J, Richard G (2017) A new analytical framework of farming system and agriculture model diversities. A review. Agron Sustain Dev 37:21. https://doi.org/10.1007/s13593-017-0429-7

van der Ploeg JDD, Laurent C, Blondeau F, Bonnafous P (2009) Farm diversity, classification schemes and multifunctionality. J Environ Manag 90:S124–S131. https://doi.org/10.1016/j.jenvman.2008.11.022

Vanloqueren G, Baret PV (2009) How agricultural research systems shape a technological regime that develops genetic engineering but locks out agroecological innovations. Res Policy 38:971–983. https://doi.org/10.1016/j.respol.2009.02.008

Wezel A, Bellon S, Doré T et al (2009) Agroecology as a science, a movement and a practice. a review. Agron Sustain Dev 29:503–515. https://doi.org/10.1051/agro/2009004

Part III
Support Methodology for Territorial Agroecological Transition Design, and Feedback from the TATA-BOX Project Experience

Participatory Methodology for Designing an Agroecological Transition at Local Level

Elise Audouin, Jacques-Eric Bergez, and Olivier Therond

Abstract The purpose of the TATA-BOX project was to develop a toolbox to support local stakeholders in the design of an agroecological transition at local level. A participatory process based on existing conceptual and methodological frameworks was developed for the design of new configurations of stakeholders and resource systems in the farming systems, supply-chains and natural resources management that were to form a new agroecological territorial system. This process, presented here, was adapted and tested on two adjacent territories in south-western France. It was structured around three main stakeholders' workshops to support the holistic diagnosis, the design of a normative vision, and the backcasting approach of the transition pathway. We describe the participatory methods and the multimodal intermediary tools used to support the collective design of the agroecological transition. We also present the main turnkey outcomes of the design process for local stakeholders, including shared diagnosis, vision for an agroecological territorial system in 2025, and a projected action plan for transition from the initial to the desired agriculture and associated governance structures. Finally, we discuss the limits of the process and the conditions that would enable stakeholders to implement the transition, by reducing remaining uncertainties.

Introduction

Agroecological transition (AET), i.e. the development of an agriculture based on diversified agricultural systems and associated ecosystem services, can be seen as an innovation process towards sustainable agriculture. It involves a complex

E. Audouin (✉) · J.-E. Bergez
AGIR, Université de Toulouse, INRA, Castanet-Tolosan, France
e-mail: elise.audouin@inra.fr; jacques-eric.bergez@inra.fr

O. Therond
LAE, Université de Lorraine, INRA, Colmar, France
e-mail: olivier.therond@inra.fr

J.-E. Bergez et al. (eds.), *Agroecological Transitions: From Theory to Practice in Local Participatory Design*, https://doi.org/10.1007/978-3-030-01953-2_9

co-evolution of technological, social, economic and institutional dimensions, and depends on appropriate management and fostering of interactions between stakeholders of the farming system, the supply chain, and natural resources management at local level (Duru et al. 2015). The latter challenge calls into question the role and organisation of research in supporting the AET. The development of action-research and transformative post-normal science is intended to address this issue. According to Cash et al. (2003), the effectiveness of scientific inputs can be evaluated against three criteria: (i) the impact of science on how issues are defined; (ii) the production of useful information for society i.e. credible, salient and legitimate information; and (iii) a strong interface between scientists and stakeholders by means of effective communication.

The complexity, management challenges and societal demands of AET led some of the scientists towards a progressive change of paradigm (Pretty 1995; Lane 1998). Emergent properties of this transformative research are: (i) a holistic approach, and (ii) changes in the researcher's position to rely more on local empirical knowledges in order to create transition dynamics adapted to and accepted by the intended actors. In parallel, civil society has called for more involvement in local policy making towards a democratic ideal, targeting more integration of local perspectives into development strategies (Pinto-Correia et al. 2006; Shucksmith 2010).

Accordingly, an increasing number of recent research projects or studies dealing with societal issues, like the AET, include participatory approaches. They generally share common concepts such as systems thinking, inter-disciplinarity, and multi-stakeholder representation.

Among them, the TATA-BOX project – based on Duru, Therond and Fares' (2015 – hereafter denoted as DTF, chapter "TATA-BOX at a Glance") conceptual and methodological frameworks – was intended to develop an operational participatory methodology to support stakeholders in thinking and designing an AET at local level. However, even if these authors developed conceptual and methodological frameworks tailored to deal efficiently with AET challenges at local level, they did not provide operational procedures and tools to support stakeholders in the design process. Yet these operational dimensions determine meaningful knowledge-sharing and collaboration between stakeholders. They have to be designed: (i) to foster creative pathways towards sustainability; (ii) to reach agreements and foster dynamics that facilitate transition; and (iii) to generate stakeholders' engagement towards change (Checkland and Poulter 2006).

In other words, the TATA-BOX project effected a transformation from AET design theory to operational and effective practices. Which methods and tools do actually support stakeholders? How do we evaluate action research outcomes?

As a methodological project, TATA-BOX aimed to: (i) provide researchers with new perspectives on procedures to support territorial actors in the design of AET; and (ii) provide feedback on the outcomes of the project's methodology.

The objective of this chapter is to provide an overview of how a conceptual and methodological framework for AET has been processed as an operational procedure with methods and tools, and what resulted from their implementation in real local case studies. Going back from practice to theory, implementation feedback is used as a basis to assess how the operational procedure could actually support the complexity and management issues inherent to AET.

Material and Methods

Theoretical and Methodological Frameworks

In the theoretical framework proposed by DTF (2015), AET is defined as a process that implies three interacting domains at local scale, each of which is characterised by specific stakeholders and material resources:

- the farming system, in which farmers manage resources like land, water, infrastructures, labour, inputs, biodiversity, and semi-natural landscape features;
- the socio-technical system consisting of supply chains in which stakeholders manage resources like stocking infrastructure, agricultural products, operating standards, and production standards;
- the socio-ecological system consisting of territorial resource management arenas in which a diversity of stakeholders, including farmers, manage natural resources such as soil, water, labour, biodiversity, natural and semi-natural landscape features, artificial infrastructures, legislation and operating standards (cf. Fig. 1)

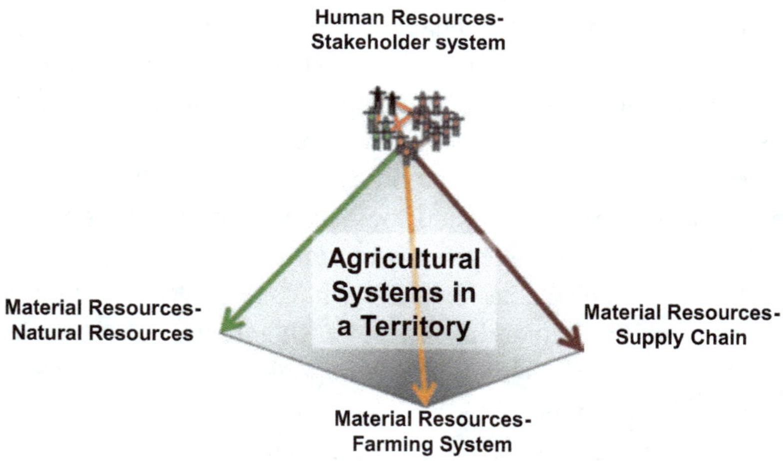

Fig. 1 Duru, Therond, Fares' conceptual framework (DTF 2015)
Local agriculture as a system of stakeholders managing three types of material resource systems through information technologies. The system of stakeholders consists of farmers and other stakeholders involved in supply chains and management of natural resources, with cognitive resources (e.g. beliefs, values, individual strategies) and whose behaviour is determined by informal norms and agreements (another type of cognitive resource) and formal rules. The tetrahedron reflects local agricultural development's reliance on interactions between its four dimensions. Each edge of the tetrahedron (double arrows) corresponds to a diversity of information technologies used to manage material resources and concrete management processes within a variety of farming systems, supply chains and natural resource management institutions

This conceptual framework highlights the critical role of the stakeholder system emerging from interactions between and the multiple roles of actors managing these three domains. The stakeholders of these three sub-systems develop specific knowledge and management strategies.

Considering this conceptual framework, DTF (2015) proposed a 5-step participatory design process to support local stakeholders in this transition design process (cf. chapters "General Introduction" and "TATA-BOX at a Glance") (Fig. 2):

(i) analyse the current situation: this step sets the transition arena, defines the issue, identifies key stakeholders and causality chains (Duru et al. 2015). For

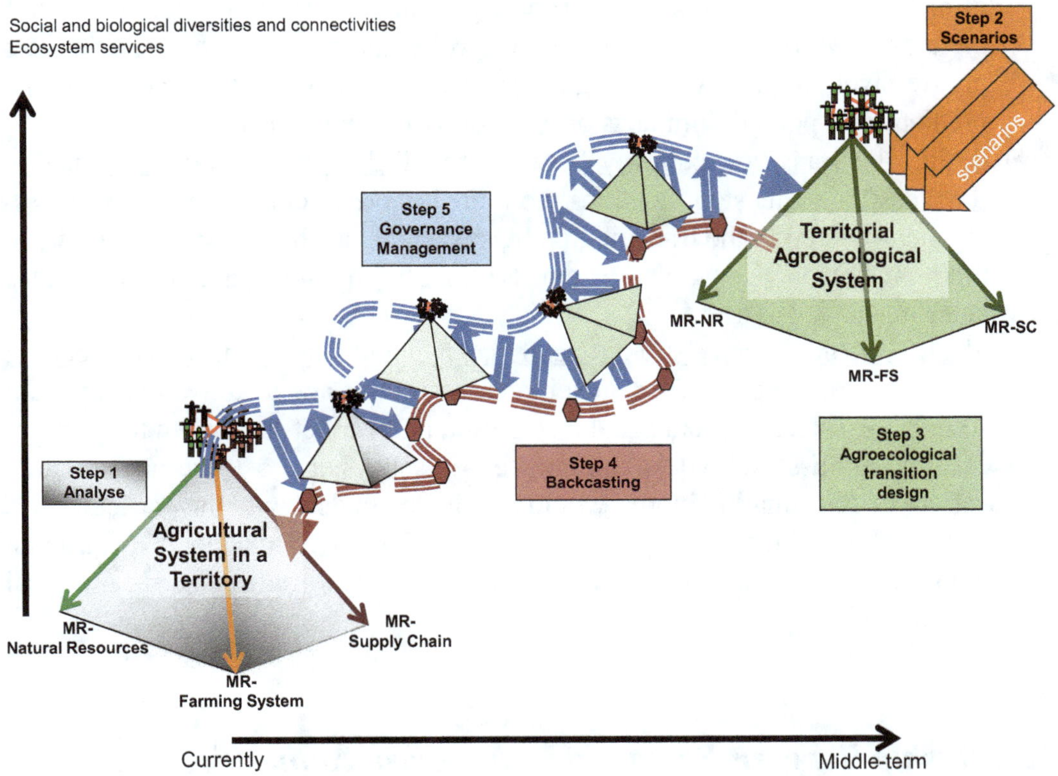

Fig. 2 DTF Methodological framework (Duru et al. 2015)
Participatory design methodology of "territorial biodiversity-based agriculture" and the transition from the current situation to this new form of agriculture. This methodology is driven by Participatory-Design Facilitator-Scientists who manage and steer a multi-stakeholder group ("transition arena") that includes stakeholders from the three management domains (farming systems, supply chains and natural resources) with key knowledge about the functioning of local agriculture. This participatory methodology is composed of five steps: (1) co-analysis of the current situation: the system of stakeholders and their material resources (MR); (2) co-identification of future changes exogenous to local agriculture, which can determine its future; (3) co-design of the expected territorial biodiversity-based agriculture; (4) co-design of the transition (pathway) from the current situation to territorial biodiversity-based agriculture (the reverse arrow indicates a backcasting approach); and (5) co-design of governance structures and adaptive management strategies enabling stakeholders to guide the transition they designed. Each step must be performed by considering and integrating interactions between farming systems, supply chains and natural resource management

this purpose, it identifies stakeholders, resources, human-driven actions, and ecological processes that have a decisive influence on the functioning of farms, supply chains, and natural-resource management.

(ii) develop scenarios of major exogenous forces (explorative forecasting): this step identifies external changes that could affect the local territory. The identification of these changes can be based on a morphological approach that addresses the potential constraints the territory could suffer from and the potential opportunities that could impact it positively.

(iii) design a desired Territorial AgroEcological System – TAES (normative foresight): this step designs a new organisation of the local agriculture that meets local stakeholders' expectations, considering current local issues and scenarios of exogenous forces. For this purpose, graphical tools (conceptual diagrams, pictures, cognitive maps) are used iteratively. Iteration fosters innovation and progressively improves the design of scenarios with cycles of propositions and prediction of potential impacts of the proposed innovations.

(iv) design the transition pathway between the TAES and the current situation (backcasting): this step identifies the most important conditions for progressing step-by-step in the transition pathway, as well as the decisive changes and their impact on the whole system. Settings of monitoring criteria are also determined.

(v) identify the governance structure and management strategy needed to steer the transition: this step identifies governance structures and adaptive management strategies to steer and manage the transition pathway. It starts with the hypothesis that multi-stakeholders and polycentric subsystems of governance with a variety of coordination modes would be adapted to deal with the particularities of the AET (Biggs et al. 2012; Duru et al. 2015) (cf. chapter "Towards an Integrated Framework for the Governance of a Territorialised Agroecological Transition").

Partnership Between Researchers and Local Authorities

Project Team and Organisation

Forty-two researchers formed the project team. They came from various disciplines (agronomy, computer science, informatics, economy, ergonomics, management science, and sociology); from six French research organisations (CNAM, ENSAT, ENSFEA, INRA, IRSTEA, UTT[1]).

[1] CNAM = National Conservatory of Arts and Crafts; ENSAT = National High School for Agronomy of Toulouse; ENSFEA = National High School of Agricultural Training and Education; INRA = National Institute for Agricultural Research; IRSTEA = National Research Institute of Science and Technology for Environment and Agriculture; UTT = University of Technology of Troyes.

The researchers were divided up into six working groups. The first three (knowledge dynamics, integrated assessment, crop-livestock farming) were set the task of providing reflective and scientific bases for AET. The fourth group was in charge of the operationalisation of the conceptual framework into process, methods and tools, along with their application on territories. This group benefited from the inputs from theoretical and empirical outcomes (from case-studies) of the first three groups, and proposals from the fifth group working on information and communication technologies (ICT) (cf. chapter "Information and Communication Technology (ICT) and the Agroecological Transition"). The last group was in charge of a reflexive analysis on the organisation of the research project ("How does the research team produce its outputs"?, cf. chapter "Towards a Reflective Approach to Research Project Management") and of the participatory methodology ("How relevant are the process, the methods, tools and the resulting products? What are the impacts on the territories?", cf. chapter "Evaluation of the Operationalisation of the TATA-BOX Process"). Accordingly, the work of the fourth group was a continuous interdisciplinary development process based on outcomes of the five other groups of the TATA-BOX project.

Case Studies and Time Scale

French government agencies for regional and rural development, named PETR,[2] were chosen as a relevant scale to implement a design process of transition towards a territorial agroecological system (tTAES). This is a public institution scale used in territorial development planning for inter-municipalities. PETR local authorities are in charge of economic, ecological and cultural development, land use planning (SCoT[3]), and ecological transition. They act at an intermediate scale: larger than the municipality but smaller than the French *département* (FADN NUTS III, http://ec.europa.eu/agriculture/rica/). Inhabitants of the PETR share a common identity but they manage heterogeneous resources requiring varying degrees of cooperation between one another.

The PETR of Centre Ouest Aveyron (129 municipalities, 2998 km^2) and the PETR of Midi-Quercy (48 municipalities, 1192 km^2) were selected because: (i) these adjacent territories share a common key water resource: the Aveyron River (upstream and downstream respectively); and (ii) they could use territories' complementarities as a catalyst for transition, e.g. to organise interactions between crop- and livestock-oriented farming systems (cf. details in Moraine et al. 2017). Their topographical, geological and landscape features are contrasted. Farming types evolve according to a gradient, from prevailing grassland-based upstream to rainfed crop-livestock systems (middle stream), to prevailing irrigated cropping and orchards downstream. Emblematic agricultural products are sheep cheese in Centre Ouest Aveyron and apple in Midi-Quercy (cf. Fig. 3).

[2] Territorial and Rural Balance Pole.

[3] Territorial Coherence Scheme.

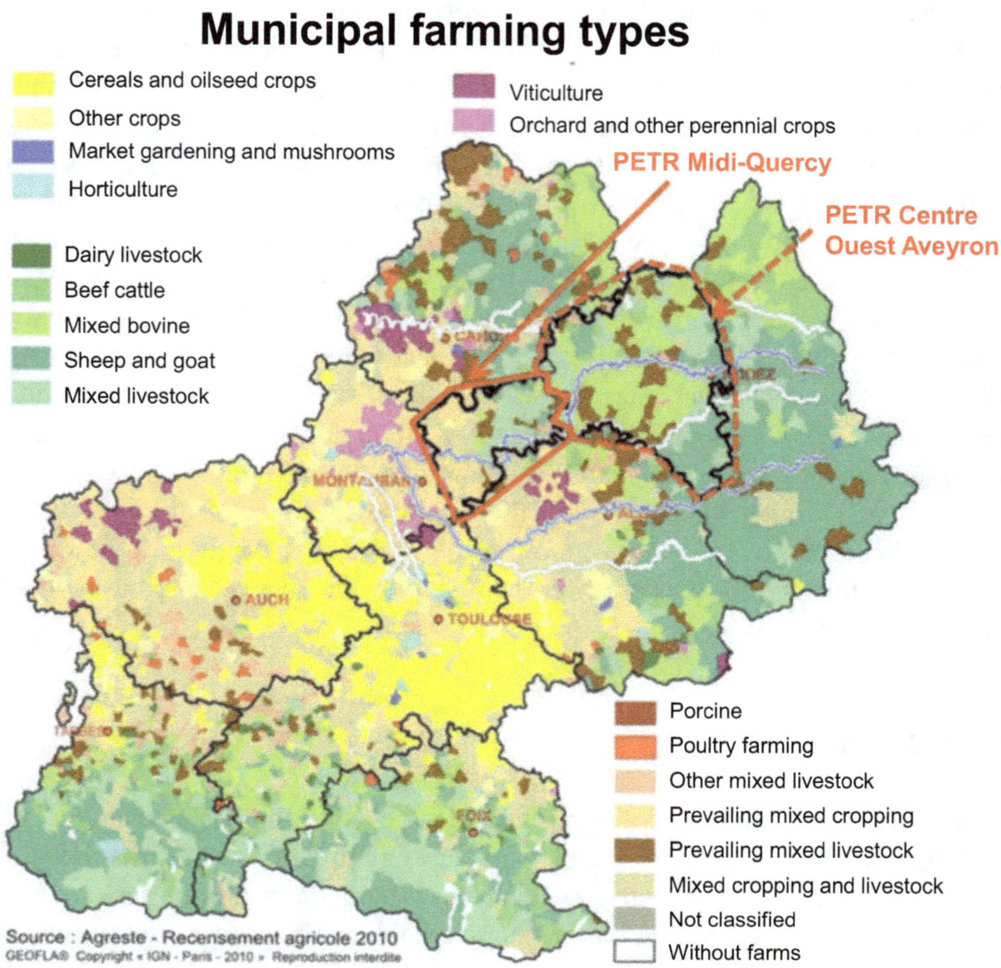

Fig. 3 Prevailing farm types map at municipal scale for Midi-Pyrénées region, AGRESTE 2010. The framed area indicates the position of the two selected PETR
Medium-term horizon scale of 10–15 years was proposed to local stakeholders to balance long-term agroecological transition issues with classical short-term issues managed by local stakeholders in their current projects

These two PETR signified their keen interest in developing a participatory process to design an AET at local level and, accordingly, became local partners for the TATA-BOX project.

Stakeholder Analysis and Involvement

In both territories, representative stakeholders were targeted to participate in the workshops. The operational goal was to constitute a "transition arena" along the process, i.e. a relatively small group of innovation-oriented stakeholders who reached consensus about the need and opportunity for systemic changes, and engaged in a process of social learning about future possibilities and opportunities. These specific stakeholders: (i) questioned the limits of efficiency-/

substitution-based agriculture; (ii) wanted to forecast which activities required changes to promote biodiversity-based agriculture; or (iii) wanted to take part in already implemented biodiversity-based agricultural systems and associated innovations (Duru et al. 2015; Foxon et al. 2009).

Representative stakeholders' identification relied on a classical stakeholder analysis (Grimble and Wellard 1997) based on key stakeholders' network mapping. The stakeholders' network was identified on the basis of the researchers' exploratory interviews with the local key actors, and the scientific knowledge drawn from their previous work in these fields. We used an interest/impact diagram to classify local stakeholders according to their willingness and their potential impact on AET (Therond et al. 2010). A third dimension taken into account in our analysis was stakeholders' position regarding AET (pro or con) (cf. Fig. 4).

This diagram was used as a decision support tool with local PETR partners to build a stakeholders database for the participatory process. In order to ensure the representativeness of the participatory process outcomes and thus to ensure its relevance, all diagram stakeholder categories were contacted: favourable, unfavour-

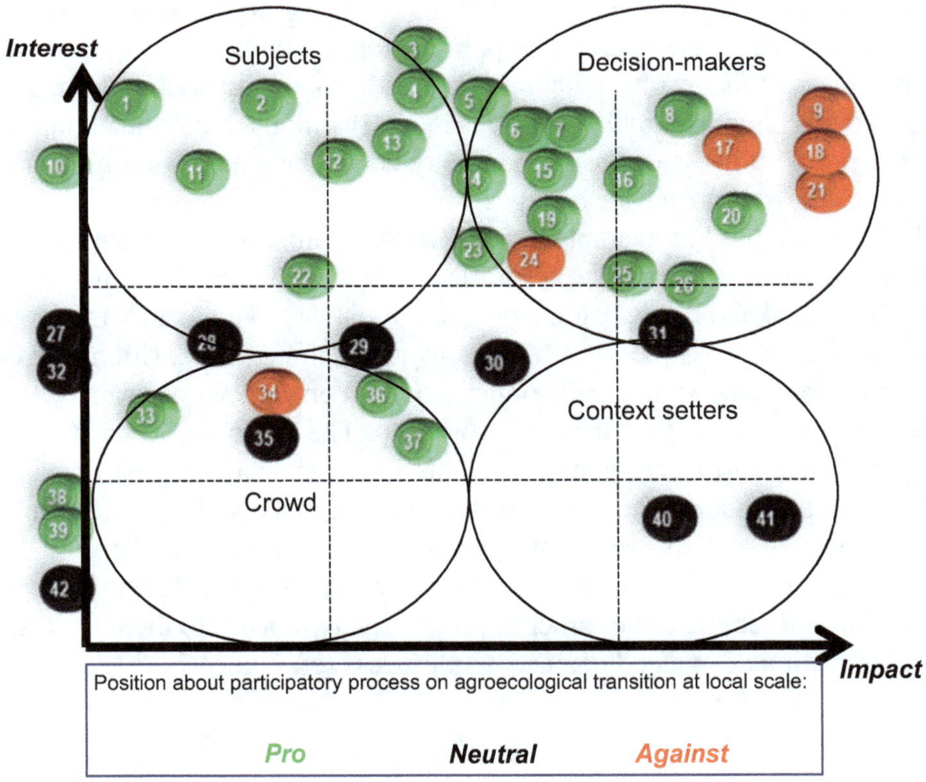

Fig. 4 Interest/Impact analysis diagram for Tarn-Aveyron watershed
Stakeholders related to local agroecological transition are listed. They are positioned according to their interest (ordinate) in or their impact (abscissa) on Territorial Agroecological Transition. Finally, their position on participatory processes of Territorial Agroecological Transition is assessed and formalised with colours: pro (green), neutral (black), against (red)

able and neutral stakeholders, with low, medium or high interest or impact. We finally gathered 57 heterogeneous workshop participants in total for both territories:

- Farmers (organic, conservation agriculture, conventional)
- Farmer groups (conventional and organic)
- Civil society (consumers or environmental associations)
- Farming advisory groups (conventional and organic)
- Territorial planning institutions
- Supply chain and retailers: conventional large chains (cooperative, suppliers), certification labels, short chains

Scientific Design of the Participatory Methodology

The participatory methodology was co-designed by the fourth group of the TATA-BOX project in 20 work sessions, from January 2014 to May 2017. We collectively discussed how to operationalise a multi-level and multi-domain participatory approach along a sequence of participatory workshops. For each workshop, the team set: specific explicit goals, a customised programme, and associated methods and tools for each step of the programme. Considering the methodological framework of DTF (2015), the operational participatory methodology was built step by step, in relation to the results, outcomes and feedback on the organisation, methods and tools of each workshop (7 workshops in total) with the local stakeholders.

The scientific process of the methodology design benefited from regular inputs from territorial experts on local issues for AET: 7 interface times with 21 local stakeholders in total. For example, the TATA-BOX process started with an immersion of researchers in the studied territories. The journey included transects, meetings with key stakeholders of agroecological initiatives to collect their testimonies and determine local needs, with a view to expanding the transition movement. Interface times were systematically organised between workshops to analyse results and new needs, and to adapt the next workshop. The outcomes of each workshop were formalised into three reports available for both workshop participants and other local stakeholders. The Scientist-Territory interface was completed by a bi-annual newsletter containing regular updates on project outputs and events.

For the last workshop, we intensified local partners' involvement in workshop organisation and facilitation in order to progressively transfer to them the responsibility of transition design and management process (cf. Fig. 5).

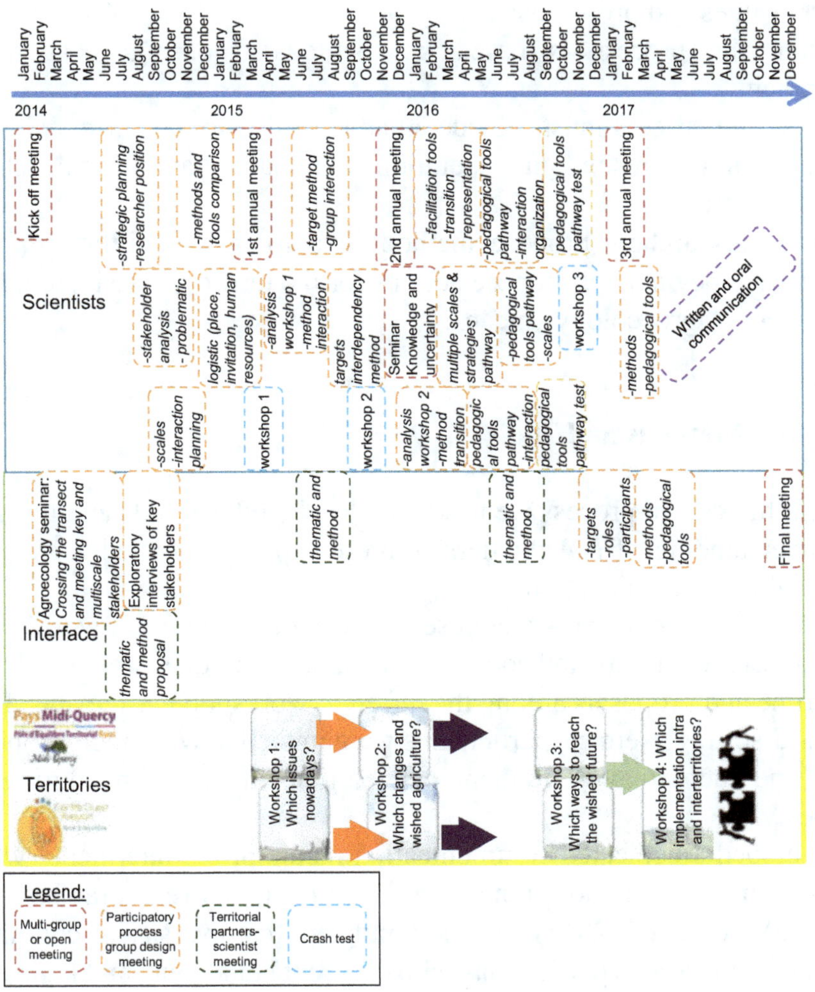

Fig. 5 Main collective reflection sessions, divided between the scientists' sphere, the territorial sphere, and the interface between them

Participatory Guidelines

Participatory Action Research Guidelines

For all workshops, we adopted 5 basic principles (Bryson et al. 2012; Vergne 2013; Jordan and Kapoor 2016):

- transparency on targets and frame: the project aims at developing and testing supporting methods for AET design. The responsibility for the implementation of the outcomes must belong to local stakeholders; in other words, it is up to the local stakeholders to implement and manage the designed transition;
- equity: the process, methods and tools are open to the integration of local stakeholders' values and proposals;
- inclusion: the process is based on methods favouring local stakeholder involvement;
- relevance: the process must provide an added-value to the current situation; the outcomes must result in operational impacts and local stakeholder empowerment;
- neutrality: researchers propose methods and tools to facilitate stakeholders' interactions; they neither propose actions for the territory considered, nor define what exactly "agroecology" means.

Participatory Methods and Tools

Throughout the design process we used and developed methods and tools to foster mutual understanding of stakeholders' representations, knowledge exchange and creative thinking and innovation.

Techniques like icebreakers were used to develop an informal work atmosphere and thus to foster creativity and cooperation. Other techniques were used to balance speaking time and proposals among the different workshop participants. For example, in a card-sorting exercise, participants were invited to write their proposals, one per card, and then to present them one participant at a time during iterative roundtables.

We developed a dedicated communication medium, i.e. intermediary/boundary objects to open representation frames and boost creativity (e.g. drawing, maps and card games; Vinck 2009, 2011). When cumulated, the diversity of communication mediums enabled various participants with different behaviours and logics to understand and therefore to contribute to the exercises. Some participants will be more confident with oral communication, while others prefer written communication and will require visual representations to express their ideas. We most often proposed a combination of representation modes to ensure that each stakeholder would find a way to participate.

In addition, some tools were developed to enable participants to materialise the detailed description of their proposals by representing all information categories that should be informed (cf. section "Intermediary tools").

Finally, in order to conserve a maximum amount of information from one step to the next, intermediary tools were developed to facilitate the use of one workshop's outcomes in the next workshop. Here, each intermediary tool corresponds to a representation at different scales of empirical and collective knowledge (Audouin et al. 2018a).

Results

Methodological Results

Participatory Process

Process

The DTF methodological framework includes an iterative cycle of 5 steps (cf. section "Theoretical and methodological frameworks"). The authors highlighted the fact that this segmentation is theoretical since the steps are interconnected and interdependent. While translating the original methodological frameworks into operational procedures, in a 4-year participatory process, researchers had to deal with several constraints and objectives:

- devote the first year to the analysis of the territorial context, the development of partnerships, and strategic planning: refining the process targets, identifying key stakeholders, etc.;
- limit the duration of the participatory process to facilitate continuous involvement of stakeholders throughout the process;
- avoid excessively frequent meetings with stakeholders, some of whom have limited availability;
- respect a minimum duration between workshops to allow time for their analysis and the step-by-step design of their methodology.

Due to these different constraints we opted for a total of three workshops on an 18-month basis. This process corresponded to a single iteration of the DTF methodological framework cycle (cf. section "Theoretical and methodological frameworks"). Each workshop was 1 day long. The specific targets of the workshops were to: (1) develop a shared analysis of the current situation and issues of local agriculture; (2) co-design a desired TAES resilient to future exogenous changes, involving the collective development of shared goals and visions; and (3) co-design the transition pathway to reach the future vision, with special attention paid to governance. Finally, at the workshop participants' request, we organised a last workshop to support territorial partners in managing the outcomes of design process and transition plan implementation. This step also resulted in the identification of territories' complementarities and potential synergies, in order to facilitate the implementation of action plans.

Methods to Foster Participants' Interaction

We organised participants' interactions at two different scales: during the process, during workshops, and during work sequences in workshops. In line with the objective of each work sequence, we grouped participants together either in diversified groups, ensuring the presence of representatives of each domain (farming system, supply chain, natural resources management, cf. section "Theoretical and methodological frameworks") or, in contrast, specialised in a domain. During the workshop and throughout the whole design process we organised interactions through four key sequences: (i) synchronisation of knowledge in plenary sessions; (ii) exploration of the space of possible options (divergence) and selection of the most interesting ones, in diversified groups; (iii) in-depth description of options by domain group (deepening); and (iv) consistency analysis of described options across domains (convergence). Divergence sessions corresponded to free expression and comparing of multiple points of view. Further reflection was held in mono-domain groups, to take advantage of participants' expert assessments in each domain. Convergence sessions corresponded to agreements and stabilisation of ideas by comparing domain insights in trans-domain groups (cf. Fig. 6).

Within a workshop, articulation between plenary sessions and group sessions, and more specifically between mono-domain and trans-domain group sessions, encouraged participants to foster a multi-domain and multi-level approach to the investigated AET. The articulation between these group session layouts allowed them to go back and forth in scale and level during each workshop.

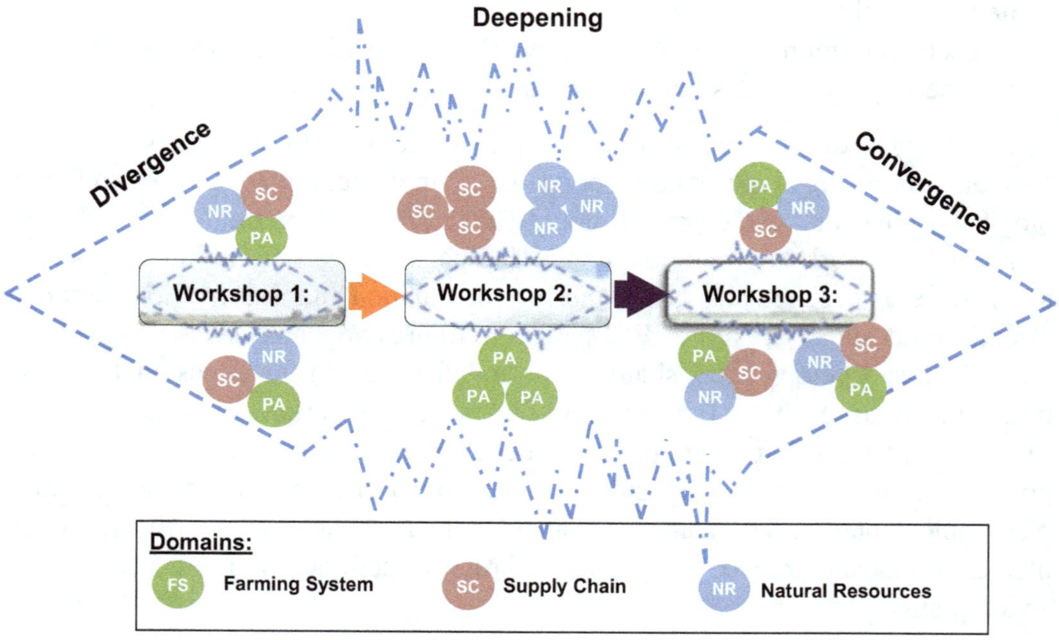

Fig. 6 Participants' interaction throughout the participatory process. The process of Divergence – Deepening – Convergence was used during the three workshops. However, depending on the goal of the workshop, this process was managed on separated or mixed domains (FS/SC/NR) to emphasise a tendency towards either Divergence, or Deepening, or Convergence

Tools for Mutual Understanding, Innovation and Cooperation

The main outputs of the fourth group of the project, dedicated to step-by-step design, are represented in the table below. Table 1 details process planning in terms of a global dynamic, the step-by-step organisation of the participants' interaction, the facilitation methods, and the intermediary tools.

When designing a workshop, in addition to organisation layout, attention was paid to creating a gradual evolution from multiple general ideas in the divergence session, to empirical knowledge during the deepening and convergence sessions.

The first workshop was dedicated to the development of a shared representation of local agricultural issues. It alternated plenary sessions with trans-domain group session. The first plenary session included researchers' presentation of local issues from previous studies. The group session was then dedicated to the formalisation of problems to solve, assets to maintain, and opportunities to take. The second plenary session was organised to share and broaden the groups' outputs. A description of the issues was then refined by a trans-domain perspective and spatial representation exercise. A plenary session shared and gathered the groups' outputs (cf. Table 1). In this workshop a shared representation of current agriculture issues was progressively developed.

The second workshop was devoted to the collective description of targets for 2025's local agriculture. It alternated plenary sessions with group sessions. The first plenary session synchronised participants' knowledge about local agricultural issues defined during Workshop 1, and the main exogenous forces that could influence the future of local agriculture. These forces were presented by the researchers and then discussed with the participants. To develop consensus about territorial targets, the first session was devoted to the identification of general targets in trans-domain groups. The mono-domain groups then broke down these general targets into specific targets, i.e. targets concerning the specific issues of these domains (cf. Table 1). In a plenary session the groups' outputs were pooled within a common share vision for local agriculture in 2025 (cf. Table 1).

The third workshop was devoted to the description of transition pathways, i.e. to the co-design of pathways from the current state (Workshop 1) to the desired future state (Workshop 2) and to the identification of the subsequent forms of governance. The workshop started by a plenary session to synchronise participants' knowledge on the desired future state developed during Workshop 2. A trans-domain session, to take advantage of participants' complementarity, then sequenced targets to develop general action plan strategies. The participants posted the pathways of each target with milestones such as intermediary states, monitoring indicators, and modes of governance (cf. Table 1). A plenary session shared and gathered the groups' outputs.

This process progressively deepens considered scales and domain specificities, evolving from "general targets" to "domain-specific targets" in Workshop 2, to target-detailed pathways in Workshop 3.

This pattern was also observed in the types of representation. Divergence sessions were based on abstract representations to foster creativity: brainstorming or

Table 1 Main results of process planning in terms of global dynamic, the step-by-step organisation of the participants' interaction, the methods and the tools

	1st workshop	2nd workshop	3rd workshop
Conceptual step of the DTF framework	1	2 and 3	4 and 5
Goal	Share representations of local agriculture issues	Co-design a desired territorial AgroEcological system resilient to future exogenous changes	Co-design transition pathways to reach the future vision with special attention paid to governance
Main dynamic	Divergence	Deepening	Convergence
Session 1: Divergence	**Plenary**	**Plenary**	**Plenary**
	Issues of 3 domains	Ongoing exogenous changes	Multi-domain, multi-scale and interconnected targets for 2025
	Presentations of researchers	*Presentations of mixed participants to build hybrid understandings*	*Rich picture*  *Interdependency* 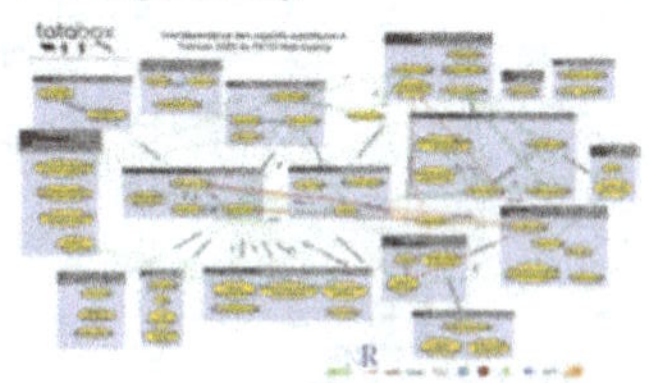
	Trans-domain groups	**Trans-domain groups**	
	Territorial issues of agriculture	General targets for 2025	
	Brainstorming 	*Brainstorming*	
		Plenary	**Plenary**
		Common feature of general targets for 2025	Which strategy to plan actions
		Oral presentation 	*Ranking of clusters of targets* 

(continued)

Table 1 (continued)

Session 2: Deepening	**Trans-domain groups**	**Mono-domain groups**	**Trans-domain groups**
	Localisation of territorial issues of agriculture	Domain's targets for 2025	Which pathway to evolve from the initial step to the desired step
	Participatory mapping 	*Brainstorming* *Participatory mapping* 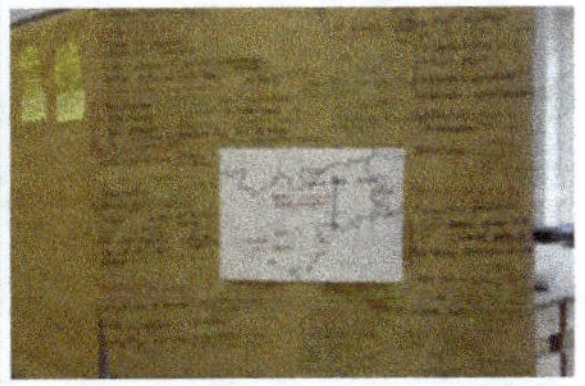	*Icebreaker*  *Nested card game:* *Cluster > envelope > target* *arrow > action, resource, obstacle, intermediary state, governance cards* 
Session 3: Convergence	**Plenary**	**Plenary**	**Plenary**
	Group diagnosis of local issues of agriculture	Group diagnosis of domain's targets and diagnosis of domain targets' independency	Group diagnosis of transition pathways and diagnosis independency
	Oral synthesis and collective analysis	*Oral synthesis and collective analysis* 	*Oral synthesis and collective analysis on timeline*

Legend:
Organisation of participants' interactions
Topic
Method/Tool

rich picturing. Deepening and convergence were based on concrete representations such as participatory mapping or nested card games reproducing a detailed action plan on a timescale with a playful and interactive format.

Intermediary Tools

To illustrate original intermediary tools designed for the TATA-BOX process, we focus on a description of Workshop 3, Session 2 (cf. Table 1), in which the objective was to produce a detailed description of the design pathway.

To reach this objective of the workshop session, we developed an original tool (cf. Fig. 7). Each target of the desired vision identified during the second workshop has been symbolised by an arrow (65 targets for Midi-Quercy, 83 for Centre Ouest Aveyron). The target arrow's colour corresponded to a domain group as identified during Workshop 2. The arrow's centre indicated what the target was about and the desired orientation feature (increase ↗, keep the same =, decrease↘). Icons were created for each target to improve its re-appropriation and handling by participants. The icons were all gathered in another intermediary tool: a rich picture providing a global overview of the desired future (cf. Section "Rich picture of a shared vision for 2025"). The ends of the arrows represented the known current state (Workshop 1) and the final desired state (Workshop 2).

Target arrows were clustered into thematic envelopes. The transition pathway was designed one envelope at a time. This work session organisation was intended to facilitate the progressive and incremental analysis of a large numbers of targets.

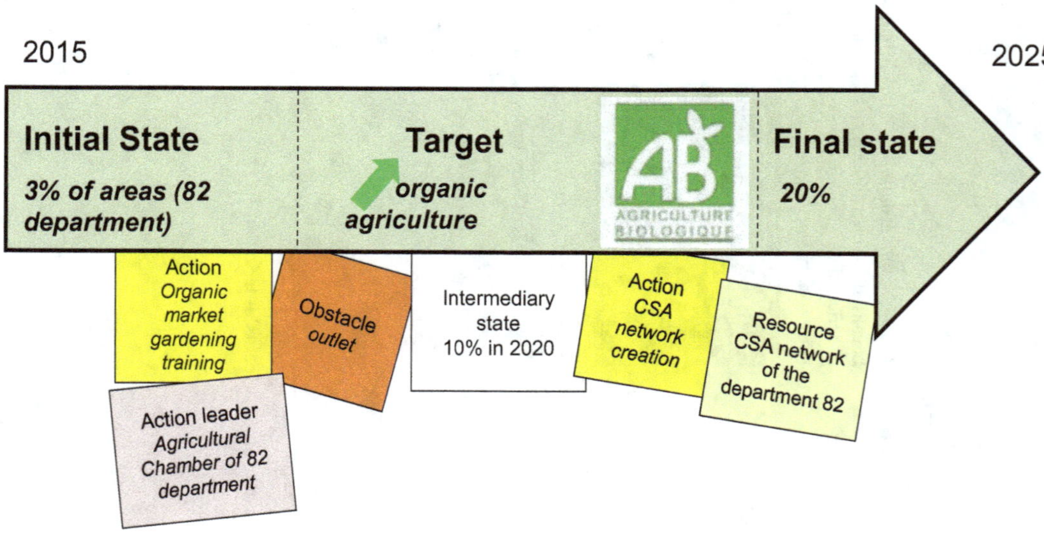

Fig. 7 Card game to examine and detail transition pathways
The transition pathway card game is made up of a set of target-arrows designed from Workshop 2 outcomes: Action, Resource, Obstacle, Action leader and Intermediary states cards used as milestones from the current to the desired state of the target

The first step of the exercise was designed to help participants to define the adapted temporal sequence for reaching the different targets through the positioning of arrows on a virtual time line, in order to reflect the expected general chronology. They were then asked to detail the strategy to reach each target, using cards of different colours to identify respectively: actions (yellow), possible obstacles (orange) to overcome, resources (green) for action, action leaders (pink), and key intermediary states to monitor between the current state and the expected future one. Participants were free to deal with the cards in any order. Information extracted from Workshop 2 outputs provided participants with predefined cards (e.g. actions, obstacles, resources) but they were free to create new cards and to rule out the predefined ones. Identified links between target arrows were materialised by a sticker, referring to the related target code number.

This exercise was performed in a trans-domain group session. A final plenary session presented the group's main outputs and initiated reflection on a global timeline (cf. Table 1). The aim was to sketch a general action plan by arranging the removable arrows and cards on a vertical support. This reflection continued by reporting a global overview of the arrows and cards on a Gantt diagram (cf. section "Action plan"). The overall chronological organisation was based on the resulting general action plan and could be adjusted on the basis of the dates indicated on intermediary state cards.

Operational Results

The application of the TATA-BOX participatory design process led to numerous varied operational outcomes. This method supported the formalisation of a shared diagnosis, original transversal targets, the action plan, and associated governance. To illustrate these outcomes, we present three operational outputs: the rich picture, the action plan, and the main effects in the field.

Rich Picture of a Shared Vision for 2025

During the second workshop the participants detailed a shared vision of local agriculture on the 2025 time line. This vision was described through the identification of 65–83 targets respectively for Midi-Quercy and Centre-Ouest Aveyron.

To synthesise and make more easily accessible the richness of these two visions, scientists developed two corresponding integrative "rich picturing" representations (cf. Fig. 8).

Apart from individual targets, the Rich Picture served to highlight targets' links. Some of them were situated during a participatory mapping exercise (cf. Table 1). The three key organisational levels accounted for by participants were symbolised by nested circles: Farm, Territory, Country (France). These levels were crossed by inflows and outflows. An initial orientation of the picture was set, and all the targets

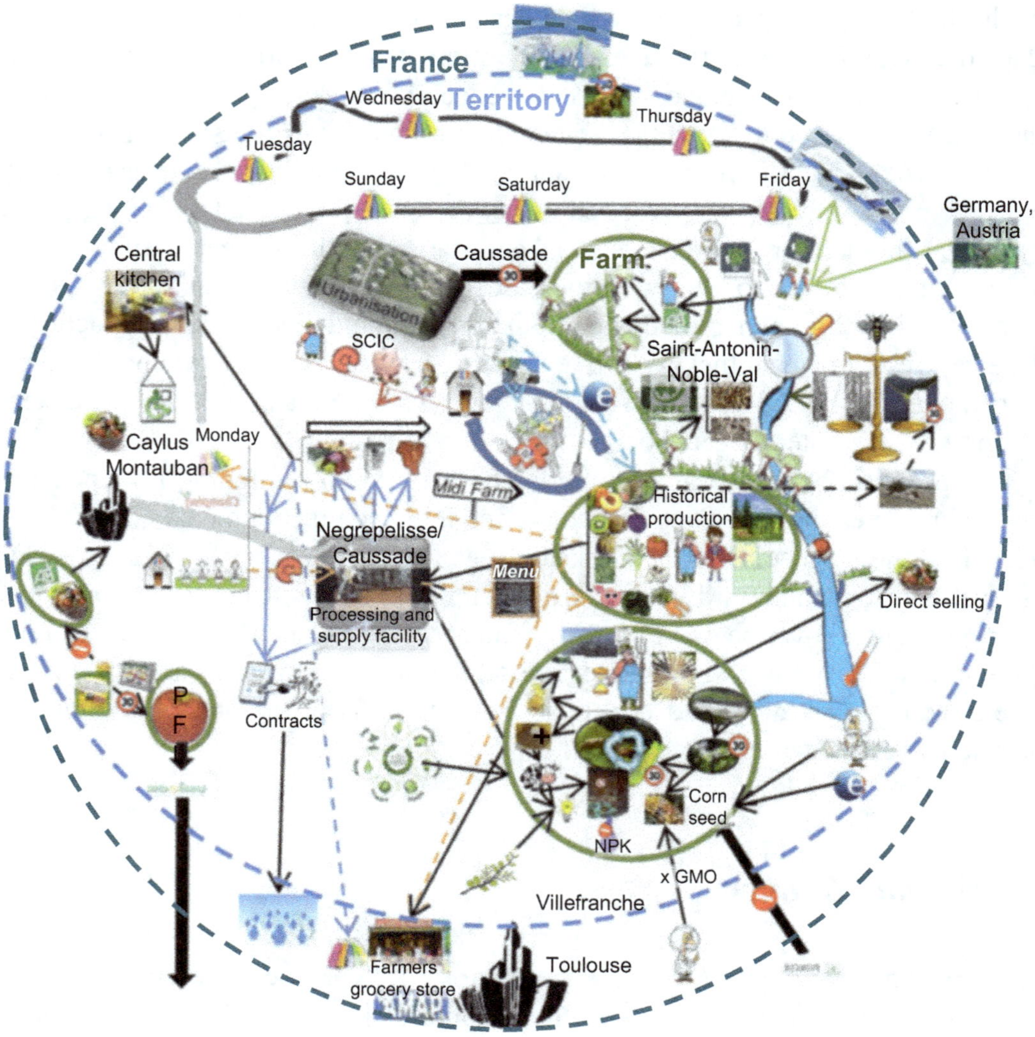

Fig. 8 Rich picture presenting the 65 targets for agriculture in 2025 in Midi-Quercy PETR

were situated on the picture to reflect their geographical embeddedness and to draft a basis for landscape planning.

This output from Workshop 2 was used during the introduction to Workshop 3, to summarise and detail all the outputs of the previous workshop. The facilitator zoomed in on each zone of the rich picture and commented on it. Participants could then react to this representation to highlight agreements, disagreements or obsolete targets since last workshop. The rich picture was displayed throughout the third workshop to keep an overview on the targeted future while detailing transition pathways.

Action Plan

In the third workshop the participants detailed transition pathways from initial states (Workshop 1) to final desired state (targets identified in Workshop 2) (cf. Fig. 9).

Fig. 9 Overview of third workshop outputs for Midi-Quercy (left side) and Centre-Ouest Aveyron (right side)
Each arrow represents targets; they are sequenced chronologically. Action plans were drawn for each target arrow with action, action leader, resource, obstacle and intermediary states cards

After the workshop, scientists translated this action plan into a Gantt diagram at short, medium and long term (Fig. 10).

This systemic normative foresight led to a large provisional action plan that included numerous axes and action scales, in accordance with the three domains considered: farming system (green arrows in Fig. 9), supply chain (pink), natural resources (blue). 83 and 100 actions were planned respectively for Midi-Quercy and Centre Ouest Aveyron counties, and distributed over the 2016–2025 period (cf. Fig. 10). 28 action leaders in charge of the governance of corresponding actions were identified in each territory respectively (in pink).

Workshop participants had free access to the action plan and could add comments or modify it online.

A second table summarised each organisation's involvement for each target, so that they would have access to information on the actions relevant to them.

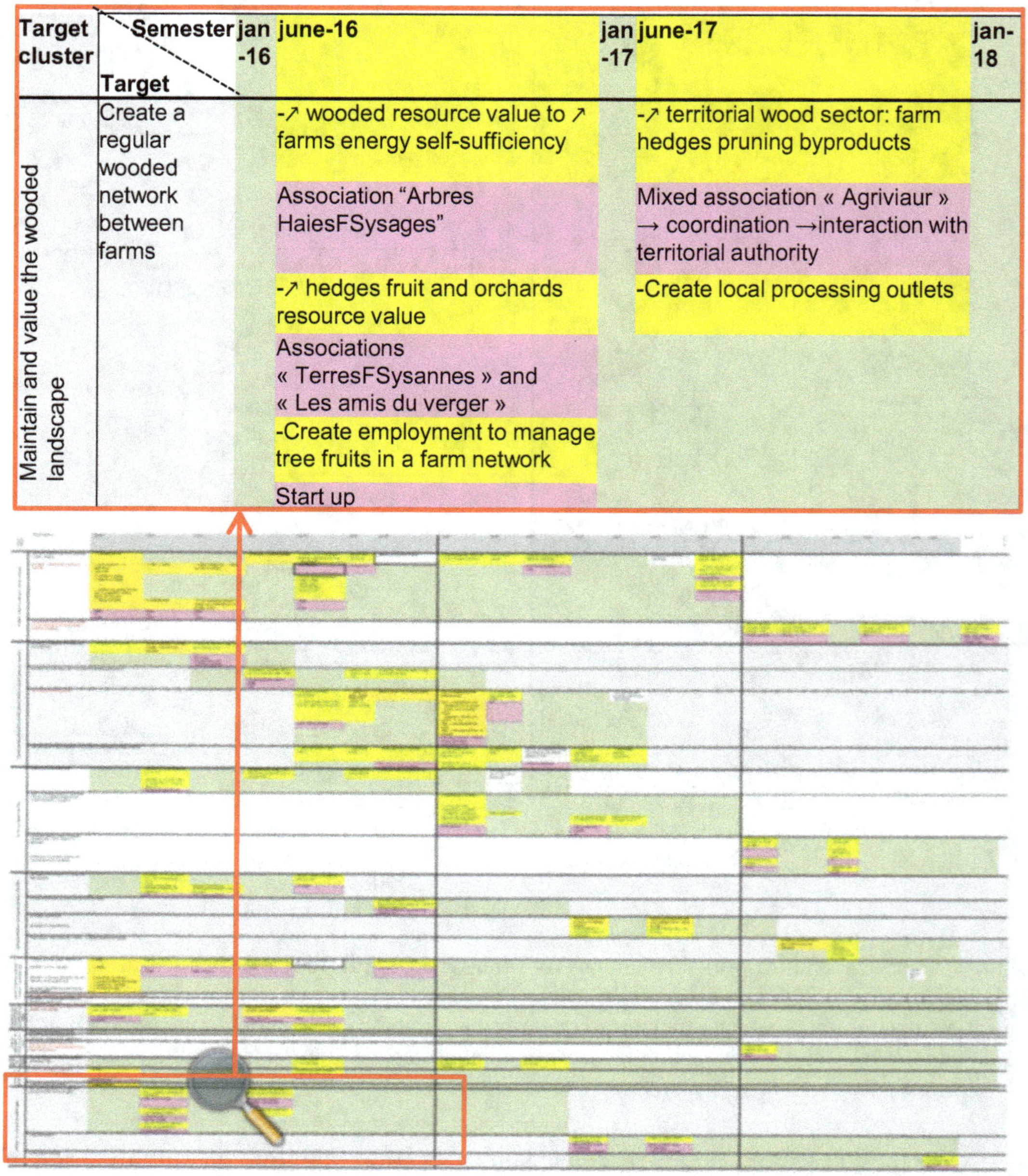

Fig. 10 Detailed action plan of Centre Ouest Aveyron, for each detailed target within each cluster target, transition pathway between the initial and the final state: actions, governance, intermediary states

Implementation of the Methodological Results and Workshop Outputs

Local partners used some of the outcomes and representation tools for local policies and their own animation activities, respectively. For example, the three illustrated workshop reports were used as a basis for the Territorial Coherence Scheme. Development agents also used representation tools such as the animated "rich picture" during meetings on the development of a Territorial Food Plan.

Local partners quickly started to implement some of the actions identified in Workshop 3. For example, Midi-Quercy PETR engaged a process to create a Social Cooperatives of General Interest dedicated to renewable energy, as identified in the action plan.

To improve action plan implementation and go further in the process of the transition management, local partners asked scientists to organise a fourth workshop. After discussions between scientists and local partners, the main assigned goals of the workshop were: (i) to share how PETR have used workshop outcomes to this day; (ii) to develop a shared analysis of the particularities and complementarities of action plans in each partner's territory; (iii) to identify possible synergies between both action plans in order to foster their implementation; and (iv) to plan collaborative actions involving both territories. This fourth workshop was entirely designed with local partners.

During this workshop, initially not planned by scientists, local partners built a common strategy, including 11 potential cooperation axes. Like the outcomes of the previous workshops, the corresponding action plan and the required resources were formalised in a formal report.

Finally, the TATA-BOX process also had some indirect impacts concerning participants' networks and their position on AET (cf. chapter "The Key Role of Actors in the Agroecological Transition of Farmers: A Case-Study in the Tarn-Aveyron Basin"). These results and the assessment of the extent to which participants' expectations had been met are detailed in chapter "Evaluation of the Operationalisation of the TATA-BOX Process".

Discussion

Can the Initial DTF Framework Be Translated into Operational Tools to Design Transition Toward a Territorial AgroEcological System?

The original conceptual and methodological framework considered three domains and five steps to address major AET issues (cf. Figs. 1 and 2) by means of redesign. This section discusses the extent to which the original frameworks were finally operationalised.

Did the Operational Process Reflect Major Agroecological Transition Issues?

For a systemic and resilient AET, the DTF framework considered key transition dimensions to be tackled in AET design. Many of the topics included in the initial expectation (Bergez et al. 2013; Duru et al. 2015) were addressed in both territories throughout the participatory process:

- **Biodiversity** was considered in the natural resource domain in both territories, with a focus mainly on conservation issues. The topic was also mentioned when dealing with issues of the farming system and supply chain domains, through targets concerning diversification of cropping, farming and landscape systems and food chains. For example, it was addressed in the final action plan through targets like "Develop agrobiodiversity", "Insert pasture within forage systems", or "Develop new chains for former outputs". The initial definition of agroecology was however deliberately broad and open, in order not to influence participants' transition design. Therefore, the process was not long enough to get into further details of expected role and effects of diversity, redundancy, connectivity and feed-backs (Biggs et al. 2012). Even if the process did not focus on specifications of interrelationships between climate change, water resources, crop and animal production, these interdependencies were explicitly tackled by the participants.
- **Energy transition** was considered as the third priority in the action plans of both territories. It was mainly addressed through the objective of farm production diversification, with a special focus on wooded resources for Centre Ouest Aveyron, and a larger range of renewable energy sources for Midi-Quercy: hydraulic, photovoltaic, solar panels, methanation, and seaweeds.
- The reduction of **anthropogenic inputs** was an explicit target in the action plan of Centre Ouest Aveyron. In this territory, the issue was closely related to the promotion of farming system autonomy by increased agrobiodiversity ("Insert pasture within forage systems"). For Midi-Quercy, it was not central but was mentioned in relation to water quality.
- Anthropogenic inputs reduction and biodiversity topics were closely related to the development of ecosystem services. However, the term "**ecosystemic services**" was not used during the first three workshops. The fourth workshop took place with local partners. During this inter-territorial interaction, the participants finally used this term and tackled **eco-economy** to insert TATA-BOX outcomes within local public policies.

Did the Operational Process Reflect the Three Targeted Domains?

The design process explicitly concerned the farm system, supply chain and natural resources domains and their interactions. In a divergence–deepening–convergence process, domains were formalised and treated separately in Workshop 2 (vision) while they were treated transversally in Workshop 3 (transition pathways) (cf. Section "Methods to foster participants' interaction".). The rate of actions transversal to at least two domains diverged from one territory to the other. In Midi-Quercy, 3/11 action plan axes mixed two or three domains while this rate reached 7/11 axes in Centre-Ouest Aveyron. Domains were materialised in order to take into account and develop a co-evolution of technological, technical, social, economic and institutional dimensions, with a view to improving the overall sustainability. We actually developed a trans-domain action plan that might favour a multi-dimensional co-evolution in the resulting transition.

Did We Develop a Functional Process, Methods and Tools for Redesign?

The TATA-BOX research process enabled scientists to develop operational tools to equip local partners for territorial planning. The choice, sequencing and application of methods and tools in the participatory process were helpful to broaden participants' innovation ability while leading them towards a concrete action plan. Compared to other participatory foresights, TATA-BOX had the particularity of enabling participants to design precisely the transition pathways to be implemented at local level. The methods and tools were designed to make this heavier operationalisation step more interactive and didactic, through a one-day workshop.

The tested participatory design process, methods and tools enabled local stakeholders to design a shared provisional action plan with potential action leaders. Additional multi-medium reports on methods were provided to local stakeholders to hand over the methodology: raw intermediary tools; a comic book on key issues of a participatory approach for AET (Audouin et al. 2018b), and a methodology guide for the whole process (Audouin et al. 2018a).

The Scientist-Territories interface could nevertheless have been intensified during the process by additional collaborative tools to develop stakeholder-stakeholder and stakeholder-researcher interaction to foster interaction and methodological adaptations (cf. chapter "Information and Communication Technology (ICT) and the Agroecological Transition").

What Is the Role of Adaptive Multilevel Governance?

Multi-form and multi-level governance features may strengthen the adaptive capacities of governance systems by increasing the ability to change and ultimately the resilience of the managed social-ecological system (Biggs et al. 2012). Polycentric governance should be multi-level in order to be resilient to change and to address the complexity of the social-ecological system (Pahl-Wostl 2009; Pahl-Wostl et al. 2010; Folke et al. 2011).

The two action plans designed by local stakeholders were based on such multi-form and multi-level governance, with a total of 28 identified organisations in charge of actions in each territory, acting at different levels, in different domains, and dealing with different issues. In other words, these action plans were based on a poly-centric governance based on different arenas dealing with specific issues at the organisational level at which they emerge. Importantly, local authorities and territorial policies were identified to drive some of the individual actions constituting the global action plan.

Some of the planned actions had to be led by various complementary organisations (33% in Midi-Quercy; 79% in Centre-Ouest Aveyron). The interaction modalities between these existing organisations and the corresponding governance system were however not completely informed. Other actions required the creation of new structures (31% in Midi-Quercy; 8% in Centre Ouest Aveyron). For example, Centre-Ouest Aveyron identified the need for a new local business owner to manage the wooded resource and derive value from fruit and hedge pruning in a farm network. Midi-Quercy identified a need for two new Social Cooperatives of General Interest (SCIC) dedicated to a local supply-chain platform and local renewable energies coordination. SCIC structures have the specificity to potentially gather farmers as well as elected representative in a same governance structure. Midi-Quercy also considered the creation of an Economic and Environmental Interest Group (GIEE) devoted to organic matter exchanges between livestock farmers and cereal farmers, as well as a bulk purchase group for water resources equipment. An inter-territorial governance organisation was proposed for water resource management.

How to Increase the Impact of the Participatory Process on Local Territories?

While DTF (2015) argued for the necessity to iterate the design process, the TATA-BOX project allowed for only one iteration. The cycle implemented during the TATA-BOX project resulted in two final action plans collectively designed by workshop participants. These action plans are at once multi-domain, multi-stakeholder, and multi-level, and involve many different biophysical and socio-economics items and variables (cf. Section "Did the operational process reflect major agroecological transition issues?").

Additional steps would be relevant to deepen action plan outcomes in order to facilitate their implementation. These workshops could: (i) deepen action operationalisation modalities and address remaining uncertainties, and (ii) address governance issues and deepen management strategies for the aforesaid actions.

While computer-based models are often used in the design process of cropping and farming systems (Bergez et al. 2010; Martin et al. 2012; Duru et al. 2015), the use of such models to deal with the full complexity of the design process presented here seems difficult or even impossible. It was not possible to use this type of tool to analyse in greater depth the impacts and trade-offs of the various transition pathways. Additional workshops focusing on specific topics and considering a reduced number of variables would however make the use of computer models possible. For example, as DTF (2015) have highlighted, agent-based models could help to address place-based interactions between human decisions and ecological processes at landscape level. GIS- and indicators-based approaches could also be used to deal with issues at field, farm or landscape level.

Adaptive management considers the iterative design-action-monitoring process in order to improve practices and policies by identifying and taking advantage of learning from implementation (Pahl-Wostl 2009). As workshop participants were not individually empowered to deal with such management strategies, additional workshops would have been necessary to enable participants to refine action plan management strategies step-by-step and to refine diversified intertwined multilevel governance systems.

More generally, while these additional workshops might empower the stakeholders in action plan implementation, new relevant stakeholders, especially operators of technologies and techniques, would need to be included throughout the additional process steps, according to the particularities of the workshop theme. New intermediary tools needed to be developed to support the stakeholder's activities during these meetings. However, this raised the question of the role of TATA-BOX's scientists in this process of transition implementation, as the researchers' withdrawal from the local process was planned at the end of the third workshop. The translation of action plans into local projects required the stakeholders to sort actions into different projects, to find project resources, and to set a final project agenda. These projects were a cornerstone between planning and implementation that actually impacted the other workshop participants. Because the PETR² were identified as key stakeholders of action plan governance in both territories, this cornerstone now entirely relies on PETR² coordination and political will (cf. chapter "Evaluation of the Operationalisation of the TATA-BOX Process"). From the beginning of the project we assumed that the PETR², as local partners, would take in charge the coordination of further steps, and we made sure to equip them with ready-to-use results and the appropriate tools to do so (cf. sections "Scientific design of the participatory methodology" and "Did we develop a functional process, methods and tools for redesign?"). We also complied with their request to organise an additional fourth workshop for exchange and coordination between the two PETR on the possible further steps they would take. Compared to the breadth of the action plan, the actual steps taken by the PETR subsequent to the workshops have however been minimal. Our final question concerns scientists' role in PETR² empowerment in further steps. Should the scientists of the TATA-BOX project wait for the PETR² to switch on their own from transition arena workshop participants to transition governance status? Or which additional tools could scientists provide them with to make this status transition? Did the researchers' withdrawal from the transition process take place at the right time or should they have carried on their maieutic support within the transition process?

What Does It Mean for Scientists?

Developing Trans-Disciplinary Research

TATA-BOX was a research project involving a group of 42 researchers. It was clearly defined that the scientific issues were methodological and consisted in testing methods with actors for support and not prescription purposes. As we have seen, the initial DTF framework considered three main areas: "farming systems", "natural resources"and "supply chain". None of the researchers were specialists in all domains. Through our training and individual research, we could provide insights and knowledge on certain themes, but not on the coherence of the whole. The facilitation methods implemented allowed us to introduce new methods and to add the researchers' scientific knowledge to the empirical knowledge of the local actors. The group of researchers wanted the participants to articulate local expertise and knowledge, consistent with a post-normal research approach (Funtowicz and Ravetz 1993). Multi-disciplinarity allowed it to delve deeper into emerging themes while mobilising the global systemic vision of the local actors of the territory during the integration of the different domains. Moreover, because of the initial transparency of the researchers' position in the process, and the purpose of the device itself, their disengagement at the end of the project was explicitly discussed and planned with the partners in the field.

"Cheating" to Propose the Project

The structure of research projects can be a real obstacle to their development in territorial co-design. In the "idealised" scheme of the National Research Agency, knowledge of a partnership and the issue to be addressed is one of the keys to successful acceptance of the project. However, for research topics like the one proposed by TATA-BOX, the partnership is built along the way, and the question – often vague (ill-structured) at the beginning of the project – is clarified over time. The same goes for the choice of methods. Here, we admit, we "cheated". We proposed a project in traditional task groups, but from the first meetings of the scientists, we modified the overall structure of the project to allow for an adaptive research schedule. Moreover, we tried an adhocratic type of governance (which allowed us to be more consistent with our wish to favour the emergence of a poly-centric adaptive governance of the territories, cf. chapter "Towards a Reflective Approach to Research Project Management"). This was difficult to maintain over time because a researcher's reasons for participating in a project vary (financial interest, management, interest in the method/thematic, network, etc.) and their availability evolves. As a result, their degree of involvement in the project varies as well. We therefore maintained a fixed trinomial of facilitators rather than a circle of facilitators in constant rotation.

Conclusion

The TATA-BOX project aimed at developing a participatory toolbox to support local stakeholders in the design of an AET at local level. Considering the objectives of agroecology, that is, the development of diversified agricultural systems providing ecosystem services that drastically reduce the use of industrial inputs, a "redesign" transition strategy was targeted rather than so-called "efficiency" or "substitution" strategies. The purpose of the TATA-BOX project was the operationalisation of the conceptual and methodological frameworks proposed by DTF (2015) for designing an AET. These authors claim that the design of an AET requires reconfiguration of the stakeholders and resource systems emerging from the interaction between farming systems, supply-chains and natural resources management strategies. Their methodological framework was designed to support local stakeholders in steering the AET of these three interdependent domains. The targeted TAES should be resilient to exogenous drivers owing to a multi-domain perspective and adaptive governance.

From theory to practice, the TATA-BOX project effectively succeeded in creating an operational and replicable participatory methodology based on the DTF frameworks. It was tested in two adjacent territories of south-western France (Midi-Quercy and Centre-Ouest Aveyron counties). The methodology is structured around three main participatory workshops enabling stakeholders to perform exploratory, normative and backcasting prospective analyses over a medium-term, 10-year period. Each step was organised through a divergence–deepening-convergence process, in which stakeholders' interactions were structured in mono-domain and trans-domain groups. The process was based on the use of different intermediary tools favouring innovative, realistic propositions, individual appropriation and exchange of information. Special attention was paid to equitable shared outputs by means of speaking-time sharing and multi-modal communication.

The workshops resulted in turnkey outputs for local stakeholders, i.e. shared agricultural diagnosis for 2015, a vision for 2025's agroecological territorial system, and a projected action plan for transition from the initial to the final desired agriculture organisation. The projected action plans included about 100 actions each, suited to the territory considered, and the associated action leaders, i.e. the governance structure.

The workshops' outputs actually reflected local particularities through various strategies and trajectories, depending on the territory considered. The analysis of current and future agricultural organisation, based on the characteristics of and interactions between three DTF domains – farming system, supply chain, and natural resources management – have proved to be helpful for stakeholders.

Stakeholders identified other indirect results such as widening networks or crossed learning.

The TATA-BOX project organised only one iteration of the design cycle: diagnostic, normative forecasting, and backcasting. As DTF (2015) expected, other iterations or additional steps would enable stakeholders to improve step-by-step transition design and adaptive governance.

Although this was a process of normative forecasting, it was large enough to adopt a free and holistic approach to transition trajectories. The problematic was too large to apply computer-based models to obtain more details on potential impacts and performances of the desired agriculture vision. More in-depth analysis on certain actions may now be investigated using such modelling tools.

The TATA-BOX process acted as a maieutic support within the transition process. Scientists supported the co-design of an action plan, but left its implementation, monitoring and management (including an iterative and continuous design process) up to the stakeholders. Partners and local stakeholders' commitment in the transition implementation and management could be encouraged by means of appropriate procedures and tools for operational adaptive governance and management of implemented transitions. The story of developing a methodology to support transition is to be continued!

References

Audouin E, Bergez JE, Choisis JP et al (2018a) Petit guide de l'accompagnement à la conception collective d' une transition agroécologique à l' échelle du territoire. Toulouse

Audouin E, Bergez JE, Therond O (2018b) TATA-BOX, Neuf métaphores des concepts clefs des démarches participatives pour la transition agroécologique

Bergez JE, Colbach N, Crespo O et al (2010) Designing crop management systems by simulation. Eur J Agron 32:3–9

Bergez JE, Sarthou JP, Soulignac V et al (2013) Programme Agrobiosphère édition 2013, Projet TATA-BOX, Document Scientifique. Toulouse

Biggs R, Schlüter M, Biggs D et al (2012) Toward principles for enhancing the resilience of ecosystem services. Annu Rev Environ Resour 37:421–444

Bryson J, Quick KS, Slotterback CS, Crosby BC (2012) Designing public participation process. Public Adm Rev 73:23–34. https://doi.org/10.1111/j.1540-6210.2012.02678.x

Cash DW, Clark WC, Alcock F et al (2003) Knowledge systems for sustainable development. Proc Natl Acad Sci U S A 100:8086–8091

Checkland P, Poulter J (2006) Learning for action: a short definitive account of soft systems methodology and its use, for practitioners, Teachers and Students. Wiley, Chichester

Duru M, Therond O, Fares M (2015) Designing agroecological transitions; a review. Agron Sustain Dev 35:1237–1257. https://doi.org/10.1007/s13593-015-0318-x

Folke C, Jansson Å, Rockström J et al (2011) Reconnecting to the biosphere. Ambio 40:719–738. https://doi.org/10.1016/j.gloenvcha.2006.04.002

Foxon T, Reed M, Stringer L (2009) Governing long-term social–ecological change: what can the adaptive management and transition management approaches learn from each other? Environ Policy Gov 2:3–20. https://doi.org/10.1002/eet.496

Funtowicz S, Ravetz JR (1993) Science for the post-normal age. Futures 31:735–755

Grimble R, Wellard K (1997) Stakeholder methodologies in natural resource management: a review of principles, contexts, experiences and opportunities. Agric Syst 55:173–193

Jordan S, Kapoor D (2016) Re-politicizing participatory action research: unmasking neoliberalism and the illusions of participation. Educ Action Res 24:134–149. https://doi.org/10.1080/09650792.2015.1105145

Lane DC (1998) The greater whole : towards a synthesis of system dynamics and soft systems methodology. Eur J 2217:214–235

Martin G, Martin-Clouaire R, Duru M (2012) Farming system design to feed the changing world. A review. Agron Sustain Dev 33:131–149. https://doi.org/10.1007/s13593-011-0075-4

Moraine M, Mélac P, Ryschawy J et al (2017) A participatory method for the design and integrated assessment of crop-livestock systems in farmers' groups. Ecol Indic 72:340–351

Pahl-Wostl C (2009) A conceptual framework for analysing adaptive capacity and multi-level learning processes in resource governance regimes. Glob Environ Chang 18:354–365. https://doi.org/10.1016/j.gloenvcha.2009.06.001

Pahl-Wostl C, Holtz G, Kastens B, Knieper C (2010) Analyzing complex water governance regimes: the management and transition framework. Environ Sci Pol 13:571–581. https://doi.org/10.1016/j.envsci.2010.08.006

Pinto-Correia T, Gustavsson R, Pirnat J (2006) Bridging the gap between centrally defined policies and local decisions: towards more sensitive and creative rural landscape management. Landsc Ecol 21:333–346

Pretty JN (1995) Participatory learning for sustainable agriculture. World Dev 23:1247–1263

Shucksmith M (2010) Disintegrated rural development? Neoendogenous rural development, planning and place-shaping in diffused power contexts. Sociol Rural 50:1–14

Therond O, Paillard D, Bergez JE, et al (2010) From farm, landscape and territory analysis to scenario exercise: an educational programme on participatory integrated analysis. In: IFSA symposium, 4–7 july 2010, Vienna, Austria. pp 2206–2216

Vergne A (2013) Qualité de la participation. In: CASILLO I. avec BARBIER R., BLONDIAUX L., CHATEAURAYNAUD F., FOURNIAU J-M., LEFEBVRE R. NC et SD (dir. . (ed) Dictionnaire critique et interdisciplinaire de la participation. GIS Démocratie et Participation, Paris

Vinck D (2009) De l'objet intermédiaire à l'objet-frontière, Vers la prise en compte du travail d'équipement. Rev d'anthropologie des connaissances 3:51–72

Vinck D (2011) Taking intermediary objects and equipping work into account in the study of engineering practices. Eng Stud 3:25–44. https://doi.org/10.1080/19378629.2010.547989

Towards a Reflective Approach to Research Project Management

Lorène Prost, Marie Chizallet, Marie Taverne, and Flore Barcellini

Abstract This chapter describes how we supported the project leaders of TATA-BOX in their task of designing a management system for the project. We did so by fuelling their reflectivity: rather than making suggestions on how to manage the project – in a normative approach –, we analysed the on-going project management and mirrored what had been done after a year. The TATA-BOX project leaders would thus be able to decide how to adjust their management and to carry on – in a reflective approach. We report on this process in this chapter: after giving some theoretical background on the concept of reflectivity and its role in helping the project leaders to manage TATA-BOX, we describe: (1) how we worked with them over 6 months, 1 year after the project began, and (2) the different methods we used to meet the project leaders' expectations. We then discuss the efficiency of these methods, their effects on the management of the project, and some lessons learned for the management of such research projects generally.

L. Prost (✉)
LISIS, INRA, EPNC, ESIEE, Université Paris-Est Marne La Vallée, Marne La Vallée, France
e-mail: lorene.prost@inra.fr

M. Chizallet (✉)
LISIS, INRA, EPNC, ESIEE, Université Paris-Est Marne La Vallée, Marne La Vallée, France

CRTD, CNAM, Paris, France
e-mail: marie.chizallet@cnam.fr

M. Taverne
Territoires, Université Clermont Auvergne, AgroParisTech, INRA, Irstea, VetAgro Sup, Aubière, France
e-mail: marie.taverne@irstea.fr

F. Barcellini
CRTD, CNAM, Paris, France
e-mail: flore.barcellini@cnam.fr

J.-E. Bergez et al. (eds.), *Agroecological Transitions: From Theory to Practice in Local Participatory Design*, https://doi.org/10.1007/978-3-030-01953-2_10

Introduction

For scholars interested in design studies, the TATA-BOX project (cf. Chap. 2) is like a concentrate: it intertwines three design processes, each of which can be studied individually. The first design process was the initial focus of the project: *designing* local agroecological transitions (AET). Acknowledging the nature and stakes of agroecological forms of agriculture, the researchers in the TATA-BOX project then felt that this design process should be conducted by the local stakeholders themselves. The project was therefore aimed at *designing* a methodology to support local stakeholders in designing their AETs. This was the second design process. The course of the project was devoted to proposing such a methodology to local actors, and to implementing its different steps with them. The idea was to iteratively adapt the methodology in order to continue its design into use (Béguin 2003) in an adaptive way. In fact, the initiators of the project had quickly established that biodiversity-based agriculture required farmers to deal with complexity and uncertainty in a process-oriented and goal-seeking approach (Duru et al. 2015). Organising a transition towards this form of agriculture consequently had to rely on specific bases, that is, on an adaptive and participatory approach. But the question was how to organise a project to achieve such a process? This was where the third design process came into play: *designing* a management system that would support the project. How could the TATA-BOX project leaders *design* the management of a project that was intended to *design* a method to support the *design* of a local AET?

As it was funded by the French National Research Agency, TATA-BOX was structured as a typical project, that is, with "*the accomplishment of a clearly defined goal in a specified period of time, within budget and quality requirements*" (Lenfle 2008), with work packages, milestones and deliverables. But its project leaders had claimed from the very beginning that what applied to the design process of AETs should also apply to the project itself. They wanted the project to be "participatory, collective, evolutionary, adaptive and adhocratic" (Chizallet 2015). This placed the project leaders of TATA-BOX in a very particular management position, in between project management and adaptive management (Holling 1978; Walters 1986), for they had to design their own management for the project.

This chapter describes how we endeavoured to support the project leaders of TATA-BOX in their task of building what we have called the third design process, that is, their design of a management system for the project. We proposed to do so by fuelling their reflectivity: rather than making suggestions on how to manage the project – in a normative approach –, we would analyse the on-going project management and mirror what had been done after a year. The TATA-BOX project leaders would thus be able to decide how to adjust their management and to carry on – in a reflective approach. We report on this process in this chapter.

After giving some theoretical background on the concept of reflectivity and its role in helping the project leaders to manage TATA-BOX, we describe: (1) how we worked with them over 6 months, 1 year after the project began, and (2) the different methods we used to meet the project leaders' expectations. We then discuss the efficiency of these methods, their effects on the management of the project, and some lessons learned for the management of such research projects generally.

Theoretical Positioning

We grounded our work in the Activity-Centred Ergonomics approach to activity, design and design process management (Daniellou and Rabardel 2005; Barcellini et al. 2014). Over the past 30 years, activity ergonomics has developed an approach to support design projects that aims to foster better interactions between project stakeholders, better integration of existing activity, and anticipation of future activity: in our case, the activity of project management. This approach acknowledges that: (1) there is often a lack of strategic management of projects, that is, not only of coordination issues but of actual management (e.g. strategic decision making); and (2) the structure of the project itself is often at fault, with a focus on the technical dimensions of the project to the detriment of the aspects related to the work of those impacted by the project, and the organisation of work and training. We were therefore interested in the project leaders' intent in TATA-BOX: based on their understanding of the design processes of a transition towards agroecology, they wanted the project management to be "participatory, collective, evolutionary, adaptive and adhocratic". In view of this position, we decided to support them by not giving them immediate design management solutions, especially since activity ergonomists have always pleaded for specific, adequate and localised interventions adjusted to the partners' demand and to their actual activities (Daniellou 1992; Guérin et al. 2006). We moreover wanted to build an intervention that would support the project leaders' learning about their activity. Activity-centred ergonomics has revealed that every work activity comprises a productive dimension directed at performing the task, and a constructive dimension that transforms workers' skills and organisation (Samurçay and Rabardel 2004). We intended to develop this constructive dimension of the project leaders' activity. Accordingly, and due to the investigative nature of the management that they wanted to explore, we chose to place them in a position to reflect upon their own project management, in other words, to be "reflective practitioners" (Schön 1983).

In this respect, Activity-Centred Ergonomics has proposed methodologies to foster the constructive dimension of work by engaging workers collectively in a reflective activity. This implies *"a critical analysis of the activity, either to compare it to a prescriptive model, to what one should or could have done differently, and to what another practitioner might have done, or to explain and critique it"* (Perrenoud 2001: our translation). This critical analysis may support the construction of "new" knowledge about work activity and related skills (Teiger and Falzon 1995). Collective reflective activity aims at learning from experience and *"switching from knowledge in action to knowledge of action"* (Mollo and Nascimento 2014: 208). Supporting a reflective activity implies helping participants to build reflection on past actions (Mollo and Nascimento 2014) through *"a social, language-based and intra-/inter-subjective activity based on actual experience"* (Le Goff 2014: 3, our translation). This particular activity must be fed by a representation of actions performed by participants, that is, what they actually did and not what they planned or intended to do (Mollo and Nascimento 2014). On the basis of the "critical analysis"of their own activity, participants may: learn about their own experience and thus develop meta-understanding about themselves and their capacity for action (Teiger and Falzon 1995), and may enhance their potential to act, and their adaptive skills. The setting built to enhance this reflective activity is of prime importance. One of its main characteristics is its anchorage in the intermediary objects (Vinck 2009, 2011) that represent actual activity – in our case, project management activity. Another characteristic is the need for interpersonal mediation as essential to the performance of reflective activity (Perrenoud 2001; Petit et al. 2007; Chaubet 2010).

The reflective intervention that we proposed was clearly inspired by activity-centred ergonomics and had two objectives. The first was to allow the project leaders to learn about their project management activity and to improve it if necessary, in order to achieve their own goals. Our second objective was to show the project leaders how to build some reflective areas by themselves, for the project people, in order to support their intention to manage their project in a "participatory, collective, evolutionary, adaptive and adhocratic" way. In fact, to support their idea of adaptive management, we had assumed that specific management tools inspired from reflective tools would be needed to adapt the course of the project over time.

Material and Methods

Construction of a First Diagnosis

Our intervention began in October 2014, 9 months after the beginning of the project. In line with the principles of ergonomics, the intervention began with an analysis of the project management and a reformulation of the project leaders' expectations. The project leaders were the two researchers who had designed most of the project: the official project leader, called "scientific coordinator", in charge of its strategic management, and a research engineer in charge of the project's coordination.

This first step involved six semi-structured interviews with the three project leaders and three researchers involved in the project. The objective of these interviews was to collect the project leaders' and researchers' representation of the project. All these interviews were recorded and transcribed. To complete them, an activity ergonomist carried out so-called "global observations" (Guérin et al. 2006) in the same office as one of the project leaders. These observations were intended to opportunistically capture real-world project issues such as gaps between project leaders' representations and actual actions, regulations performed by the project leaders to cope with unexpected events during the project, and so on. She also sat in on three project meetings (one involving the three project leaders, one involving two project leaders, and one involving researchers and the three project leaders). Finally, the ergonomist performed an analysis of documents related to the project (TATA-BOX project proposal, main publication of researchers, such as Duru et al. 2014, 2015) to reveal the project leaders' initial representations of the project: the stakes (economic stakes, production stakes, stakes related to work activity), whether these were made explicit or not, and the project structure that had been implemented.

Reflective Intervention: An Exploratory Building Process

The diagnosis described in Section "Construction of a first diagnosis" was the first input to begin the reflective intervention with the project leaders. The set-up of this intervention was then iteratively built from one meeting to the next according to the outcomes of the meeting and the development of the project leaders' thinking about their project management.

Global Framework of the Intervention

A first meeting called "Intervention Proposal"was organised (January 2015) between the TATA-BOX project leaders and the members of the "Reflectivity Group" (RG; the four authors of this chapter). During this meeting, the first modalities for the implementation of reflectivity were established. The following were agreed: the time to develop a reflexive activity (a two-hour reflectivity meeting would be convened once a month), the space (which office), the roles (the TATA-BOX project leaders would be the reflective practitioners, the reflectivity group – RG – would be in charge of the facilitation). All the participants also agreed on the idea of adapting the next meetings based on discussions, reactions and requests of the current meeting. Finally, they agreed that, for each meeting, a time of contribution by the RG would be coupled with a participatory exercise at the end of the meeting. The exercises would be proposed by the RG to the practitioners in order to encourage the emergence and evolution of their reflective activity. These workshops would place the project leaders in a reflective exercise on a particular aspect of the project or its organisation. They would encourage discussions among practitioners with a sharp reduction in the facilitator's intervention.

Description of the Reflective Intervention

After this first meeting, the construction of the intervention was carried out in an exploratory way during a series of four RG meetings that consisted of: (1) determination of objectives for the meeting; (2) search for research material to feed the meeting and achieve its objectives (theoretical frameworks, researchers' activities, project elements, etc.); (3) preparation of the facilitation of the meetings (slideshows, exercises); (4) the meeting itself; (5) assessment at the end of the meeting with the TATA-BOX project leaders and the RG; (6) debriefing of the meeting within the RG and analysis; and (7) determining new objectives for the next meeting based on these analyses.

This led to the design of the reflective intervention described in Fig. 1. All these meetings were held with the project leaders. Table 1 provides details about the goal of each meeting, the inputs that were used to build and lead the meetings, and the exercises that were done with the project leaders.

Some objectives of the intervention required the extension of the initial diagnosis to an analysis of on-going events of the project. That was particularly the case of Meeting 3: "Feedback on methodological seminar". During this meeting, the RG decided to focus on a specific seminar that had been organised a few months earlier, and during which all the TATA-BOX researchers had been asked to design the organisation of the project. Feedback about this seminar was drafted by the RG based on:

- An analysis of written and audio tracks of the seminar: initial intentions of project leaders, actual object discussed during the seminar.
- Seven semi-structured interviews with participants at the seminar. The interviews took place 8 months after the seminar. The duration was 1 h each. A reminder of the seminar programme was read by the ergonomist to the researcher interviewed at the beginning of the interview.

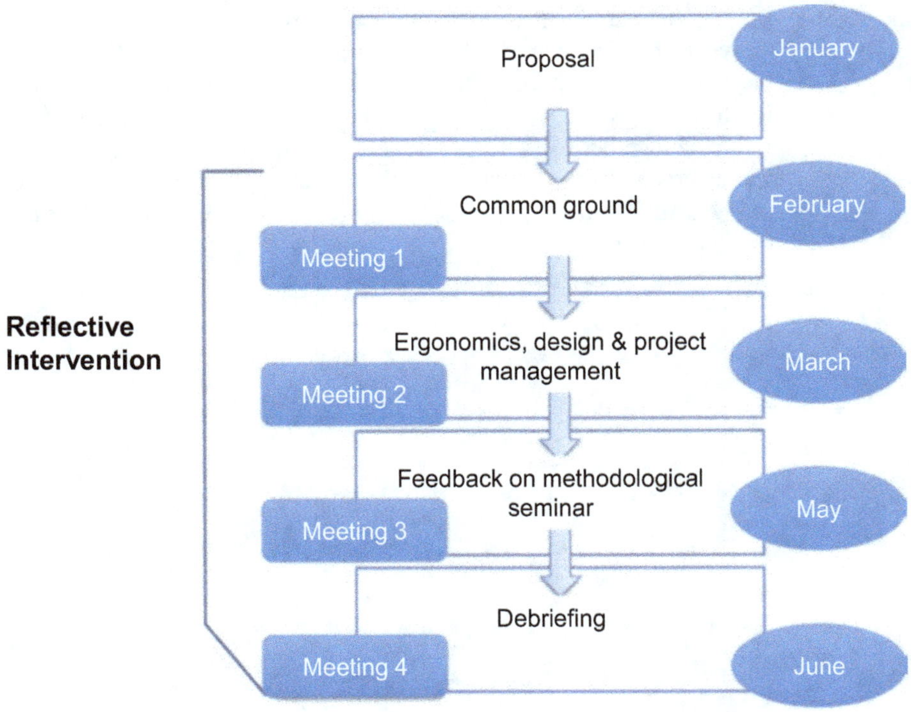

Fig. 1 Synthetic view of the reflective intervention

Table 1 Goals, inputs and methods used during the reflective interventions

	Goals	Inputs presented by the ergonomist	Participatory exercises
Meeting 1 – February 15 **Building a common ground**	Provide an external vision of the project and share a common vision. Discuss the objectives of the project and its difficulties. Define the roles of the project leaders and distribute them.	Diagnosis built by the ergonomist	**Role sharing** *Instructions*: Please, ask yourself: What does this role, or this task, entail? Who carries out this task? *Material*: Four sheets of different colours, one for each of the three researchers of the nucleus, a fourth for the other researchers of the project. A list of roles and tasks on labels: Some reported from interviews, others added by the ergonomist. Other blank labels are available for other roles or tasks that do not appear here, and to multiply certain roles or tasks that would be distributed to several researchers.
Meeting 2 – march 15 **Ergonomics, design and project management**	Bring in concepts of ergonomics. Build links between the concepts proposed by the ergonomist and the project. Reflect together on what a collective and participatory conception implies. Examine together the characteristics of the project: Adhocratic, participative and collective project.	Bibliographic search on collective design, participatory design, project management, adhocracy Diagnosis built by the ergonomist	**Appropriation** *Instructions*: Please explain how you would apply the general concepts brought by the ergonomist to TATA-BOX. *Material*: Each slide presented by the ergonomist had to be taken up by the project leaders and adapted to the specific case of the TATA-BOX project.

(continued)

Table 1 (continued)

	Goals	Inputs presented by the ergonomist	Participatory exercises
Meeting 3 – May 15 **Feedback on methodology seminar**	Show important moments of the seminar on methodology (during which the organisation of the project was rebuilt). Ask about these moments. Consider these moments in different ways. On the basis of this feedback, ask about the evolutionary, adaptive and collective characteristics of the project.	Analysis of the methodological seminar	**Rethink the methodological seminar** *Instructions*: Please think about how you could do otherwise if the seminar methodology had to be reorganised, for each key situation of the methodological seminar. *Material:* Post-it notes
Meeting 4 - June 15 **Debriefing**	Have a return on the reflective intervention by the project leaders. Examine the current organisation of the project.	Analysis of all the meetings of the intervention	**"Build a common representation of the organisation of the project"** *Instructions*: Please reconstruct the current organisation of the TATA-BOX project by using labels with the names of the workgroups and arrows. *Materials*: Arrows and labels with the names of the different project workgroups

Table 2 Elements of the coding scheme of reflective activity (Adapted from Chizallet 2015)

Places	Activity
Withdrawal	A participant avoids a problem or does not answer
Testimony	A participant clarifies or explains his/her view on the basis of an experience
Clarification	A participant asks for details/explanations
Questioning	A participant outlines a difficulty
Proposal	A participant imagines another way to do something

Characterising the Reflective Activity

For Meeting 4: "Feedback on project management and reflective intervention", the three previous meetings were analysed as part of the reflective intervention. All these meetings had been recorded and transcribed. Table 2 presents the coding scheme defined to reveal reflective activity in interaction (Jorro 2005; Chizallet 2015). In this scheme, the reflective activities were classified from the least reflective (withdrawal) to the most reflective ones (proposal).

Results

In this section we show three types of results. First, we provide information on the RG's diagnosis of the project management at the beginning of its intervention. This diagnosis produced questions and assumptions that were debated with the project leaders to build the following intervention. Second, we browsed through the results of the four meetings organised with the project leaders. Finally, we discussed the efficiency of the reflective intervention by following the development of traces of reflectivity throughout the intervention.

Diagnosis of the Project Management

From a Structured to an Adhocratic Project?

During our first interviews, the project leaders regularly used five adjectives to characterise the project: "participatory, collective, evolutionary, adaptive and adhocratic" (Chizallet 2015).

In the diagnosis they tried to collect information to understand how these characteristics emerged. The various interviews helped us to trace the history of the project: (1) from the emergence of a first intention of the project, (2) through the design of its first version and (3) a redesign of the project.

(1) The intention of the project initially emerged with the question of "How to support an AET?", raised by a team leader who envisaged an adhocratic and participative project emerging from the team. It was then taken up by three other researchers (an economist and two agronomists). Faced with the complexity of the AET concept, they decided to spend time on the conceptual framework that would help to define and support an AET. This brought together different disciplines, theoretical frameworks, and views of agroecology and of transition and research postures. There have been many debates, mainly on how to represent local agriculture and think the transition dynamically. This complexity within the project was discussed at length and the construction phase of the project was long. It resulted in a conceptual framework and a five-step methodology that structures the TATA-BOX project.

(2) A proposal for the TATA-BOX project was then drafted specifically to obtain funding from the French National Research Agency, and therefore did not correspond to the project that the researchers had in mind. For instance, initially there was not supposed to be a project manager, as the project leaders wanted an adhocracy (i.e. organic governance), but the normative frameworks of the ANR did not allow that. The director of the unit consequently took the lead with two other people: an agronomist from the small initial group and a full-time engineer on the project. This leading trio kept its effectiveness in the organisation of the project. To stick to the ANR requirements, the project proposal submitted to the ANR was also divided into five work packages, although the leaders assumed that this organisation would be modified by the project researchers themselves, with a view to building an iterative and more collective project.

(3) To help the project researchers to appropriate the newly-funded project and to reorganise the project collectively, the project leaders decided to begin the project with several seminars dedicated to first contact and knowledge sharing. These seminars were built to allow the researchers to collectively take ownership of the project, agree on what the project was, what the field was, and discuss misunderstandings. A kick-off meeting marked the launch of the project in January 2014. All the researchers presented their research activities so that each of them could locate themselves in relation to one another and to the project. This allowed the researchers to build bridges between their various research activities. In addition, the seminar highlighted the existence of several representations of agroecology. A field seminar was organised 4 months later, in Aveyron, to discover the terrain and discuss the concept of agroecology. This seminar also allowed the researchers to meet some of the stakeholders. From there, it was decided to re-organise the TATA-BOX project. Following their idea of rebuilding of more polycentric project after its acceptance by the ANR, the project leaders organised a "methodological seminar" in September to reorganise the project work. Seven groups emerged from this seminar. There were about ten researchers in each group, some of whom were present in several groups, and of whom had joined voluntarily. The groups were not supposed to be fixed, but rather to be reorganised during the course of the project.

Identification of Project Management Issues

Apart from setting out the history of the project design, this diagnosis allowed us to highlight various issues that might be improved to enhance the functioning of the project. In this chapter, we detail only detail those three that relate to the project management.

Firstly, our diagnosis underlined a first project management issue related to the objectives of the project. It revealed that there was a lack of synchronisation among the participants, with regard these objectives. The project leaders shared the same idea that the project was not intended to support the AET but rather to design a methodology to support the actors in building their own AET. However, when asked about the objectives of the project, the participants were not so clear. There seemed to be some confusion and discrepancy between them, mainly concerning the researchers' intention to support the territorial AET or not. We thus assumed that, after 9 months and in spite of the different seminars, not all project researchers had managed to share a common vision of the project. This is a well-known difficulty of project management. Many studies on design processes have revealed that the objectives of a project are often "ill-defined" and that a synchronisation activity between the project participants is deeply needed. Several studies have focused on this highly important but time-consuming activity in design meetings (e.g. Falzon and Darses 1996; Détienne 2006; Visser 2009).

Secondly, following the same idea of a lack of synchronisation, we characterised a second project management issue dealing with the conceptual and methodological framework of the project. We have explained above that after extensive debate about what an AET could be and how to analyse and support it, a framework was proposed in Duru et al. (2014, 2015), distinguishing five steps for designing an AET. The interviews showed that while some researchers had taken up this framework, others had more difficulties with it. For some of them, the framework was too different from their own conceptual and methodological backgrounds. Others pointed out that it seemed contradictory to claim that the project needed to be adaptive and evolutionary and, at the same time, to set quite a rigid framework from the outset. There was thus a lack of synchronisation about the conceptual and methodological framework of the project. Note nevertheless that the project leaders claimed not to set up definite concepts. Acknowledging the complexity of the subject, they thought that some vagueness was needed to allow the participants to work together. As soon as the framework would be too definite, it would exclude some participants. There was then a balance to find between too much and too little framing.

From these two elements of difficulty, we built two assumptions. Firstly we assumed that the fact that the project had been built by several little collectives successively did not favour its quick take up by all the project researchers. Secondly, we assumed that the various transformations of the project were not sufficiently thought out and that not all project researchers adhered to these transformations. Some of them did not adhere to the transformation from the initial project into an ANR normative project. Others had difficulties with the idea of detaching themselves from the normative aspect of the project to re-create a new dynamics for the project. As a result, a two-speed project was appearing: one with the participants who applied the methodological framework, and the other with the researchers who fed the project but outside of its main dynamics. This may be considered as a success – albeit partial – with regard to the project goals: the project leaders had indeed succeeded in creating a collaborative dynamics among some researchers who applied the methodological framework, whereas research projects often consist in gathering competencies without building a collaborative dynamic.

Finally, a last project management issue appeared. From the diagnosis, it appeared that the three project leaders had significant decision-making power in the project, which was partly contradictory with their own will of building a participatory, adhocratic project. Moreover, the interviews showed that the distribution of roles between these researchers was not clearly formalised and that the roles each of them assumed sometimes impinged on the role of their colleagues.

On the basis of this first diagnosis outlining an on-going redesign process of the project and these three project management issues, we began our work with the three project leaders. Our common objective was to build a reflective intervention in order to help the project leaders to redesign the project on an on-going basis by following their participatory and adhocratic intention, and in order to deal with the project management issues identified. In addition, we opportunistically adapted the intervention to the demands expressed by the project leaders (e.g. Meeting 2, to be equipped with conceptual issues regarding project management from an ergonomic point of view), and to the events of the project (e.g. Meeting 3, feedback on a seminar conducted during the intervention). We now first review the dynamic aspects of the intervention, and then consider its efficiency in relation to the enhancement of reflectivity.

Dynamics of the Reflective Intervention

The first meeting was used to build a common understanding of the project management and its stakes. Starting with our external understanding of the project, developed during the diagnosis, we intended to discuss and clarify this common understanding. We thus presented the main results of our diagnosis to the project leaders: the way we had understood the building of the project over time, the difficulties the researchers of the project had to define the precise objectives of the project and their current role in the transformation of the agricultural practices in the field, and their questions about the structuring of the project and the need to organise the work of the three project leaders. All these points were discussed with the project leaders. We then organised a participatory exercise to support the project leaders' thinking about their respective roles. This enabled them to clarify these roles, as shown in Fig. 2.

At the end of this first meeting, the project leaders asked for more information about the conceptual frameworks that we used to analyse their management. The second meeting was therefore intended to open up possibilities for the project leaders by giving some bibliographic elements about conceptual frameworks in relation to design, project management and participation. A focus was put on the notion of "adhocracy" which was often used by the project leaders. A discussion was then initiated on the transposition of these frameworks to the TATA-BOX project. This discus-

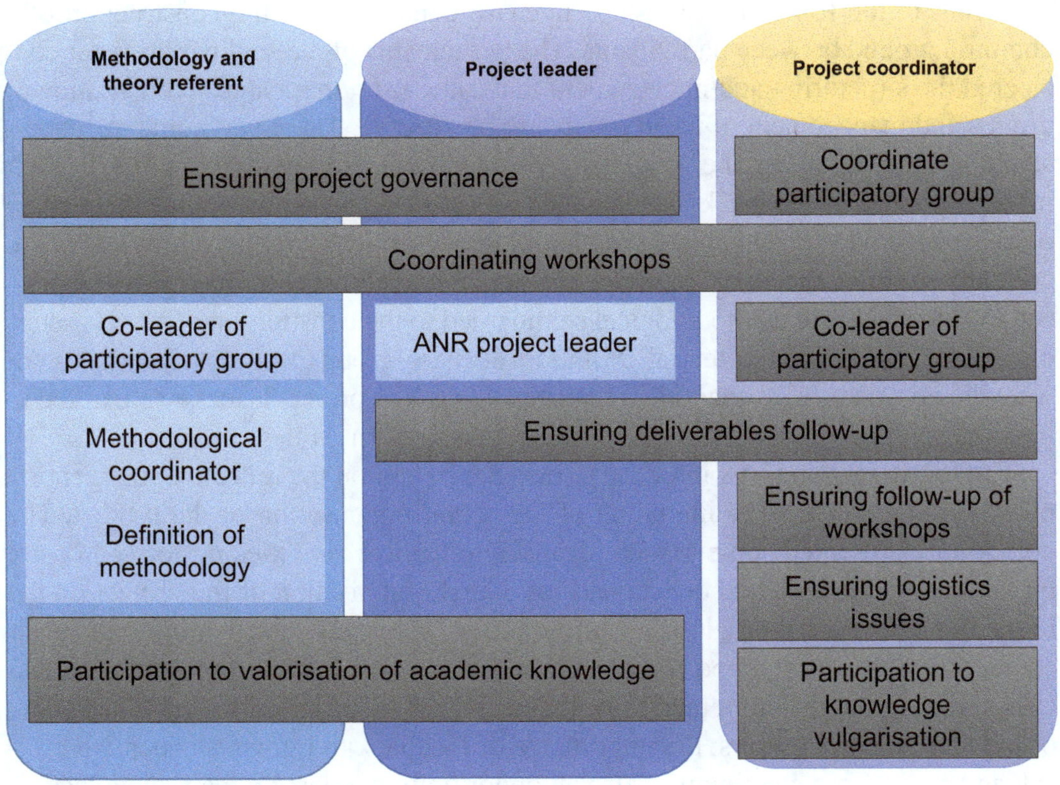

Fig. 2 Representation of project leaders' roles built by themselves

sion showed that, through the use of adjectives to qualify the project, like "participatory, evolutionary, adhocratic, adaptive and collective", the project leaders wanted to find a way to involve all the project participants. Their intention was to generate a dynamic, constantly re-designed that incorporated learning generated through its implementation. So what they wanted was responsiveness, collegiality and shared management. We argued that this type of management may be supported by a continuous reflective appraisal. As the project leaders still had difficulties in imagining how to implement such ideas, it was decided that we would demonstrate the use of feedback to feed reflection about their management and ways to make it evolve. We then decided to provide some feedback about the methodological seminar. This feedback was based on interviews with some of the participants and an analysis of the audio recording of the seminar, between the second and the third meeting.

The third meeting provided feedback on the "methodological seminar" of the project that was key to organising its management. This feedback was meant to allow the project managers to imagine how they could have run this seminar differently. Based on the ergonomist's analysis of the seminar, the project leaders could experience the gap between their initial intention, the actual execution of the meeting, and the participants' feedback. The seminar was structured around the three main topics planned by the project leaders: (1) positioning the respective research activity of each researcher; (2) building work groups of researchers; and (3) beginning to work in groups. The interviewed participants evaluated the interactions occurring during the first part as the most useful ones, for they helps them to develop a better understanding of each participant's objectives and tasks. Discussions and interactions occurring in the following two steps likewise contributed to the construction of a common ground between participants. The participants of the seminar however considered the structuring and management that came from these steps (constitution of groups and management of these groups) to be "fuzzy". There was some misunderstanding about the way the groups were constituted and some ambiguity in their management. This feedback was intended: (1) to be a probe to stimulate the design of alternative ways of organising methodological seminars in the future; and (2) to critically examine the actual participatory and adhocratic way of managing the project. For instance, the methodological seminar led to the constitution of seven groups of researchers, but the potential evolution of these groups was not discussed, nor were the criteria used to adapt the project *en route*. The project leaders were asked to imagine solutions to deal with these project management issues.

The fourth meeting was intended to be a debriefing on the reflective intervention proposed and a critical examination of the actual organisation of the project. The ergonomist presented some results regarding reflective activities performed or not by the project leaders, and questioned the actual organisation of the project on the basis of a synthesis of previous meetings and participants' contribution to the project feedback. It was an opportunity once again to go over the various elements discussed during the different meetings: the need for synchronisation, the functioning of the three project leaders, the importance of feedback to inform an adaptive project, and so on. This meeting was also an opportunity to question the project leaders about the intervention itself: what they had learned about managing the project, what they would have done differently, what they would do in the future about the organisation of the project, and what kinds of tools they would use to do so. We discuss these elements in the following section by combining the project leaders' comments during this fourth meeting with our own analysis of the efficiency of our intervention.

Efficiency of the Reflective Intervention

The objective of our intervention was to enhance the reflective postures of the project leaders regarding their project management activity. To evaluate in which way our methodology was successful in reaching this objective, we analysed the actual activities performed by the project leaders in the three meetings we organised, using the coding scheme presented in Table 2. Various analyses were carried out: we analysed the weight of the different degrees of reflectivity for each meeting and over time; we compared the profiles of each project leader and their evolution over time; we also worked on the links between the position of the ergonomist (what she does or asks when she intervenes) and the reflectivity it provokes. Finally, we collected the opinions of the three participants during Meeting 4.

To illustrate the analysis carried out from the transcriptions of the different meetings and the coding of the degrees of reflectivity, we discuss the comparison of the participants' postures between Meetings 1 and 3. The results of this analysis were consistent with several of what Mollo and Nascimento have called "Golden Rules of reflective practice". These rules do not precisely describe how to implement reflectivity but they draw the boundaries within which reflective methods "*may be deemed constructive*" (Mollo and Nascimento 2014). Four of them are defined as follows: "*focusing on the real aspects of work activity*"; "*a regular and perennial collective*"; "*the joint elaboration and evaluation of solutions*"; and "*the involvement and commitment of the hierarchy*" (Mollo and Nascimento 2014). The first two have an interesting illustration in the comparison of the participants' postures between Meetings 1 and 3. Figures 3 and 4 represent the distribution of postures in these meetings and show that the project leaders were actually in reflective postures in the sense of Jorro (2005) but to differing degrees. In both cases, "testimony" and "clarification" were the most important postures. Interestingly, the "proposal" posture – the highest degree of reflectivity – was more important in Meeting 3: "Feedback on methodological seminar", and no attitude of withdrawal – the lowest degree of reflectivity – was observed in this meeting.

There is thus a global improvement of the reflectivity from Meeting 1 to Meeting 3. Various factors explain these differences.

- We can assume that there was more trust between the participants of the reflectivity meetings as the process progressed, and that the project leaders had a better understanding of what the RG was trying to build with them. This could be linked with the rule of having a "regular and perennial collective".

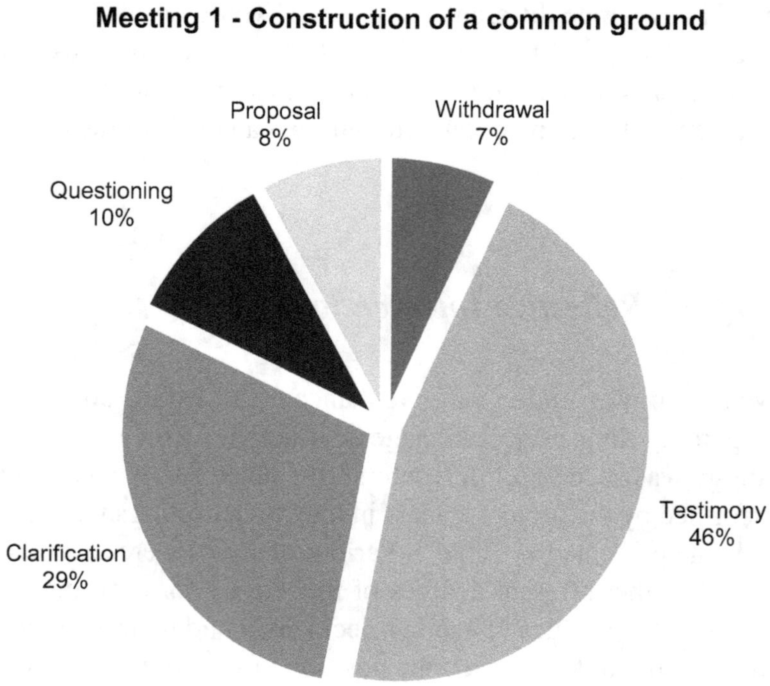

Fig. 3 Postures of participants in meeting 1 (% of contributions of the three project leaders)

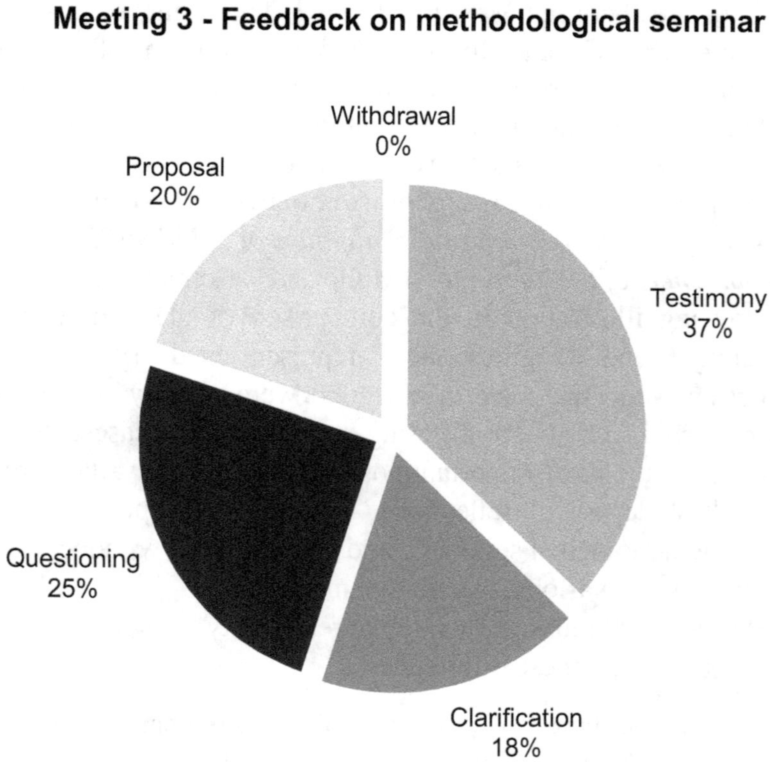

Fig. 4 Postures of participants in meeting 3 (% of contributions of the three project leaders)

- More discussion around "proposals" can also be linked with the objective of Meeting 3, which was to support the project leaders in imagining alternative ways of running methodological seminars in the future. Actually, the ergonomists' inputs were different. Meeting 1 was based on their diagnosis of the project. This meeting had been defined as an opportunity to clarify this diagnosis, which was questionable and encompassed the entire project. It was therefore logical that the project leaders were more in a posture of testifying and specifying, in order to feed the project diagnosis and foster mutual understanding with the RG.
- Finally, Meeting 3 was built on feedback from an extremely important seminar for the organisation of the project. This feedback was by the ergonomist, based on the minutes of the seminar and interviews with the participants. It was thus less questionable as it was based on actual observations, activity analysis and testimonies of their colleagues. This feedback thus contributed elements to help the project leaders in building a factual opinion about the methodological seminar. Figure 4 suggests that this is helpful to generate ideas to "do otherwise". This may outline the importance of grounding the proposition of alternatives ways of managing the project in discussions based on actual past experiences. This is in line with the Golden Rule "focusing on the real aspects of work activity" of reflective practice: "*The object of reflective practices must be work activity in the real world. To avoid 'drifting' towards a general discussion about work and life in the organisation, this practice may be supported by films, pictures or accounts of situations that emphasise the real conditions in which the work is carried out*" (Mollo and Nascimento 2014): 215–216).

Ultimately, this reflective intervention was an interesting opportunity for our RG as it allowed us to test different methods and tools to support a reflective process while trying to base our intervention, as much as possible, on the actual work and activity, in line with the principles of Activity-Centred Ergonomics. But what was the efficiency of our intervention from the participants' point of view? From the discussions in Meeting 4, we can see that their opinions were mixed. They appreciated having an opportunity to discuss the management of a project that was not easy for them. The fact of having secured monthly meetings on that subject was considered to be very positive as they would not have taken this time otherwise. They also explained that these meetings forced them to express some difficulties: "*you have identified a number of points, dysfunctions, problems, points to improve, and for that you guided us in the discussions, or you have even implemented workshops to formalise things that were very implicit in our mode of operation (…), I often had the impression that you led us to explain the implicit*" (one of the project leaders). However the project leaders expressed a lack of effective help for managing the project, as the following very interesting discussion in Meeting 4 illustrates:

- *I have the impression that we've taken dedicated time for discussions between us and that you've equipped us to discuss matters between us, and you've supported us in discussions. This is always very positive and constructive, but I didn't feel that I was adequately equipped with tools to lead the project. (project leader).*

- *I'm surprised that you say there are no tools, especially methodological. (...) We used many different methods to animate this reflective device and our idea was that these methods should also help you to continue to animate the work of organising the project with other researchers. (ergonomist)*
- *I didn't think at all that the methods you used to get us to discuss and to be reflective about the project were potentially methods to animate the project. And it seems that you thought we were aware that the methods you put on the table to animate these meetings were potentially methods that could have been used to animate the project (project leader).*

This interaction suggests that our second objective of showing the project leaders how they could, on their own, build some reflective areas for the project people in order to support their intention of managing their project in a "participatory, collective, evolutionary, adaptive and adhocratic" way was largely missed. Whereas we had the impression of having expressed this objective clearly during the Proposal Meeting and through the participatory exercises, this was obviously not the case. As such, even though our intervention was an interesting step for the project leaders, its continuity was compromised.

Discussions and Perspectives

The TATA-BOX project leaders definitely had a challenging task of managing a research project in a "participatory, collective, evolutionary, adaptive and adhocratic" way – an intention they had expressed to be consistent with their understanding of transition processes in agriculture. A crucial question is how to manage a project to make it innovative. In fact, traditional project management has been criticised extensively when it comes to innovative design that "*render[s] its hypothesis (i.e., the ability to identify a clear objective, to plan the work, etc.) irrelevant*" (Lenfle et al. 2016). And the particularities of the transition processes in the agricultural world show that these hypotheses are currently largely irrelevant. Acknowledging this, the TATA-BOX project intended to have an innovative management by involving all the researchers of the project in the decision process and by being adaptive. Although several studies have advocated the use of adaptive and iterative modes of design management in agriculture, they are mostly conceptual (Le Gal et al. 2011; Meynard et al. 2012) and fall short of proposing methodological tools to support action. Or when they do describe how to implement this type of management (Giller et al. 2011), they are most often focused on only one iteration (Dogliotti et al. 2014; Falconnier et al. 2017), which raises the question of the management of such processes over the long run and the ways to support it. TATA-BOX was precisely a project intended to address such questions. But our diagnosis of the project management after just 1 year highlighted the fact that the participants had not sufficiently discussed the project objectives and the conceptual and methodological framework together. In terms of management, our diagnosis showed a discrepancy between the intention of the project leaders and their actual possibilities. Confronted with the need not only to produce deliverables and to report to the ANR, but also to manage a project whose objectives and framework were not totally clear for every participant, they had to take a lead in a way that was at odds with their intended adhocracy. The reflective intervention we built with them over 6 months was intended to give them an opportunity to discuss these discrepancies and how to solve them.

As such, the reflective intervention proposed was one of the first attempts, as far as we know, to implement Activity-Centred Ergonomics proposals around the development of reflective intervention, which is part of a more general project of conceptualisation and operationalisation of enabling intervention (Barcellini 2015, 2017 for a synthesis, and Arnoud and Perez Toralla 2017). Although our results show that our intervention succeeded in provoking reflectivity among the project leaders, it partly failed to enable them to manage the project further. This reveals that methods supporting reflective activity *per se* are not sufficient to support the development of a redesign activity in project management. It would have been necessary to couple them with methods explicitly targeting a more projective activity (e.g. Chizallet et al. 2018) submitted), that is, an activity dedicated to re-designing the activity at stake, using tools such as organisational simulation (Barcellini and Van Belleghem 2014). In this sense, the workshops organised would have benefitted from more applied exercises to help the project leaders in addressing very real difficulties of project management. It may have helped to transform this reflectivity into a more projective activity. In other words, our intervention supported the constructive activity of the project leaders but not through to the end. If we consider that our intervention was expected to support the design of a project management, we can see that it supported the first typical steps of a design process, that is, synchronisation and grounding between the project leaders regarding the actual situation and the goals to reach, but that it stopped before totally supporting the generative step of design processes. This was attempted with the exercises that concluded each meeting, but they were obviously not linked enough to the actual management issues of the project leaders. This may suggest that moments of reflective and projective activities should have been distinguished over time in a longer intervention. Moreover, due to the financial constraints of the funding programme, the RG was provided with funding for only two 6-month interns. Looking back, this funding was largely insufficient compared with the ambition of the RG. An ergonomist recruited specifically to monitor the project leaders over the long run would have been necessary to support them throughout the project and to carry out additional analyses of the researchers' actual work in the project and with the stakeholders (e.g. Chap. 11). This would have been more in line with the objective.

As a last point of discussion, we would like to come back to the tension we evoked between the idea of clarifying the objectives and framework of the project, and the idea of leaving them vague enough so that participants are not excluded. We have the feeling that these two approaches are not necessarily mutually exclusive. If openness is needed for complex concepts, to build certain interdisciplinary work, this does not mean that such openness should be experienced as ambiguity. It would be better to collectively acknowledge it so that it becomes a resource for the collective and not a source of confusion. This is in line with the idea of integrating uncertainty into the management of "exploratory" projects rather than managing to progressively reduce uncertainties (e.g. Lenfle 2016). Our proposition to use reflectivity to manage the project precisely aimed at making this uncertainty visible, in order to make it manageable. The reflective tools did allow for a representation of actual actions performed by participants when confronted with actual situations, which made it possible for the project leaders and the project researchers to adapt, correct, and reorient their action. In that sense, the reflectivity was used to feed the adaptive character of the project. More broadly, the points that we discussed with regard to the design of the TATA-BOX project management are also significant for other design processes in agriculture. What was at stake in the TATA-BOX management was the project researchers' ability to build by themselves an organisation that would be efficient and adaptive. There was then a challenge to articulate a design direction or intention (that is to say, an intention for the future, a goal that directs the design project) to a continuous adaptation of the actual situation. Looking at the design processes of agricultural systems, we can find the same stake and the same challenge (Prost et al. 2018). Given the complexity and uncertainties of designing agricultural systems, farmers should be reconsidered as designers of their own production systems. The role of research agronomists in the design processes is consequently being called into question: their role is seen more as a support for farmers' design activity than as a substitute for it. These researchers are therefore confronted with the same challenge as the project leaders of TATA-BOX: they have to find how to support the farmers in the design of their own agricultural systems. To do so, they can feed a design direction by helping the farmers to be innovative. At the same time, acknowledging the nature of design processes, they also need to assess and show the effects of the design solutions implemented in actual situations, in order to allow the farmers to iteratively adapt their design processes. This would require identification of the appropriate tools.

References

Arnoud J, Perez Toralla M-S (2017) L'intervention capacitante: quels enjeux pour la pratique de l'ergonome? Activités. https://doi.org/10.4000/activites.3042

Barcellini F (2015) Développer des interventions capacitantes en conduite du changement. Comprendre le travail collectif de conception, agir sur la conception collective du travail. Université de Bordeaux

Barcellini F (2017) Intervention Ergonomique Capacitante: bilan des connaissances actuelles et perspectives de développement. Activités. https://doi.org/10.4000/activites.3041

Barcellini F, Van Belleghem L (2014) Organizational simulation: issues for ergonomics and for teaching of ergonomics' action. In: ODAM 2014 (11st Human factors in organizational design and management conference)

Barcellini F, Van Belleghem L, Daniellou F (2014) Design projects as opportunities for the development of activities. In: Falzon P (ed) Constructive ergonomics. CRC Press, Boca Raton, London, New York, pp 150–163

Béguin P (2003) Design as a mutual learning process between users and designers. Interact Comput 15:709–730

Chaubet P (2010) Saisir la réflexion pour mieux former à une pratique réflexive: d'un modèle théorique à son opérationnalisation. Éducation Francoph 38:60–77

Chizallet M (2015) Conduire l'activité réflexive pour soutenir la conduite de projet: une étude exploratoire. Master Sciences Humaines et Sociales, Mention CNAM, Travail et Développement, Spécialité Ergonomie

Chizallet M, Barcellini F, Prost L (2018) Supporting farmers' management of change towards agroecological practices by focusing on their work: a contribution of ergonomics. Cah Agric 27:35005. https://doi.org/10.1051/cagri/2018023

Daniellou F (1992) Le statut de la pratique et des connaissances dans l'intervention ergonomique de conception. 1992. Université de Toulouse-Le Mirail

Daniellou F, Rabardel P (2005) Activity-oriented approaches to ergonomics: some traditions and communities. Theor Issues Ergon Sci 6:353–357

Détienne F (2006) Collaborative design: managing task interdependencies and multiple perspectives. Interact Comput 18:1–20. https://doi.org/10.1016/j.intcom.2005.05.001

Dogliotti S, Rodríguez D, López-Ridaura S et al (2014) Designing sustainable agricultural production systems for a changing world: methods and applications. Agric Syst 126:1–2. https://doi.org/10.1016/j.agsy.2014.02.003

Duru M, Fares M, Therond O (2014) A conceptual framework for thinking now (and organising tomorrow) the agroecological transition at the level of the territory. Cah Agric 23:84–95. https://doi.org/10.1684/agr.2014.0691

Duru M, Therond O, Fares M (2015) Designing agroecological transitions; a review. Agron Sustain Dev. https://doi.org/10.1007/s13593-015-0318-x

Falconnier GN, Descheemaeker K, Van Mourik TA et al (2017) Co-learning cycles to support the design of innovative farm systems in southern Mali. Eur J Agron 89:61–74. https://doi.org/10.1016/j.eja.2017.06.008

Falzon P, Darses F (1996) La conception collective: une approche de l'ergonomie cognitive. Coopération et conception, Octarès, 123–135

Giller KE, Tittonell P, Rufino MC et al (2011) Communicating complexity: integrated assessment of trade-offs concerning soil fertility management within African farming systems to support innovation and development. Agric Syst 104:191–203. https://doi.org/10.1016/j.agsy.2010.07.002

Guérin F, Laville A, Daniellou F, et al (2006) Comprendre le travail pour le transformer, la pratique de l'ergonomie. Lyon

Holling CS (1978) Adaptive environmental assessment and management. John Wiley & Sons, London

Jorro A (2005) Réflexivité et auto-évaluation dans les pratiques enseignantes. Mes évaluation en éducation 27:33–47

Le Gal PY, Dugué P, Faure G, Novak S (2011) How does research address the design of innovative agricultural production systems at the farm level? A review. Agric Syst 104:714–728. https://doi.org/10.1016/j.agsy.2011.07.007

Le Goff JL (2014) La réflexivité dans les dispositifs d'accompagnement: implication, engagement ou injonction? ¿ Interrog? 11

Lenfle S (2008) Exploration and project management. Int J Proj Manag 26:469–478

Lenfle S (2016) Floating in space? On the strangeness of exploratory projects. Proj Manag J 47:47–61. https://doi.org/10.1002/pmj.21584

Lenfle S, Le Masson P, Weil B (2016) When project management meets design theory: revisiting the Manhattan and Polaris projects to characterize 'radical innovation' and its managerial implications. Creat Innov Manag 25:378–395. https://doi.org/10.1111/caim.12164

Meynard JM, Dedieu B, Bos AP (2012a) Re-design and co-design of farming systems: an overview of methods and practices. In: Darnhofer I, Gibbon D, Dedieu B (eds) Farming systems research into the twenty-first century: the new dynamic. Springer, Dordrecht, pp 407–431

Mollo V, Nascimento A (2014) Reflective practices and the development of individuals, collectives and organizations. In: Falzon FD-P (ed) Constructive ergonomics, pp 223–238

Perrenoud P (2001) Développer la pratique réflexive dans le métier d'enseignant
Petit J, Querelle L, Daniellou F (2007) Quelles données pour la recherche sur la pratique de l'ergonome? Trav Hum 70:391–411
Prost L, Reau R, Paravano L et al (2018) Designing agricultural systems from invention to implementation: the contribution of agronomy. Lessons from a case study. Agric Syst 164:122–132
Samurçay R, Rabardel P (2004) Recherches en didactique professionnelle, Octarès. Toulouse
Schön D (1983) The reflective practitioner: how practitioners think in action. Basic Books, New York
Teiger C, Falzon P (1995) Construire l'activité. Performances Hum Tech Hors Série 34–40
Vinck D (2009) De l'objet intermédiaire à l'objet-frontière, Vers la prise en compte du travail d'équipement. Rev. d'anthropologie des connaissances 3:51–72
Vinck D (2011) Taking intermediary objects and equipping work into account in the study of engineering practices. Eng Stud 3:25–44. https://doi.org/10.1080/19378629.2010.547989
Visser W (2009) Design: one, but in different forms. Des Stud 30:187–223. https://doi.org/10.1016/j.destud.2008.11.004
Walters CJ (1986) Adaptive management of renewable resources. MacMillan Pub Co, New York

Evaluation of the Operationalisation of the TATA-BOX Process

Marie Taverne, Sarah Clément, Lorène Prost, and Flore Barcellini

Abstract The chapter evaluates how the TATA-BOX process supported the collective design of an agroecological transition. In order to carry out this evaluation, we interviewed a panel of 24 participants about their experience of the process and their opinions on it. In this chapter we set out the results in relation to three questions: How did the workshops go? What characterised the outputs? What effects were identified? On these bases, we discuss some possible improvements in the TATA-BOX process and the ways in which this process supported the design of an agroecological transition. We show in particular that the TATA-BOX process successfully initiated a collective design process as it allowed the participants to establish a common ground, define a range of goals to meet, and identify actionable means that could help to reach these goals. The process will nevertheless have to be continued through actual implementation. Various actors will most likely take responsibility for limited actions, rather than for the territorial agricultural transition project in its entirety. They will select the design solutions they need and might revise them. The TATA-BOX participatory process thus appears to be one step in the process of designing the territory's transition.

M. Taverne (✉) · S. Clément
Territoires, Université Clermont Auvergne, AgroParisTech, INRA, Irstea, VetAgro Sup, Aubière, France
e-mail: marie.taverne@irstea.fr; sarahclement34@gmail.com

L. Prost
LISIS, INRA, EPNC, ESIEE, Université Paris-Est Marne La Vallée, Marne-la-Vallée, France
e-mail: lorene.prost@inra.fr

F. Barcellini
CRTD, CNAM, Paris, France
e-mail: flore.barcellini@cnam.fr

J.-E. Bergez et al. (eds.), *Agroecological Transitions: From Theory to Practice in Local Participatory Design*, https://doi.org/10.1007/978-3-030-01953-2_11

Introduction

Many researchers today argue that the issues facing French agriculture call for an in-depth redesign of farming systems and territories (Martin et al. 2012; Meynard et al. 2012, 2017; Duru et al. 2015; Prost et al. 2017). This would require the development of capabilities around design in the agricultural world, from both a methodological and a theoretical point of view. In this regard, the TATA-BOX project (cf. chapter "TATA-BOX at a Glance") is original in two ways: its subject of research, namely how to support the design of an agroecological transition (AET) across a territory; and its strategy, which is to wager on "learning by doing". In order to produce scientific knowledge around this research topic, the project's researchers chose to implement a proposal for a design support process, on their own. This is demonstrated by the reflexive work carried out throughout the project, described in chapter "Towards a Reflective Approach to Research Project Management". It also implies being able to evaluate *a posteriori* the process that was implemented.

This process can be defined through two main features: it is participatory, and it is intended to support design. The evaluation of participatory methods is the subject of an abundant literature in the agricultural world, in the fields of agricultural development and the impact of agronomic research (Alvarez et al. 2010; Perez et al. 2010; Joly et al. 2015; Thornton et al. 2017). This literature emphasises the variety of dimensions of evaluation on contrasting scales. The evaluation of design methods is also a subject that has been addressed extensively, particularly in the field of Design Studies (Moultrie et al. 2006; Clevenger and Haymaker 2011; Détienne et al. 2012; Hatchuel et al. 2016). Depending on the case, the scope of this evaluation covers the product designed (value, originality, etc.), design as a process (quality of collaboration, team performance, effectiveness, etc.), or the learning derived from these processes.

Taking into account the objective of the TATA-BOX research project and the issue of developing design capabilities, we chose to evaluate the TATA-BOX process by applying design process evaluation frameworks. We accordingly focused both on the course of the process and on its outputs. This type of choice raises a number of methodological questions, for these frameworks were not originally intended for designing objects such as an AET. Therefore, wondering about how to evaluate the design process and its results afforded us the opportunity to rethink the methods, in order to carry out this evaluation. How did the workshops go? What characterised the outputs? What effects were identified? To answer these questions and to evaluate how the TATA-BOX process supported the design of an AET, we chose to survey a large panel of participants. Interviews were thus held subsequent to the design process. The intention was to evaluate participants' experience of the process, while remaining as true as possible to their statements in order to be able to pass these statements on to project collaborators and promote their reflection *a posteriori* on the process proposed.

Materials and Method

The TATA-BOX Participatory Process (cf. Also Chapter "Participatory Methodology for Designing an Agroecological Transition at Local Level")

The participatory process implemented as part of the TATA-BOX project is a variant of the methodological framework of Duru et al. (2015). It is composed of three workshops with different goals. These goals are defined as follows in management documents of the TATA-BOX process:

- Workshop 1: "Development of a common understanding of the situation", "Co-construction of an inventory of farming issues";
- Workshop 2: "Construction of a shared vision of forms of farming to develop, and of the organisation of a Territorial AgroEcological System", "Creation of forms of farming to develop";
- Workshop 3: "Design of intermediate states of the transition and identification of monitoring indicators", "Design of local farming governance for this transition", "Creation of transition pathways and governance to develop to achieve these desired forms of farming".

These three workshops were implemented in two fields of study, PETR (Territorial and Rural Balance Pole) of Midi-Quercy and Centre-Ouest Aveyron.

Material Collected

Twenty-four people were interviewed at the two fields of study for the project: 14 in Midi-Quercy, and 10 in Centre-Ouest Aveyron (cf. chapters "TATA-BOX at a Glance" and "Participatory Methodology for Designing an Agroecological Transition at Local Level"). The people who had participated in the 3 workshops of the process were targeted as a priority (we met with 13 of them). Interviews were also held with 7 people who had participated only in workshop 3, with 3 people present in workshops 1 and 3, and with 1 person present in workshops 1 and 2 (Fig. 1).

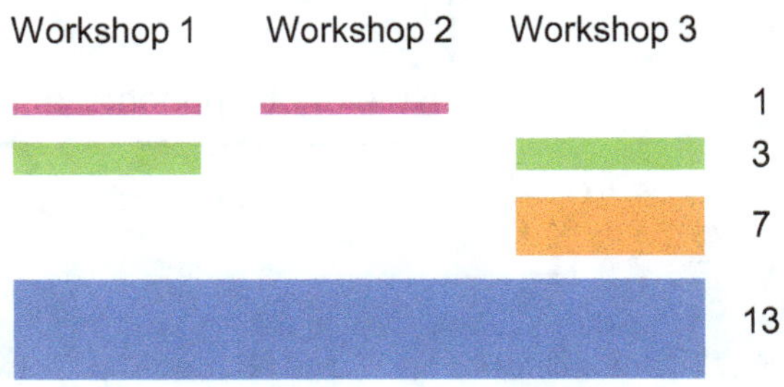

Fig. 1 Among the 24 interviewees, the number of participants in the different workshops

The data were collected during a semi-structured face-to-face interview with each of these people, with five main components: their expectations regarding workshops, how the workshops went, the results, the position of the process compared to others, and its influence. Photos of the different sequences of days were shown to them, to help them to remember.

Data Analysis

This data were analysed based on three main components: (i) the course of the participatory process, (ii) its direct results, and (iii) its effects (Fig. 2).

First, analysing the course of the workshops was intended to identify the real-life experience of the interviewees with regard to the process and the forms of interaction that took place between participants. Discourses were examined both qualitatively and quantitatively. On the qualitative level, the components of the process

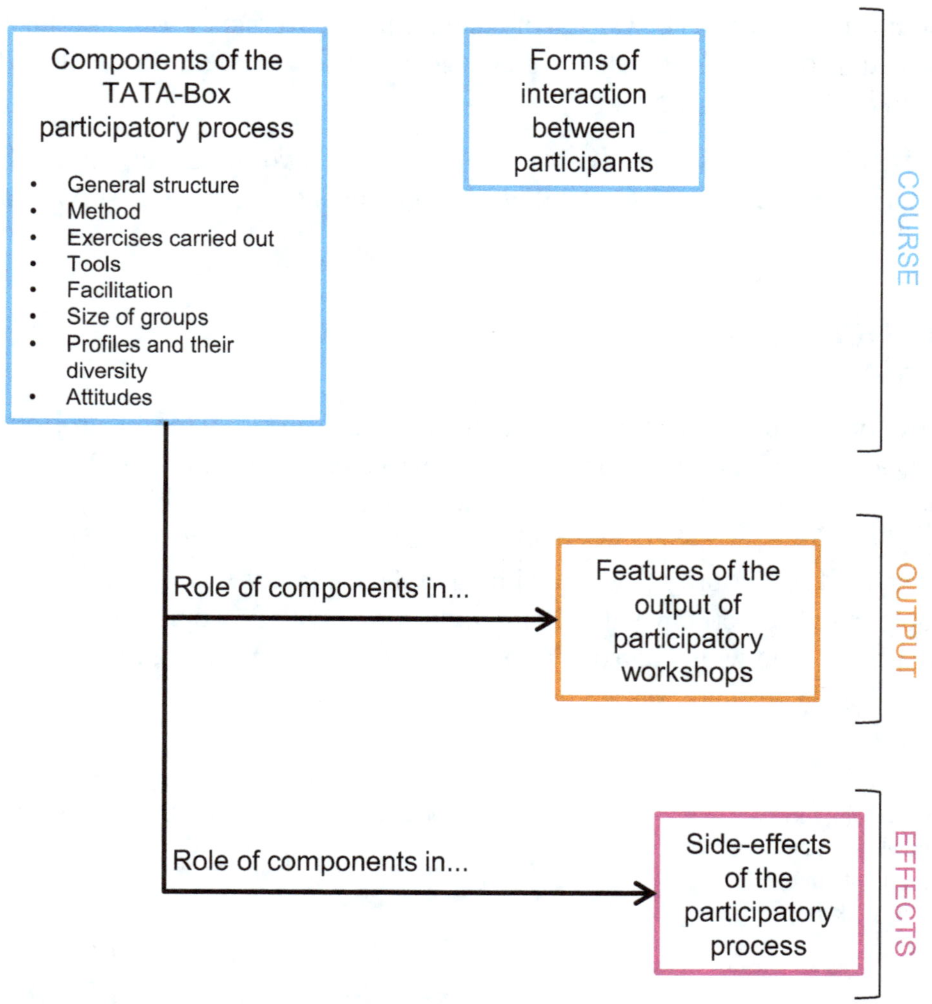

Fig. 2 Summary of objects and relationships between the objects analysed

discussed by the interviewees were initially detected by the analyst inductively: its general structure, the method, the different exercises carried out, intermediary work objects, the facilitation, the size of work groups, the diversity of participant profiles, and the participants' attitude.[1] The underlying meaning of the statements, whether positive, negative, or mixed-undetermined, was then defined. Likewise, discourses on the forms of interaction between participants were detected and categorised inductively (exchanges of viewpoints, articulation of viewpoints, discussions, consensus building), and then distinguished based on the meaning of the words. NVivo 10 software was used to codify the different categories of statements and then to quantify the number of contributing interviewees. For each component of the process and each category of interaction between participants, a second qualitative analysis "on paper" was then carried out inductively to describe the discourses in more detail. This at times led to adjustments in the coding carried out by NVivo.

The second thrust of analysis aimed at recording interviewees' points of view regarding the features of the outputs of the process. The interview guide contained targeted questions on the exhaustiveness, accuracy, originality, and degree of completion of outputs. These dimensions stemmed from interaction with the designers of the participatory process. They reflected these designers' effectiveness criteria for the process, concerning the quality of the result. The interviewees sometimes also addressed other features of the outputs. Interviews were processed quantitatively and qualitatively. NVivo software was used to: (1) allocate spoken elements to the categories of "exhaustiveness", "accuracy", "originality", "degree of completion", and "other"; (2) categorise these speech elements based on the meaning of the statement, whether positive, negative, or mixed-undetermined; (3) quantify the number of contributors to these different statement categories. Like previously, each statement category was then qualitatively analysed (inductively) to describe the statement in more details, which sometimes led to readjustment of the coding in NVivo.

Lastly, the third analysis component consisted in identifying the categories of effects mentioned by the interviewees and in describing their experience and expectations regarding these side-effects. Interviews were processed qualitatively. The content of statements was inductively grouped into thematic categories.

[1]As participants' attitude often appeared in their statements, it was treated like the other components of the process. It pertained to more than just the process, but this process explicitly aimed at allowing each person to express him- or herself without judgement, with balanced turn-taking to speak.

Results

The Course of the Participatory Process

A Process Deemed to Be Positive Overall

Table 1 shows that the method, exercises carried out, tools, facilitation, size of groups, and participants' attitudes are more often the subject of positive statements by the interviewees than negative statements. The exercises carried out and the work tools nonetheless present a significant number of "mixed, undetermined"-type statements. Statements on the diversity of participant profiles appear to be mixed. Those regarding the general structure of the process, which encompasses various aspects, appear to be negative somewhat more often. While this initial analysis provides a general idea about interviewees' feelings, it hides a wide diversity of experiences. The detailed analysis of the corpus for each component allowed us to identify the main ideas put forth and to group them together.

Diversity of Experiences Among the People Interviewed Regarding Each Component of the Process

Regarding the General Structure of the Process

General Feelings

There were few general expressions of feelings regarding the process, but they were positive (*"left thrilled", "only good things", "it went really well"*, etc.) – although they did not stop the people in question from making some criticism or pointing out difficulties, which are presented below.

Table 1 Number of interviewees with a positive, negative, or mixed or undetermined discourse regarding each of the components of the process; the higher the number, the darker the colour

	Number of interviewees that discussed the component	Positive	Negative	Mixed, undetermined
0: General structure of the process	22	11	18	10
1: Method	18	14	4	2
2: Exercises carried out during the day	20	15	10	9
3: Tools	18	14	9	6
4: Facilitation	17	13	5	3
5: Size of groups	8	7	1	0
6: Profiles and their diversity	24	17	21	14
7: Participants' attitudes	12	10	3	1

Institutional and Territorial Framework

The interviewees' statements highlight the original nature of the process framework, which is out of the ordinary and contains some degree of neutrality. The invitation from a research institute (INRA – French National Institute for Agricultural Research) facilitated the inclusion of people with different sensitivities, and participants contributed on an equal footing, which is not the case of typical hierarchical or institutional frameworks (two interviewees). However, some of them found it unfortunate that the process was not designed in closer conjunction with the territory, and in particular other initiatives underway, to better mobilise actors and to ensure the continuity of the work initiated by the TATA-BOX process afterwards. One interviewee was positively surprised by the attendance of so many INRA engineers to put local actors to work (*"at a place that we know, a local place"*).

Overall Dynamics of the Process

Three subjects were brought up in relation to the overall dynamics of the process. First of all, the process dimension was seen in a positive light (3 interviewees): the fact of defining a point of departure, a point of arrival, and reflecting on the pathway to reach it seems to have been particularly important. Next, the interrelation between the three workshops (understanding of the logic, appropriating and re-using the results from one workshop to another, iterations carried out on the results) was perceived in a variety of ways. One participant who joined the process during Workshop 3 spoke of the difficulty of understanding the goal of the undertaking, whereas another one insisted on the clarity of the overall interrelation of the workshops. Five participants mentioned difficulties in appropriating and re-using the results because the summary was not available before the following workshop, although for some this was offset by the way that the previous work was summed up during meetings (e.g. rich picture and cognitive map). The progression between the outputs of work sessions and those of the research team from one workshop to the other was viewed positively, even though the interviewees felt that the visibility of this work between workshops and the nature of the conclusions drawn from it could be strengthened. To conclude, the length of the process (1.5 years) and the intervals between workshops (6 months to 1 year) partially explain difficulties in appropriating and re-using results. Several interviewees saw this as a hindrance to memorising the work carried out during the previous workshop and to maintaining the involvement of the same actors throughout the entire duration of the process (especially when some of them had changed jobs in the meantime). Some people suggested the creation of an "intermediate session" to keep participants in the process between workshops (3 interviewees).

The Meeting Days

Statements on meeting days concern three aspects. First, several interviewees felt that the duration of single-day meetings was too long, taking into account availability or the intensity of the work. Workshop 3 was nevertheless perceived as being too short to achieve the established goals. Second, the calibration of the time dedicated to the different exercises was viewed positively. The diversity of the types of exercises within a single day allowed people "*to not get bored*". However, it also sometimes limited in-depth exploration of certain subjects (1 interviewee). These dynamics also had differing effects on people. For example, two interviewees said that they finished the days "*fried*" or that it was "*gruelling*", whereas another one said, on the contrary, that she "*felt great*" at the end thanks to the format of these days. Third, the interrelation between these activities was experienced very positively ("*it flowed*") and certain people even specified that this fluidity was not typical in the meetings in which they usually participated.

The Method

A few statements regarding the method remained general and were positive. Three commonly addressed topics stand out. First, the comprehensive nature of the exercise set it apart from other processes. This was due to the diversity of the subjects addressed and the sheer range of conditions of the AET that had to be specified in the design of the pathway (2 interviewees), as well as the structuring of reflection in three domains (agricultural production, natural resources, and agricultural supply chains) (2 interviewees). Next, according to several participants, the interactive dimension allowed a variety of people to work together in a livelier way than with more classic approaches (1 interviewee), and helped them to envision future scenarios (1 interviewee) – a tricky activity that was nonetheless desirable in territorial reflection. Two interviewees nonetheless expected a more analytical and in-depth posture: one thought it unfortunate that neither a more in-depth quantified inventory of the state of local agriculture, nor forecasting scenarios, were done; the other emphasised the role of the organising scientists in the production of the results, allocating them a role of data analysis and the creation of proposals based on the work during workshops, which appeared to be reduced to data collection. Lastly, the method was deemed to be effective: it was productive (emergence of ideas, concretely specifying them in writing, effective resumption) without losing time due to its reflexive structuring and framework, a structure perceived as being non-restrictive (2 interviewees: "*It didn't feel like we were working hard, I mean it wasn't a constraint*"), contrary to other processes.

The Exercises Carried Out During the Workshops

At the time of the interviews, the majority of the interviewees no longer remembered the details of the different workshops. The photos presented by the interviewer helped, but not all of the exercises were systematically addressed. The interviewees gave the most details on Workshop 3, which was more recent (Table 2).

Table 2 Number of interviewees that had positive, negative, or mixed or undetermined statements on each workshop

	Number of interviewees who gave their opinion of the workshop	Positive	Negative	Mixed, undetermined
1: Workshop 1	9	8	0	1
2: Workshop 2	7	4	1	6
3: Workshop 3	17	13	8	6

The Main Exercises

Most points of view on Workshop 1 were positive and mainly concerned the benefit of building a common and structured representation of the current situation (Fig. 3). Only one interviewee had a mixed opinion (he felt that he was barely involved due to the general nature of the exercise). Statements regarding Workshop 2 are often vague (Fig. 4). Several people had only a vague memory of the course of that workshop and the ideas originating from it. Concerning Workshop 3, the statements of eight interviewees demonstrate their interest in the method for designing the transition pathway (Fig. 5). The originality of this method compared to the previous workshops and what the interviewees were accustomed to practising was often highlighted. Several interviewees, and sometimes the same people, nonetheless noted that this exercise was complicated or difficult, and spoke of a long start-up phase to understand the meaning of the tools and what had to be done. Difficulties persisted at times: (1) in finding the meaning of the ideas resulting from previous workshops embedded within tools; and (2) in providing the elements requested when the group did not have the required capabilities. Lastly, a lack of time was often mentioned for Workshop 3. For this reason, two interviewees would have liked to have limited the number of goals defined during Workshop 2.

Additional Exercises

The "prioritisation" of issues (Workshop 1) or of goals (Workshop 3) with the help of coloured stickers was often spontaneously mentioned as something that participants liked (Figs. 3 and 5). The same goes for the icebreaker implemented during Workshop 3, except for one person who claimed to have been unsettled by it (Fig. 5) ("*I don't know. She gave us those cards, and you had to imagine something or rather, and I didn't really understand what the point was, or if there even was one*"). Some participants said they did still wonder, at the time of the interview, if there was any connection between this icebreaker and the remainder of the exercises carried out during the day. Lastly, two interviewees made statements regarding plenary meetings. The first, intrigued by the consensus obtained during Workshop 2 despite

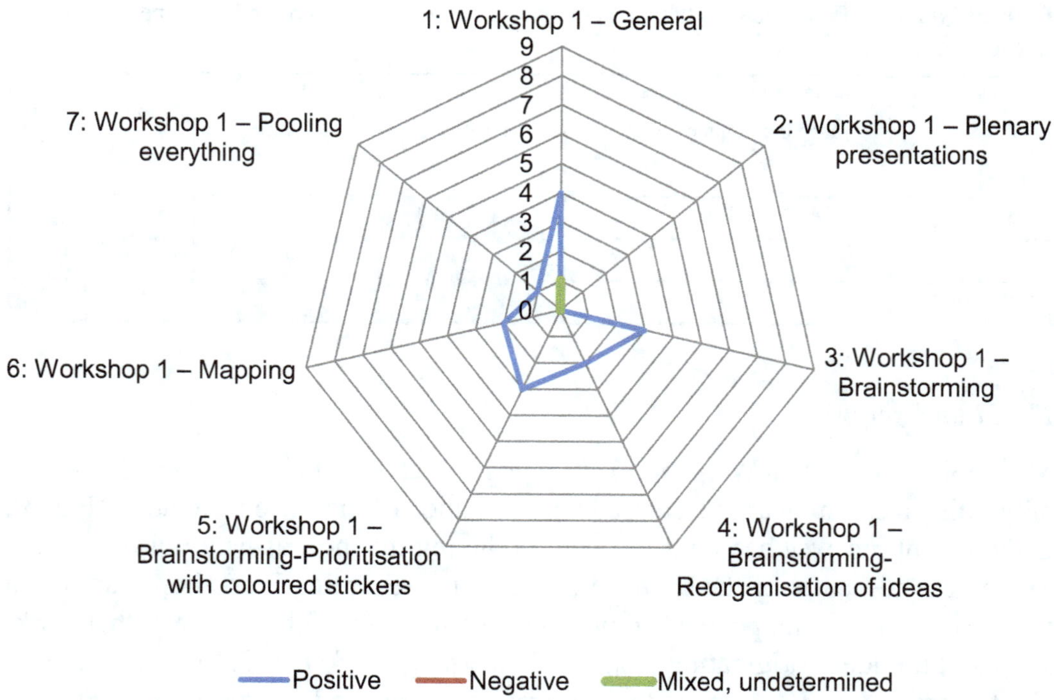

Fig. 3 Number of interviewees who gave their opinion on the exercises of Workshop 1

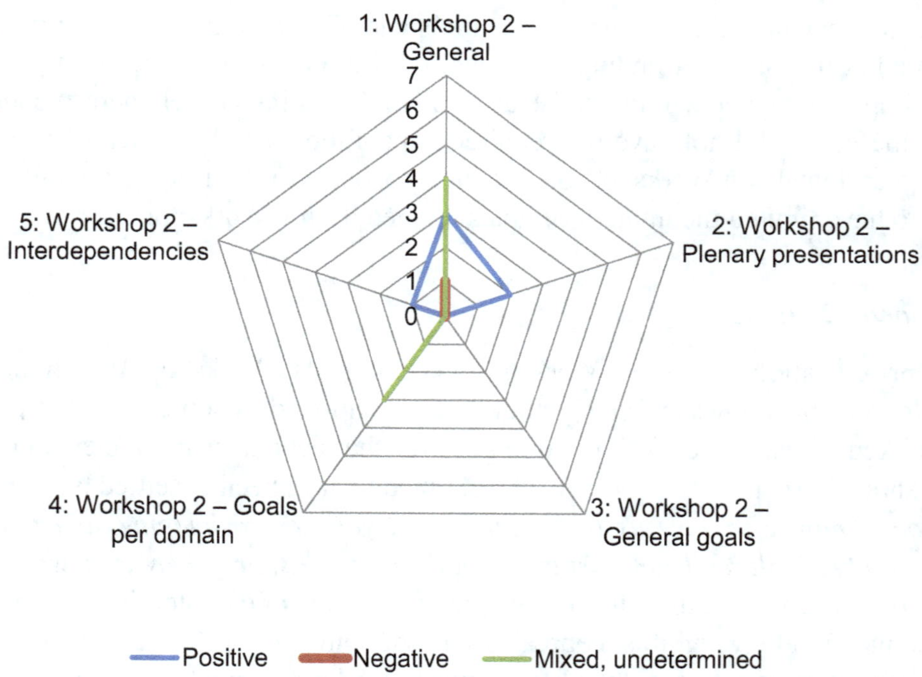

Fig. 4 Number of interviewees who gave their opinion on the exercises of Workshop 2

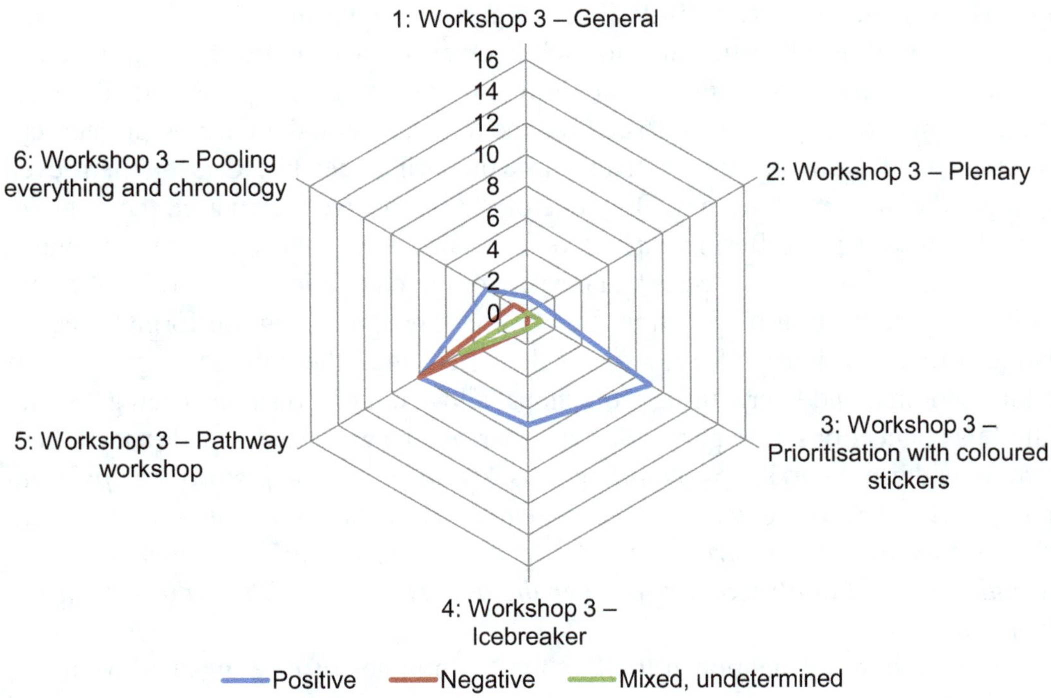

Fig. 5 Number of interviewees who gave their opinion on the exercises of Workshop 3

the diversity of participant profiles, wondered about the role that could have been played by the introductory presentations during the plenary meeting (cf. chapters "TATA-BOX at a Glance" and "Participatory Methodology for Designing an Agroecological Transition at Local Level"). The second person found the way of carrying out the plenary meeting during Workshop 3 to be highly original and interesting: the meeting was short and graphic tools were used to summarise the results of previous workshops.

The Tools Used

The tools mentioned during interviews are mainly those associated with Workshop 3.

Tools for Summarising the Results of Previous Workshops

Seven of the interviewees found the rich picture[2] very helpful, although two of them felt that it needed to be supported by a presentation. Various qualities were attributed to it: very clear, explicit, complete ("*it's all there. [...] I didn't see any*

[2] Broad range of terms used to refer to the rich picture: "the diagram with all the relationships", "that sort of sphere", "the little potato with circles", "that sort of big map with the connections to everything", "that nice overview diagram", "the diagram presented at the beginning of the meeting", "that famous map that makes me dream".

omissions"), interesting, well-designed, figurative, educational, great, "*It really suits me*". It allowed participants to look back on what had already been done (cf. section "Regarding the general structure of the process"), presented all interrelations, made people reflect, structured the reflections on the farming project represented, and participants effectively used it during the day. One of them even thought that he would reuse it. Only one of the interviewees criticised the rich picture, because it put differing aspects on the same level, and contained too much information. Three interviewees had positive or mixed opinions concerning the cognitive map presenting the relationships between goals in diagram format: despite being difficult to interpret, they found that it clarified the choices of goals to be addressed first, and later, during workshops, allowed them to situate each goal with all of its connections. One participant summarised the way that he made use of these two tools during workshops: "*it allowed us to remember [the last workshops], that picture with the main guidelines that were adopted, with the images, and we would say to ourselves, 'that's true, we did say that, we said that', and afterwards, we would say 'but I don't really remember the details very well', and there, bam, we had that map*".

Six interviewees mentioned the fact that no minutes of meetings had been distributed before workshops, and five of them pointed out that they would have liked to have them as a reminder or to use prior to the following workshop (cf. section "Regarding the general structure of the process").

Tools Used During Workshops

Only two interviewees mentioned the post-it notes used during Workshops 1 and 2. Their use appears to be common yet was seen positively: they were used wisely to share ideas and detect similarities and differences. Five interviewees mentioned the coloured cards (and other similar materials) of pathway design workshops (Workshop 3), describing them as being more elaborate than post-it notes, visual, playful, and as good "starters" for discussions. Some highlighted the fact that the take up of these tools was not immediate and required the facilitator's mediation. Another interviewee recognised that these tools did a good job of framing the work, but felt that they also held it back. He mainly noted the benefit of organising pathway elements over time, with the help of arrows and the line that was used to position the different subjects: this produced an overview of the work and made it easier to talk about afterwards.

External Data

Two interviewees mentioned data that were unrelated to the participants' discourses. The first one felt he lacked the data to feel legitimate enough to make alternative suggestions and argue in favour of their feasibility. He would have needed a knowledge base (techniques in particular) in common with other participants to be established. Another one noted that a large amount of information was provided at the beginning of the second day (including the presentations by scientists) and wondered whether these presentations had an influence on the direction of workshop discussions.

Facilitation

On the whole, the majority of interviewees congratulated the facilitators or praised their work. Their statements also contained details on the facilitators' activities, and each positive aspect was systematically addressed by 3 or 4 interviewees. (1) They highlighted the facilitators' ability to explain exercises, set things in motion to make progress, and reframe things. Nonetheless, two interviewees perceived a lack of methodological guidance by facilitators during Workshop 3. One experienced difficulties in learning to use the colour codes of tools. The other felt that his group was confused and a bit lost in terms of the different concepts used to define the transition pathway. One person spoke of tension between the group's difficulty in appropriating the content of certain goals, and the facilitator's engagement to make the pathway defining exercise progress. (2) The facilitators' ability to make participants express themselves and to establish dialogue between different mindsets in a respectful context was also highlighted. (3) The interviewees perceived facilitators as people who listened and took all viewpoints into account without stigmatising them. Two interviewees (from different groups in Centre-Ouest Aveyron) nonetheless noted that the facilitators slipped into participation in the discussions. They related this to the small group size, which caused facilitators to feed discussions. One person disliked the fact that the co-facilitator used the various tools extensively, rather than having a withdrawn position of listening and synthesising, which would have been more conducive to reflection. (4) The relevance of facilitators' interpretation and summarising of participants' statements was also apparent in the interviews. One person in particular noted: "*a real ability to respond to what is said, to try, when a sentence isn't clear, to seek clarity by asking questions, by responding, by re-formulating*". Lastly, two interviewees noted that the method was based on the facilitator's quality. Moreover, they perceived a level of participation and production that differed to some degree, depending on the facilitator in Workshop 3. One of them therefore predicted that the reuse of this method in other contexts by other people would require training of the facilitator.

Group Size

The interviewees described the groups formed for the work sequences as small groups. Their statements regarding this small size were essentially positive and touched on what that allowed them to do in terms of interactions between participants. One interviewee also related working in a small group to the ability to "*work well*" and with accuracy. During Workshop 3, the excessively small size of certain groups in Centre-Ouest Aveyron was mentioned as a factor exacerbating the weight of individual ideas. While only one person highlighted this direct effect of the small size of certain groups, its effect via the low diversity of participant profiles was addressed much more broadly (cf. section "Participant profiles and their diversity").

Participant Profiles and Their Diversity

The diversity of participant profiles was seen in a positive light (13 interviewees), even if it was still inadequate or unequal, depending on the workshop (lower for Workshop 3). Fourteen of the 24 interviewees considered farmers' limited presence to be unfortunate, as there were fewer farmers than elected officials, administrative officials, and representatives of organisations ("*The 'real people' have to be at the heart of things*"), and some of the farmers were no longer working. Around ten interviewees also regretted the lack of stakeholders with an influence on changes in the agricultural world (elected officials from chambers of agriculture, agricultural syndicate representative, cooperative director, etc.). Other missing categories were highlighted more occasionally: food processors (butchers, manufacturers, etc.), state representatives, public project financial backers, banks and insurance companies, agricultural education and training actors, the water agency, river syndicates, as well as people working on hedges, bocages, and the development of riparian forests. On the whole, the interviewees considered there to be less diversity Workshop 3 than during the previous two workshops. This hindered the dynamics of the groups concerned ("*there were too few people to really have a debate*") and their production ("*when it comes to that topic, which we have to address, all of a sudden we don't have the right people to address it, and then we're a bit cornered*").

Two interviewees mentioned disadvantages of grouping together people with diverse profiles. For instance, the lack of common ground allowing for debate can make it difficult to call a dominant solution into question.[3] Furthermore, the promotion of one's institutional and political position can inhibit one's creativity.

Furthermore, three people said they felt they lacked the legitimacy to express themselves at times. Two interviewees thus questioned the justified nature of certain participants' statements around farming ("*you have to have first-hand experience, so my little criticism would be that people allow themselves to give advice without being familiar with the profession*").

Participants' Attitudes

Six of the interviewees spoke of participants' attitudes demonstrating respect and open-mindedness. Moreover, three of them were surprised by the discourse of certain participants, which did not reflect the mindset of the type of structure with

[3] Maize farming, for example.

which they were affiliated ("*with certain organic farmers, for example, things are really black and white, […] so-called conventional farmers aren't used to having such positive discussions with organic farmers*"). This is related to the constructive position aimed at promoting reflection, which appears in the statements of five of the interviewees. Yet, according to two interviewees, one person came to hold a more dominant position than the others in the group, which tended to hinder the discussions.

A Leading Place for Exchanging Viewpoints

Statements on interactions between participants contained strong connotations of the exchange of viewpoints (cf. Table 3): the concepts of collecting ideas, allowing everybody to express themselves, freedom of speech, a diversity of viewpoints, a wealth of ideas and exchanges, predominated. The few hesitations in this respect concerned Workshop 3 in Centre Ouest Aveyron, in which the small group size limited the diversity of viewpoints shared. The articulation of these viewpoints was also stressed. By contrast, disputes (arguments, debates, etc.) seem to have occupied little space. The lexical field related to them (11 interviewees) consisted mainly of simple phrases and nothing more ("*we talked*", "*we stuck the post-its up and then debated*"), which are difficult to interpret. Some interviewees indicated that there were no significant controversies or debates around subjects that were nonetheless controversial, or even that people voluntarily contained their arguments and avoided conflict, as a mark of respect. Some interviewees moreover mentioned consensus building as exchanges progressed, whereas others mentioned similarities in viewpoints from the beginning, with the former being slightly more numerous than the latter.

Table 3 Number of interviewees that addressed the different forms of interaction between participants

	Number of interviewees contributing to the group of ideas	Yes	No	More or less	Not easy	Don't know
1: Exchanges in viewpoints	23	22	6	3	0	1
2: Interrelation of viewpoints	11	9	1	0	0	1
3: Disputes	16	9	8	2	0	0
4: Consensus or consensus building	12	10	2	0	0	0
Average	*15.5*	*12.5*	*4.3*	*1.3*	*0.0*	*0.5*
Standard deviation	*5.4*	*6.4*	*3.3*	*1.5*	*0.0*	*0.6*

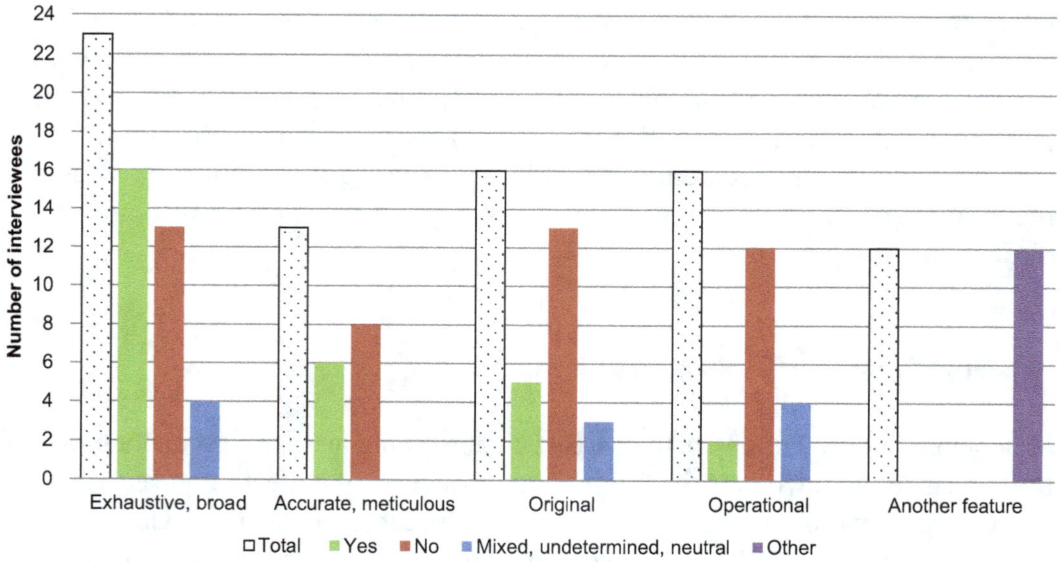

Fig. 6 Numbers in the categories of evaluations made by the interviewees regarding the different features of workshop outputs

The Features of Workshop Outputs

Figure 6 quantitatively represents the different categories of statements detected. To avoid increasing the number of categories of features, "exhaustiveness" was broadened to include all statements regarding the scope of that which was produced. Regarding exhaustiveness and accuracy, there was a similar number of interviewees with positive or negative statements. Regarding the originality and operationality of the outputs, more interviewees had negative opinions than positive ones. The exact nature of statements was subject to qualitative analysis below.

Exhaustiveness and Degree of Accuracy

Ten people stated that the outputs were generally either exhaustive or at the very least substantial, valuable, or wide-ranging. One of them said that it was the most complete exercise that he had participated in up until then. Another emphasised that this made it difficult to select a priority issue. Some comments nevertheless nuanced the scope of the outputs.

It was often highlighted that the forms of farming considered during workshops were the reflection of the people present, which included few farmers and operators in conventional supply chains, and at times few participants (in particular in Centre Ouest Aveyron for workshop 3). The interviewees thus questioned the representativeness of these workshops with respect to the supply chains and forms of farming in the territory. In particular, conventional supply chains without certification labels were rarely addressed, even though the interviewees stressed that they were the main source of income of many farms and that it was also necessary to consolidate them and to reflect on the transformation of the systems in question ("*we might have not done enough thinking on transforming systems that are embedded within big industry today*"). Other supply chains existing in the territory were not considered either (such as milling wheat production). Even though a variety of agricultural models were mentioned in the two territories and discussions stressed their complementarity and necessary cohabitation, certain orientations were prioritised. One interviewee in Centre Ouest Aveyron noted that interactions were marked by agroecology. Moreover, even though other forms of farming were not excluded, those towards which the comments were oriented seemed, according to several interviewees,[4] to be targeted at more diversified farming developing local resources, family farming with autonomous systems to create value, and short supply chains.[5] This gave some people the impression of addressing niches or somewhat idealised agriculture, rather than the "overall truth". Moreover, the agricultural orientation that established a consensus between the participants surprised one of the Centre Ouest Aveyron interviewees, who thought that the representatives of certain structures were defending others, considering their usual positions. Still with regard to the lack of representativeness but in the opposite sense, one interviewee in Midi-Quercy found it unfortunate that his group revolved around the preservation of maize farming, without investigating alternatives.

Likewise, while several interviewees considered that the definition of issues and goals was relatively complete,[6] some mentioned subjects that were barely covered or not addressed during workshops, such as supply chain difficulties (in Centre-Ouest Aveyron) and the isolation of land parcels due to urban development (in Midi-Quercy), which was only addressed from the landscape point of view. Participants' experiences across a single territory were sometimes different in this respect: in Midi-Quercy, one of them felt that water was one of the main topics, whereas another person from another group felt that it was only alluded to. With regard to the impasses in Workshop 3, they were often the result of a lack of time to address all of the goals defined in the previous workshops.

Statements regarding the degree of accuracy of outputs were often less explicit. In particular, five interviewees simply answered yes or no to the question. Four believed that the subjects addressed were generally considered in depth. In particular, all of the important components of the pathway were covered during Workshop 3: the initial state, goals, the final state, governance, key actors, means of acting, financing. Seven of the interviewees presented a more mixed opinion on the degree of accuracy of the work carried out. Four made general statements on the shallow-

[4] Belonging to the two territories

[5] One interviewee in Midi-Quercy, on the other hand, found it regrettable that organic agriculture was not discussed during Workshop 3.

[6] "*when I saw the signs and I saw those diagrams [rich picture and cognitive map] I told myself 'it's all there'*"; "*Yeah, in terms of natural resources, I think that we got to the bottom of what we – we might have forgotten, but I think that we did structure the workshop well, yes… We didn't forget much, at least I don't think so? From memory, we left quite satisfied with what we had done. For the issues as well, I don't remember having seen a major issue that could have been forgotten…*"

ness of the subjects, due to the alternation between activities, the absence of "*well-defined proposals*", the additional time necessary to completely address all of the goals, or the lack of numerical data during the diagnostic. One interviewee doubted that Workshop 3 went further than Workshop 2, which was already quite detailed. Lastly, one interviewee highlighted the fact that transition pathway indicators could not be defined properly due to a lack of sufficient expertise and time.

Originality

The interviewees almost unanimously agreed that new paths for the territory or novelties for certain actors had been identified, but noted that the majority of them corresponded to the spirit of the time or existed elsewhere. Only five interviewees said that they had detected new or innovative ideas, without qualifying their statements. One of them gave an example of this[7] but it is difficult to know what the other four meant by "new ideas": new for them or for other actors? new for the territory or truly innovative? Three other interviewees stated that in any event, the paths of action defined during the workshops already existed elsewhere or were in line with a general trend. Eight even indicated that the ideas collected, whether all or the majority of them, were already being considered in the field, or were even being implemented. A few people moreover mentioned conditions that according to them were unfavourable to innovation during workshops: "*there was no emulation that day during the workshops that would have allowed us to say 'Yeah, we can do that!'*"; it is difficult for "*the people at the bottom*" to not stick strictly to what they have learned[8]; the inadequate diversity of the profiles limited the emergence of innovative ideas (statement originating from Centre Ouest Aveyron), or the diversity of participants, on the contrary, limited capacities to debate alternative solutions (lack of shared culture) or hindered the boldness of innovation to the benefit of political ideas related to the requirements of representation (statement originating from Midi-Quercy). Beyond that, two of the interviewees in Midi-Quercy highlighted a situation of lock-in[9] in certain current solutions that nonetheless do not appear to be sustainable. In this case, the protection of existing supply chains (maize, melon) was retained over the possibility of making them change quickly.

Following another line of thought, one interviewee was surprised by the positions taken by certain participants, which seemed distant from the typical positions of their organisation. He explained this in terms of their "field" activity as opposed to representation ("*at these meetings, I felt like it wasn't the real world, and that's interesting because there were people there, and not necessarily representatives, they were technicians or people working in the field that came to provide their point of view*").

---

[7] The idea that certification labels are no longer seen as an end in itself.

[8] This interviewee considers the research world as being the most capable of proposing innovations.

[9] Due to a lack of alternative solutions or a lack of common culture to overcome blockages.

Degree of Completion

A few interviewees talked about elements produced during workshops that could be put to use directly in the field, in particular the identification of key actors in transition pathways. However, only one of them said that all of the aspects necessary for what he called the operationalisation of the project were addressed in Workshop 3. The statements of 11 other people indicated that the outcome of reflection on implementation would require follow-up to the process. Five of them moreover insisted on the importance of concrete completion ("*if we just stick with what we've done, we won't make any progress*"; "*something needs to produce a result, otherwise... not much will be left over*"; "*now we need to move on to the project phase, to the phase of initiating something that meets the needs*"). As regards follow-up, several interviewees stated that they were unsure of how the transition to action would take place. One of them suggested that "*we now need to find a methodology: 'how do we transit from defining needs to action'*". A few conditions and mechanisms to implement paths stemming from reflection were nonetheless mentioned, such as a very local scale, knowledge of capacities for financing actions, sending of the final report to financial backers, dynamic and influential relays among decision-makers, publication of outputs online (such as on Territorial Rural and Balance Pole websites), experimentation and the demonstration of practices, the identification of constraints and means for concretising the actions defined,[10] the fourth workshop of the process foreseen by workshop organisers (cf. chapter "Participatory Methodology for Designing an Agroecological Transition at Local Level").

Other Features of the Product Mentioned

During the interviews, the interviewees mentioned other features of the workshop outputs. Each of these features was mention by only one or two people.

An initial aspect was the realism and degree of justification of some goals. One interviewee thought and liked the fact that the TATA-BOX project "*[has] been a federating space, for sharing data to be able to still manage to construct something realistic*". He and another person from Midi-Quercy nonetheless questioned the possibility of reducing water consumption, while a third person found, on the contrary, that future constraints, such as the lack of water, were not adequately taken into account during workshops. One interviewee in Centre Ouest Aveyron also saw a bit of utopianism in orienting toward the re-localisation of food and supply chains, taking into account their operating conditions and the current economic environment. More generally, two interviewees heard statements that they thought were disconnected from farmers' practices and constraints, and considered it unfortunate that the operation of farming was not developed to a greater extent, so that each participant might be able to better understand existing practices and that which was possible or reasonable to do given the constraints. One participant nevertheless

[10] Nonetheless partially carried out during Workshop 3.

liked the fact that the method led to awareness of constraints without stopping at them: "*For me, in that respect, the method was very timely, that's to say, it took into account these constraints [...] but we couldn't fix them, in my opinion, it's kind of risky to bridle a hosting technique with this reality principle*".

One set of ideas addressed the relations between the elementary components of the outputs. By indicating that the subjects identified were indissoluble, one of the interviewees emphasised the systemic nature of the workshop outputs. Another interviewee, by revealing that some of the forces at play collided with one another, highlighted antagonisms ("*managing the water resource while protecting the maize supply chain: they don't go together*"). He and another person moreover defined the final result as establishing a common basis of the knowledge that each person individually brought to the table.

Some interviewees also spoke of the form of the output: ideas were "*posited*" in writing (this facilitated the identification of implementation conditions) and organised (prioritisation of subjects, chronology of goals), something was "*efficient*", "*it was subtle*", voting with coloured stickers was "*speaking without speaking*", the project was "*illustrated*", "*for agricultural production the goals were pretty clear (linking production and consumption)*", pathways were more "*complicated*".

Lastly, due to the profiles of the participants in Centre Ouest Aveyron, who were mostly sensitised to agroecology or organic farming, the outputs converged towards these forms of agriculture. This was less pronounced in Midi-Quercy, where the participants originated from a more conventional farming environment (cf. section "Exhaustiveness and degree of accuracy").

The Spin-Off of the Process for the Participants and the Territory

Individual Learning

Around half of the interviewees stated that they learned things during these meetings and that they reused this in their work. This learning concerned six different aspects: (1) reasoning logics or actors' sensibilities; (2) farming and its issues; (3) the initiatives emerging from territories; (4) reference points for acting or changing ways of acting; (5) new reflection; and (6) the detection of potential partners. The insert below contains a selection of *verbatim* quotes that illustrate the six categories of individual learning.

Some interviewees said they had become aware of issues during workshops, and that they intended to try to take that new knowledge into account in their territorial development or farmer support activities. Others had got useful ideas during workshops (on problems and key actors), that would help them to make progress on subjects that they did not previously know how to address.

Selection of *verbatim* Quotes Demonstrating Individual Learning

- *I arrived with that preconceived idea, which was finally knocked down (1)*
- *that let me size up the obstructions, to see a bit of those sensibilities, which I was aware of but I hadn't measured in terms of magnitude [...]. that let me truly see sensitive points regarding the territorial agriculture question, and therefore maize, and that in terms of organic, we have to stop seeing two opposing sides of the bad conventional farmers on the one hand and the people who grow organically and are saving the planet, on the other. (1)*
- *that allowed us to become aware of a new position, another sector of activity, another supply chain, and all of a sudden to take it into account in our way of seeing things. (1) and (2)*
- *the agroecological transition is a subject that I had never addressed [...] I was able to draw from all of this information that I didn't have before and which today feeds my thoughts. When I have projects that are emerging, I can make connections, and in the background, it also really helped me. (2)*
- *there's a whole aspect around methods that I learned a ton about, about production methods, supply cycles, the ties that will be established between farmers and their ecosystems [...] that really let me immerse myself in farmers' ecosystems, and that also let me immerse myself in their techniques and concerns. (2)*
- *that lets people become aware of certain issues that they don't know about. (2)*
- *Mr. XXX [...], who was against the Sivens dam [...], against water reservoirs, for irrigation-free farming, he didn't know much about it [...]. So, when I said when I had to say, as a professional... modest but stating things as they are, the guy completely changed position. [...] TATA-BOX was a meeting place that allowed to... in the end, today, with those same people we are having a different discussion around that project. (2)*
- *we weren't aware of all of the initiatives [...]. It wasn't that I discovered a supply chain that I wasn't aware of or something like that, but at times it allowed me to say, well, there's this or that that's being done. (3)*
- *we talked a lot about short supply chains. [...] And in them, the need is essentially having a farming activity that supports the supply chain or even a peri-urban belt. For the time being, we aren't taking this into account. [...] So, all of that will be a lot more present in my mind when I do a file analysis. (4) and (2)*
- *that allowed for a stage today during which we no longer have doubts around the desire to create this legumery. (4)*
- *the usefulness that I see is that it consolidated our common view of things and our forecasting of concrete action in territories, as we saw it [...]. We*

(continued)

are truly embedded within that logic of action promoting sustainable agriculture, with an entrance into agroecology. (4)

- *by reminding me bit by bit, there were reflections that I hadn't had that came out in the workshop [...], the story of labels, for example. (5)*
- *I participated in the meeting but after that, it stopped there; we don't have time to work on those problems. [...] on the other hand, I would be curious to read the summary. [...] it will answer questions that I never thought about, that I haven't had the time to explore in detail; it might consolidate some ideas that I have or that I've heard and for me, that's personal enrichment, that will expand my viewpoint of things, of the agricultural world. (5)*
- *It was the first time that I had heard of GIEE (Economic and Environmental Interest Groups) [...], it enabled me later to get into contact with that structure again, to inform myself, and in the end, to talk to people involved in that project, and today, we are in the middle of creating bridges with these people. (6)*

Learning legend: regarding (1) reasoning logics or actors' sensibilities, (2) farming and it issues, (3) the initiatives emerging from territories, (4) reference points for acting or changing ways of acting, (5) new reflections, (6) detecting partners.

Reusing the Method

Around ten of the interviewees talked about reusing the TATA-BOX process themselves or its reuse by other people. The majority thought this method could be adapted to a variety of situations and scales.

Among the nine people who mentioned reuse by themselves or by an organisation to which they belonged, we find a gradient of intentions. Quite naturally, the people who had already reused the principles derived from the method at the time of the interview, or who definitely planned to do so, were essentially the people whose task was to host groups (groups of farmers or groups on the territorial level). The methodological elements mentioned were: the three workshops of the process, the method and organisation of Workshop 3, collective representation on a virgin map of the territory (farming systems, constraints, and conflicts), and the representation of ideas in the form of a rich picture diagram. Other interviewees did not exclude the possibility of reusing elements of the method, but had either not yet defined the situations for which they would do so, or saw obstacles to this, such as obtaining funding to gather a large public and then concretely implementing projects, the duration of the process, or the lack of hosting capabilities. At the time of the interview, a latter group did not plan to reuse the method.

Moreover, two people foresaw others reusing the method. One particularly suggested reusing it in the "territorial food project" in Midi Quercy. He also expressed the desire for a rather broad reuse, such as by the Chamber of Agriculture or other structures in the context of the process for setting up young farmers.

Use of the Results

Several interviewees stressed the importance of the outputs of workshops resulting in actions (cf. section "Degree of completion"). In this respect, four pointed out the issue of disseminating the results to decision-makers so that they could take them into account in their strategies (financial backers, elected officials, and directors of key organisations running projects). The success of this transfer of the results did not however appear to be a given, considering the questions formulated by some participants with regard to: (1) the possibilities of the agricultural world taking up the results, considering that it had little representation at meetings; (2) the lack of directors of influential organisations in the agricultural world and the lack of project financial backers at meetings; (3) the ability to mobilise key actors, including those in the agricultural world, with this mobilisation already appearing difficult in projects underway; and (4) the human resources available to conduct projects, as certain organisations that would be legitimate in this activity had very few resources. In view of the latter point, the interviewees hoped that the participants in workshops would create relays by taking charge of certain actions themselves.

During the interviews, a number of interviewees foresaw how results might be transferred in the context of projects underway on the territorial scale. Several of them noted connections to be made with the Territorial Coherence Scheme (SCoT, *Schéma de Cohérence Territorial*), in the context of diagnostic or dialogue processes. In Centre Ouest Aveyron, the results of TATA-BOX workshops were sent to the study bureau responsible for SCoT diagnostics, but discrepancies between calendars nonetheless limited their use in this context. One interviewee moreover mentioned a limit in making use of TATA-BOX outputs in the context of a SCoT, as the latter had to draw support from numerical data and not only from the feelings and suggestions of the actors.[11] Usage in the context of different territorial projects was also mentioned. In Midi-Quercy, this was the case of the "territorial food project" (*projet alimentaire territorial*). The ranking of the goals defined in TATA-BOX (cf. colour sticker placement exercise during Workshop 3) and the rich picture diagram appeared to be usable in this context, in particular by putting them online on the PETR website. Some people even wondered if the work carried out during TATA-BOX had not already played a role in the importance granted to certain actions of this "territorial food project", and in particular the legumery. In Centre-Ouest Aveyron, different projects were mentioned, such as those on the circular economy.

[11] Note that in the synthesis of Workshop 1 (diagnosis), statements made during workshops are supported by quantitative data added *a posteriori*.

At the time of the interviews, the reuse of some of the workshop outputs was foreseen in the reflection around these projects.

Lastly, the guidelines adopted during workshops consolidated some participants' viewpoints on that which their own activity should promote in the field. Some of them would have liked to be able to reuse the rich picture in their work.

Increased Familiarity with One Another

Lastly, several interviewees noted that the fact of having met certain actors during these meetings would facilitate partnerships for establishing projects. That would simplify establishing contact, and participants had gained better knowledge of each other's respective tasks ("*The PETR, agents [...] call upon the Chamber for consular missions [...] but they don't necessarily have knowledge of all tasks, and in particular the project support task [...], so that also makes it possible, through this type of informal exchange, to also inform actors around other tasks for which capabilities are available and which aren't necessarily known, and then to prepare [...] what comes next for another collaborative framework that might be a bit broader than that for which the Chamber is known*").

Discussion

In this section, we will first highlight the features of our analysis before returning to the feedback of actors regarding the TATA-BOX process, and drawing conclusions on its ability to generate a collaborative design process.

Feedback on the Features of Data Collection and Analysis

Data collection via interviews had both strengths and weaknesses for evaluating the TATA-BOX process. The interview guide allowed us to collect stories of the participants' experiences during these workshops and their projection for the future. Participants nonetheless experienced difficulties in remembering the course of the workshops during interviews, especially with respect to Workshops 1 and 2, which had taken place 6–18 months earlier. Moreover, the characterisation of the interactions between participants, based on these interviews, draws on two successive interpretations: the interviewee's interpretation, as well as the interpretation of the interviewee's statement by the analyst. To complete this work, a direct analysis of interactions based on a corpus of complete transcriptions of the discussions is underway. Concerning the outputs of workshops, the interviewees tended not to expand on their answers, and without a more open question upstream, the questions asked may have not provided access to what they saw as priorities. Lastly, the quantitative analyses presented in this chapter have the advantage of giving a general snapshot of the statements of the survey. This snapshot is nonetheless oversimplified and should be used with caution, for two reasons: (1) each discourse category for which the numbers were calculated hides a broad diversity of content, and (2) the number of interviewees contributing to a single category of ideas does not indicate the quantity of ideas stated.[12] This picture must also be used with caution because allocating statements to different categories (NVivo coding) implies choices and the way of coding is refined as the analysis takes place. It is therefore necessary continually to return to that which was coded to ensure the homogeneity of the coding, although discrepancies can easily go unnoticed. For all of these reasons, qualitative analysis constitutes an important additional component, along with putting the quantitative analysis into perspective.

What Improvements to the Process Did the Participants Suggest?

As we have seen throughout this chapter, most of the participants perceived the TATA-BOX process in a positive light, both overall and with regard to more specific areas of analysis. We also saw that the process produced effects, whether these were the outputs of each workshop, or the individual learning that several interviewees mentioned. In this section we go back to discussing those aspects on which participants had a more nuanced assessment.

The Representativeness of Participants in Workshops

A diversity of agricultural actors participated in the workshops, yet this diversity was not as great as the effective diversity characterising the two fields of study. This imposed a limit on the exhaustiveness of the exercise and, according to some interviewees, on the originality of the solutions proposed. The implementation of the actions defined could be expected to be affected by this – and this was mentioned by the participants –, due to the lack of enrolment of key actors. From the first phases of the creation of the TATA-BOX process, its designers nonetheless took into account the issue of representing the diversity of viewpoints. To design the AET of local agriculture with the help of their methodology, Duru et al. (2015) recommended bringing together the stakeholders of different management processes involved,[13] and offered methods for identifying them. The detection of agricultural stakeholders in the Tarn-Aveyron basin was thus carried out during the first phases of specifying the TATA-BOX process, prior to organising the first workshops. While

[12] For example, a person who makes three different positive comments on an aspect of the process does not count for more than a person who expresses only one criticism.

[13] Agricultural production, supply chains, and natural resource management

this detection in itself seems to have left room for improvement (by adding categories, such as financial backers, or by specifying the diversity within certain categories, such as farmers), the lack of representativeness seems primarily to have originated from the mobilisation of actors in the different categories identified. To improve this mobilisation, different pathways could be explored, such as mobilising more relays that have an influence on certain categories of actors in order to send them invitations, organising shorter but more numerous meetings that are closer together in time to facilitate the availability of participants, or embedding the participatory process in existing devices (projects underway in the territory, regular meetings of CUMA, etc.). The latter pathway, which would disperse the process across different groups, bringing together only some of actors, would nonetheless require an intermediary process to articulate the work of the different groups.

Tension Between the Innovation and Realism of Solutions

Beyond the observation of the low level of originality of the results, some of the interviewees criticised the difficulty of extracting themselves from current solutions. Additionally, other interviewees criticised the lack of realism of some of the positions adopted by certain participants, and of certain collectively defined paths. This raises the question of the balance to be sought between these two tendencies and the mechanisms of establishing a dialogue between them. The protagonists on either side – if we can call them that, considering that they were not in open conflict –, did not change their positions during the workshops. As the purveyors of a certain vision of agriculture, they remained frustrated at not having been able to provide enough arguments in favour of this vision, and therefore having had no effect on the choices made. The solution mentioned by some interviewees, of establishing common ground upstream to enable people with contrasting positions to engage in a discussion, needs to be examined. While better knowledge of the logics underpinning the opposing sides' vision can effectively allow one to move beyond the first level of discussion, it also presents the risk of reducing innovative ideas to their obstacles and thus preventing any departure from set ways of thinking. The same goes for the idea of increasing the diversity of participant profiles: according to the majority of interviewees it would enable more creativity, but according to others it would have the opposite effect, due to the institutional positions that certain actors must maintain with respect to the outside world.

The TATA-BOX Process, a Collective Design Process?

In the Introduction we mentioned the issues in evaluating the TATA-BOX process using design process evaluation frameworks. Given the originality of the design goal, we applied a framework specific to the TATA-BOX process to achieve this. Our analysis allows us to detect, in the process, design stages that are well described in other fields of design studies. This is what we discuss in this section. It will then allow us to conclude that the TATA-BOX process did, in fact, provide an adequate toolset for a collective design process, although this design process still remains incomplete today.

Establishing Common Ground

The interviewees' discourses reflect the importance granted to exchanges of viewpoints and their articulation in verbal interactions during the process. Disputes, in the sense of the confrontation of alternative ideas and arguments, appear less present. It therefore appears that verbal interactions allowed above all for common ground to be established around knowledge.[14] In collaborative design research, we speak of a cognitive synchronisation activity. Acknowledged as a mandatory step in design processes, this activity ensures a shared understanding and representation of both the design issue (what is at stake) and the knowledge that is known or missing. The statements highlighting the significant part of reflection emerging in the field among the ideas gathered during the workshops also follow this line of thought. But what about the more original ideas that appeared, and the ways of articulating ideas as a whole during the different workshops? Are these creations limited to building common ground? How does the TATABOX process foster generativity and creativity, which are also critical in design processes? Generative design activities[15] are based on argumentation, and it is the argumentative process that allows convergence towards a collectively-acceptable solution (Barcellini 2008). The apparently modest place of contrasting alternative ideas and arguments can lead one to think that the specification of the object to design (the AET of the territory) was limited in scope. The analysis of interactions based on the full transcription of workshops will allow us to put this hypothesis to the test. We can however already reflect on the role of argumentation, taking into account the particularities of the process.

Argument as a Way to Converge Towards a Solution?

The interviewees spoke of a consensus around the workshop outputs. How can this be explained, considering how little alternative ideas and arguments were compared?

Perhaps the participants agreed overall from the beginning. Several interviewees spoke of consensus building during interactions, while according to others, the consensus was immediately established thanks to the similarities in participants' ways of thinking. In addition to the bias related to differences in interpretation and inter-

[14] Even though a few interviewees pointed out limits to this activity: the development of ideas limited by time constraints and the scope of ideas limited by the diversity of participants.

[15] In particular, these cover the creation of new solutions, the identification of alternative solutions, and the evaluation of solutions.

viewees' choice of words to describe verbal interactions, this dissonance can be explained by different situations in different groups, as some converged without requiring lengthy discussions because of similarities in viewpoints from the beginning, whereas other groups effectively built a consensus based on initially divergent viewpoints. The use of argumentation appears less crucial in the former. Unfortunately, the collection of information on the diversity of participants' viewpoints in the various groups was not sufficient to be able to verify this hypothesis.

Even with participants whose viewpoints were divergent at the beginning, the features of the object to design may have required few arguments and disputes. Specifically, the role played by argumentation has been highlighted in design processes oriented at a single, fully specified solution, constituted, for example, by the plans for a new building. During TATA-BOX workshops, even though the goals and the pathways to achieve them had to be defined, the stance adopted in favour of the coexistence of different agriculture models may not have required the exclusion of many opinions. Likewise, the pathway defining exercise may have been carried out more in a spirit of detecting all favourable conditions than in one of selecting the means that it will be necessary to effectively activate. In this case, collective creation was most likely directed more at identifying the goals of the different participants, explaining their articulation, and defining actionable means, than at trade-offs between goals and between means to implement. Argumentation therefore appears to be less crucial to the completion of the collective undertaking: consensus is established on the basis not of a trade-off between competing solutions (competing goals and pathways) but rather of the relevance of allowing several solutions to coexist. Certain workshop sequences, such as defining a chronology for different pathways to enable a transition, would nonetheless imply trade-offs. Was there more argumentation? The information collected does not allow us to analyse this.

Is the Term "Design" Appropriate for Discussing the TATA-BOX Participatory Process?

The overall goal of the TATA-BOX project, based on the methodological framework of Duru et al. (2015), was to develop methods and tools allowing the actors in a territory to collectively design an AET on the local scale (Galvez et al. 2014). The goals of the three workshops were very different (cf. section "Materials and method"). Given the objective of drawing up an inventory and common understanding of Workshop 1, it is likely that interactions allowing common ground of knowledge (cognitive synchronisation) to be established were at the heart of this workshop. This workshop consisted more of the articulation between the different sets of information, than of "typical" generative design activities. Assuming that the options foreseen by the participants in terms of the forms of farming to develop and pathways to achieve them are not all reconcilable, Workshops 2 and 3 suggest that proposals were drawn up and choices made. We can therefore expect that the participants took part in generative design activities as such, backed by arguments. However, will identifying these different types of activity, which are characteristic of a collaborative design process, be enough to say that the participatory process resulted in the design of an agricultural transition project? First of all, if there was design, the process that allowed it to be achieved may have had to include the research team's work of shaping workshop outputs, across the three workshops. The way of representing the results, even though it aimed at remaining as close to possible to the workshop materials, was the result of a creative process. For example, the way that ideas were articulated on the rich picture diagram constructed from material from Workshops 1 and 2, was not directly accessible in the raw data. Certain participants' passion for this diagram moreover clearly attests to the fact that it contained something more than this raw data. Second, designing a new building involves plans which, even though they can be adjusted during execution to deal with unforeseen constraints, will nonetheless be applied in their entirety by the project manager. The situation is different in this case, where it does not appear that a project manager is willing to implement all of the goals and paths of action that have been defined. The implementation of the results will therefore most likely be distributed among different actors, who will take responsibility for limited actions rather than the territorial agricultural transition project in its entirety. The work of Béguin and Rabardel (2000) shows however that design continues into use. The field actors who will take responsibility for transferring the results of the participatory process will certainly *a minima* select the ideas to convey. They may also appropriate them for themselves and change the goals and the pathway to achieve these goals. We can see therefore that the design of the Tarn-Aveyron basin's agricultural transition was not complete at the end of the TATA-BOX project. It continues and will be continued via the use that actors in the territory make and will make of the workshop outputs. The TATA-BOX participatory process thus appears to be just one stage in the process of designing this territory's transition. The hypothesis of implementation distributed among different actors is moreover a source of uncertainty as to the complete and coherent use of that which this process has produced. Specifying the status of the process implemented during the TATA-BOX project will require the description of its place and role in the design and implementation of a transition "steered" at territorial level.

Conclusion

The purpose of this study was to evaluate participants' experience of the design process and their opinions on its outputs and effects. It intended to help the method's designers to take a step back from it, in particular with respect to potential improvements.

According to the interviewees, the workshops resulted in a proliferation of shared information, possible pathways, and elements of the transition pathway envisaged. The content of the outputs, on the other hand, appeared unoriginal to them in absolute terms, even though ideas that were new for the territory or for certain actors emerged. The mechanisms for transferring outputs into concrete actions remained to be defined and seemed to be the main issue at this stage of the design process. In addition to questions related to appropriation in the field and to financing, the limited availability of human resources that could legitimately coordinate the implementation of the transition project in its entirety makes it difficult to foresee such implementation. On the other hand, the benefits of transferring some of the results to existing dynamics in the two territories were expressed. The interviewees furthermore also provided evidence of the significant effects of the TATA-BOX participatory process for the individuals who participated in it. The discourses of participants regarding the participatory process in itself were globally quite positive. However, for each component of this process considered in detail, it is often hard to outline a common or dominant viewpoint among the interviewees. This can be attributed to their respective profiles and sensibilities: in itself this diversity is a result.

Some of these results appear in the methodological guide intended for future users of the TATA-BOX process. They thus constitute a potential resource for encouraging and improving future implementation of the method.

This study, based on participants' discourses, will be completed by a direct analysis of interaction during workshops. The latter analysis can be expected to produce a more detailed characterisation of the interaction, and allow for a better description of the collective dimension of design work supported by the TATA-BOX process.

References

Alvarez S, Douthwaite B, Thiele G et al (2010) Participatory impact pathways analysis: a practical method for project planning and evaluation. Dev Pract 20:946–958

Barcellini F (2008) Conception de l'artefact, conception du collectif. Conservatoire national des arts et métiers

Béguin P, Rabardel P (2000) Designing for instrument mediated activity. Scand J Inf Syst 12:173–190

Clevenger CM, Haymaker J (2011) Metrics to assess design guidance. Des Stud 32:431–456. https://doi.org/10.1016/j.destud.2011.02.001

Détienne F, Baker M, Burkhardt JM (2012) Quality of collaboration in design meetings: methodological reflexions. CoDesign 8:247–261. https://doi.org/10.1080/15710882.2012.729063

Duru M, Therond O, Fares M (2015) Designing agroecological transitions; a review. Agron Sustain Dev 35:1237–1257. https://doi.org/10.1007/s13593-015-0318-x

Galvez E, Bergez JE, Therond O (2014) TATA-Box – Concevoir collectivement une Transition Agroécologique au sein du territoire (flyer de présentation du projet)

Hatchuel A, Le Masson P, Weil B et al (2016) Multiple forms of applications and impacts of a design theory: 10 years of industrial applications of C-K theory. In: Impact of design research on industrial practice. Springer, Cham, pp 189–208

Joly PB, Gaunand A, Colinet L et al (2015) ASIRPA: a comprehensive theory-based approach to assessing the societal impacts of a research organization. Res Eval 24:440–453. https://doi.org/10.1093/reseval/rvv015

Martin G, Martin-Clouaire R, Duru M (2012) Farming system design to feed the changing world. A review. Agron Sustain Dev 33:131–149

Meynard JM, Dedieu B, Bos AP (2012) Re-design and co-design of farming systems: an overview of methods and practices. In: Darnhofer I, Gibbon D, Dedieu B (eds) Farming systems research into the 21st century: the new dynamic. Springer, Dordrecht, pp 407–431

Meynard JM, Jeuffroy MH, Le Bail M et al (2017) Designing coupled innovations for the sustainability transition of agrifood systems. Agric Syst 157:330–339. https://doi.org/10.1016/j.agsy.2016.08.002

Moultrie J, Clarkson PJ, Probert D (2006) A tool to evaluate design performance in SMEs. Int J Product Perform Manag 55:184–216. https://doi.org/10.1108/17410400610653192

Perez P, Aubert S, Daré W et al (2010) Evaluation et suivi des effets de la démarche. In: Etienne M (ed) La modélisation d'accompagnement – Une démarche participative en appui au développement durable. Quae, pp 153–181

Prost L, Berthet ETA, Cerf M et al (2017) Innovative design for agriculture in the move towards sustainability: scientific challenges. Res Eng Des 28:119–129. https://doi.org/10.1007/s00163-016-0233-4

Thornton PK, Schuetz T, Förch W et al (2017) Responding to global change: a theory of change approach to making agricultural research for development outcome-based. Agric Syst 152:145–153. https://doi.org/10.1016/j.agsy.2017.01.005

Part IV
New Prospects and Cross-Cutting Perspectives

Information and Communication Technology (ICT) and the Agroecological Transition

Lola Leveau, Aurélien Bénel, Jean-Pierre Cahier, François Pinet, Pascal Salembier, Vincent Soulignac, and Jacques-Eric Bergez

Abstract The development of information and communication technologies (ICT) has to meet the needs of farmers and sustainably support the competitiveness of agriculture in a rapidly changing digital world. Under certain conditions of use, digital tools could facilitate the application to agriculture of the historical, methodological and socio-economic principles defining agroecology. This chapter is composed of four sections. In the first section we define a framework to study agricultural IC tools. The second section considers how ICT should be used during the design phase of the territorial agroecological transition – an example of which is the TATA-BOX project –, before its actual implementation. The third section sets out the four types of IC tools that can usefully be applied during this transition, and provides several examples. Finally, the last section shows the various barriers that ICT specialists will have to overcome in order to provide effective support to food systems. It also discusses the contradiction that can exist between high energy-consuming technologies and an agroecological production paradigm in which a drastic reduction of the reliance on fossil energy is essential.

L. Leveau (✉) · F. Pinet · V. Soulignac
TSCF, Irstea, Centre de Clermont-Ferrand, Aubière, France
e-mail: lola.leveau@gmail.com; francois.pinet@irstea.fr; vincent.soulignac@irstea.fr

A. Bénel · J.-P. Cahier · P. Salembier
Institut Charles Delaunay -TechCICO, Université de Technologie de Troyes, Troyes, France
e-mail: aurelien.benel@utt.fr; jean-pierre.cahier@utt.fr; pascal.salembier@utt.fr

J.-E. Bergez
AGIR, Université de Toulouse, INRA, Castanet-Tolosan, France
e-mail: jacques-eric.bergez@inra.fr

J.-E. Bergez et al. (eds.), *Agroecological Transitions: From Theory to Practice in Local Participatory Design*, https://doi.org/10.1007/978-3-030-01953-2_12

Introduction

The numbers of Information and Communication Technologies (ICT) available are constantly increasing, and so are their applications in the agricultural sector. As explained in the French ICT #DigitAg project[1] reference text, *"Digital agriculture may have started more than 40 years ago, with the first civil satellite programmes enabling earth remote sensing, followed by the computer capacity boom in the 80s, which made it possible to digitalise crop models, to build expert systems and to introduce precision agriculture. Subsequent technological breakthroughs produced mobile phones and smartphones, satellite communications, GPS, more accessible wide-ranged satellite data, and now connected objects and Internet of things. Agricultural engineering research has benefited from these offers. Concerning sensors, research has been dedicated to exploiting satellite images, developing new sensing techniques for specific properties (e.g. quality, disease). The latest developments are in wireless sensor networks in fields, and in phenotyping, as the lack of phenotyping sensors is a bottleneck of genomic research."* (#DigitAg 2017).

Under certain conditions of use, digital tools could facilitate the application of the historical, methodological and socio-economic principles defining agroecology as conceptualized by Altieri, the SAD department of INRA or the GIRAF group (Stassart et al. 2012). For example, ICT could support the development of multi-criteria guidance for agro-ecosystems, the construction of participatory research frameworks, the creation of networks promoting public debate and knowledge diffusion, and the re-localisation and co-management of food systems by both producers and citizen-consumers. It could moreover highlight the diverse forms of knowledge to take into account in the construction of a problem and the research contributions to solving it.

In this chapter, we discuss this issue of ICT's role in agroecology: are the different IC tools parts of the game and how can they be used to facilitate the agroecological transition (AET) of a territory? The chapter is composed of four sections. In the first section we define a framework to study agricultural IC tools. The second section considers how ICT should be used during the conception phase of the territorial AET – an example of which is the TATA-BOX project –, before its actual implementation. The third section sets out the four types of IC tools that can usefully be applied during the practical implementation of this transition, and provides several examples. Finally, the last section shows the various barriers that ICT specialists will have to overcome in order to provide effective support to food systems, and

[1] #DigitAG is a French research coordination in which 360 scientists are working on digital agriculture, new digital tools and services for the agriculture of the future. Sensors, connected objects, smartphones, satellite images, drones, Internet of Things, big data, high-performance computing systems, as well as decision support, territorial management, frugal innovation, ethics, confidentiality, and management are some of the keywords to meet the challenges of food security and sustainable development. The main goal is to develop information and communication technologies (ICT) to meet the needs of farmers and sustainably support the competitiveness of agriculture in a rapidly changing digital world (http://www.hdigitag.fr/fr/)

discusses the contradiction that can exist between high energy-consuming technologies and an agroecological production paradigm based on fossil resources sobriety (Altieri 1995).

Setting the Framework

ICT: Support or Slave?

A big issue when dealing with connected agriculture is how it is defined: either as a model or as a tool for agriculture. This positioning is critical to understand antagonisms, complementarities and utilities in digital agriculture with regard to agroecology.

- If **connected agriculture is considered to be a new model of agriculture**, it raises fundamental questions about the positioning of this model in relation to other agricultural models, and in particular to agroecological agriculture. This new mainstream digital model imposes on farmers and society a techno-centric point of view which may not be compatible with the principles of agroecological agriculture. In this connected agriculture configuration, there is a risk of transforming agroecological agriculture into techno-centric precision agriculture, transforming biological regulation and knowledge networks into ICT fully-equipped agriculture seeking optimisation at all costs. The solutions here stem from a new top-down approach, placing the sacrosanct trilogy "Certified Seeds, Fertilization, Phytosanitary Coverage" into a new form of dependence on technologies owned by multinational firms. The implications of the change of local-territorial scale, fundamental to the agroecological agriculture approach, are revisited here from the perspective of a territory connected by multiple sensors. Additionally, although the governance of agriculture and its data is still to be imagined, it could easily escape from the farmers' control.
- If **connected agriculture is seen as a tool for agriculture**, it could constitute a valuable opportunity to objectify agroecological agriculture by providing technologies for better qualifying it and sharing it (Bergez et al. 2016). Through the use of multiple data in predictive models, a connected agriculture with moderate instrumentation could support complex agroecological farming systems by offering coherent decision-making alternatives for innovative practices. It also could provide a geo-localisation of biological phenomena (species, pests, crops, soil type, etc.), allowing differentiated reasoning practices. In addition to this better understanding of biological processes, other IC tools could allow voluntary and chosen knowledge to be shared between stakeholders for better farm and territory management. Connected agriculture could also be seen as a vector for networking in agroecological agriculture, between farmers, between farmers and society (citizen, industry), and between farmers and researchers.

In this chapter, like Therond et al. (2017), we consider ICT not as underpinning any one agricultural model, but as a range of tools that are usefully applied to various agricultural models. In agroecological agriculture, ICT can be used to support the application of its different principles without distorting them, and without taking the decision-making power away from the farmers. The main issue is then to define the specific ICT needs of agroecological agriculture.

From the Conception to the Practical Implementation of a Transition

The TATA-BOX project focused on the primary step of an AET: the design, by the local actors, of both a desirable territorial agroecological system (TAES) and the pathway to reach it i.e. the transition to the TAES (tTAES). The methodology developed to conceive this agroecological system was mainly based on face-to-face discussions and "low tech" devices (paper maps, post-it notes, audio recordings, etc.). Some software tools could nevertheless be used to optimise the interactions during future applications of the methodology (Cahier et al. 2016). This postulate will be discussed in section "ICT recommendations to support the conception of a transition: objectives and architectural guidelines" of this chapter.

Apart from the IC tools that are useful during the design of a territorial agroecological system, we decided to widen the scope of this chapter to the digital technologies that could be used during the practical implementation of the designed system. Although these tools are not used during the stages studied by the TATA-BOX project, being aware of their existence while designing a desirable transition path can support choices that differ from a conception process based on the assumption that ICT will not be part of the game. For example, communication tools are interesting to discuss here since communication between people that do not usually work together was an important stake to maintain the momentum created by the project. The IC tools that are useful to a practical agroecological system implementation will be presented in section "ICT to support the implementation of a territorial agroecological transition".

From a Connected Farm to a Connected Food System

An agroecological transition concerns not only a farming system, but also the whole food system (*sensu lato*) of which it is part (from farm to table), as well as its relationship with society (Stassart et al. 2012; Duru et al. 2015a, b). The TATA-BOX project, which brought together a wide range of stakeholders from the territorial food systems it studied, is a good illustration of the importance of taking these multiple dimensions into account in order to identify all the mechanisms involved, their interdependences, and the obstacle or stepping stone they could represent during the transition. We have therefore chosen not to present IC tools specially conceived for a "connected farm" or for a "connected research laboratory", but rather IC tools that create networks between the actors of a food system, linking farmers to farmers, farmers to researchers, researchers to policy makers or technical advisers, farmers to consumers, and so on.

ICT moreover naturally break down the spatial and temporal boundaries characterizing agriculture, allowing interaction between actors and comparison between situations that would not have happened without them. Consequently, even if we focus on the territorial scale, some technologies will first be described at a larger spatial level so that their local consequences can be explained.

ICT Recommendations to Support the Conception of a Transition: Objectives and Architectural Guidelines

Based on the TATA-BOX Aveyron meetings, we propose in this section an outline of an ICT platform, combining several IC tools to support the practices of a tTAES. We call a TPD community (agroecological Transition with Participatory Design) any local group engaged in a tTAES. A TPD community is thus to be understood as a complex multirole community, including local stakeholders as well as facilitators, analysts and all the other roles necessary for the successful design of the tTAES. A TPD community, like many communities of practice or communities of action (Wenger 1998; Zacklad 2005; Warner 2006), needs to grow and develop, to describe and organise itself in terms of roles for collaborative work (Herrmann et al. 2004), and constantly to modify content shared among members. In terms of task organisation, the TATA-BOX method involves multiple roles, and an ICT platform must provide an infrastructure for a community portal that accommodates each of them. Each role is endowed with different prerogatives and possibilities of action on the contents.

The main purpose of the platform we propose is to make available to this community the informational content that it consults, creates and modifies through its discussions during the design process. These content-related services are intended to facilitate this provision by respecting the division of labour and confidence requirements necessary for the community design of the tTAES.

In terms of trust and security, an ICT TPD community in a given territory must protect its contents and retain complete control over them. As the technical underlying framework is constituted by Web technologies, the Community needs to have strong fence and effective confidentiality throughout its design and discussion work on the options, verbatim records and content that it develops, annotates and evaluates through the platform. This is particularly important to promote free expression of views internally at design meetings. Within this framework, the digital services of the platform, set out below, are essentially intended to serve the constitution, collective manipulation, and "capitalizing" of the traces of design discussions.

Proposed as a foundation for building and supporting socio-technical community systems in the particular TPD field, the platform has to be open and modular, and to avoid structuring too much of the TPD community's activity. In fact, the TATA-BOX method aims to stay within the frameworks to which rural actors are accustomed, without upsetting them. It incorporates the possibility that, from one TPD community to another, the conditions of the territory, as well as many human or material factors of participatory work, may influence its implementation. A digital platform must therefore adapt to local cultures and constraints, while in each case providing the different TATA-BOX roles with the best opportunities for interaction around shared content.

Technically, the core of the targeted platform is a documentary system based on Web technologies and supported by "Web Services" technologies. In addition to the flexibility of adaptation, the most important expected functionalities considered during the TATA-BOX design process were: supporting a large number and diversity of participants (present or distant); reducing costs and the number of staff involved in organising and running these workshops; ensuring the efficiency and semantic neutrality of the content tracking system; ensuring the simplicity of implementation; and facilitating appropriation of all the roles involved.

Based on these criteria, we recommend an open infrastructure, of the Social Semantic Web type, capable of easily and flexibly hosting the documentary system and these services. The services are illustrated below, in order of priority, and limited to the most essential ones needed to provide a basic solution that a TPD community can apply quickly to become socially operational.

With the exception of the MM-Record component, (discussed in priority 2 below) which was tested during the last series of TATA-BOX workshops in 2017 (cf. Fig. 2), the tools mentioned below were not tested in the project itself. However, since these tools are generic, they have already been tested in other social situations (cf. e.g. Bénel et al. 2010). These experiences with a variety of research fields have made it possible to problematise the notion of a social semantic Web infrastructure platform (Cahier et al. 2013). This social validation work would still have to be carried out on scale one throughout the whole process of implementing the TATA-BOX method, and would require major ICT developments and arrangements.

Priority 1: A Collaborative Basis for All TATA-BOX Roles and Actors

According to criteria previously listed (cost, workforce, efficiency, etc.), it is necessary to give priority consideration to the ICT platform as it is used, not only by the "field participants" of rural areas, but also by all other roles necessary to the design methodology. We observed during the TATA-BOX workshops that *basic participants'* roles are only the tip of the iceberg.

Reflection on tools must also consider the immersed part as equally important, that is, the roles inter alia of community organisers (managing invitations and staff, meeting agendas, their own agendas, etc.), facilitators, transcribers, observers and analysts (from several scientific disciplines), managers of certain technical tools (video, sound), graphic designers and producers of "rich picture" artefacts (who play a major role in the methodology).

The platform must meet a twofold challenge: to help all these actors to work on content, according to their role in the structure, but without the structure of the roles being prescribed rigidly by software workflows. Therefore, as a matter of priority, the platform requires a foundation of basic community functions, accommodating the different possible roles and supporting their mutual trust and design interactions. This is the prerequisite for access to a first stage of basic services (cf. sections "Priority 2: using topic maps to give access to verbatim records and TPD items" and "Priority 3: making visible and affordable the diversity of the viewpoints") offering all these roles (at least) read-only access to the traces of discussions, whatever their forms (audio, video, text). For example, this will allow everyone in the community to retrieve documents after the fact or in case of absence, and thus stay in touch with the community, one of the great challenges in such community design that spreads over months to years (cf. chapter "Evaluation of the Operationalisation of the TATA-BOX Process").

In the TATA-BOX method, the many roles give rise to varied interactions. Since TPD requires a complex collaborative design, the interactions between these roles and the objects and arguments under discussion are also complex. They are considered here in semiotic terms as digitisable traces. The challenge for the platform is to assist the creative or manipulative interactions of these contents by the various roles. On one hand they are interactions between actors sharing the same role. Of particular interest, for example, are the "cross-reading" scenarios between participants, in which each participant, rereads a transcribed discussion, highlights critical items, annotates and tags passages in the margins, and interacts with their peers through chained annotations. On the other hand, there are interactions between actors of different roles. For example (cf. Fig. 2) the "secretary" actor responsible for the audio recording of a discussion lasting several hours, annotates (on-the-fly or off-line) the audio stream with tags. These markers will be used: (i) by the transcriber to find his/her place quickly when he/she listens again; (ii) to construct useful markers for the group; and (iii) to indicate to the transcriber which fragments he/she has to process (this will save time and money in the overall process), etc.

From an IT point of view, this means that the platform must propose an infrastructure of participation architecture facilitating a coherent registration of actors into well-defined roles (Zaher et al. 2007; Merle et al. 2012; Tosi and Bénel 2017). Its basic services must enable the TPD Community on its own to finely regulate functions such as directory and member sponsorship,[2] role endorsement, role access

[2] Open Source tools integrated with the Hypertopic suite (see below and http://hypertopic.org) could be used for these functions, such as AAAforREST and DoLoMite ("Directories Led by Members", to federate digital identities management for multiple CSCW tools).

rights to various possible actions on documents (as in the examples provided above), and so on. Each of the actors thus benefits from prerogatives associated with their role. For example, for everyone there would be commentary rights; for some (managers, experts, etc.), the right to create tags; for other roles, more specific rights of creation or modification of contents (such as the transcribers' right to deposit texts), and so on. This type of basic infrastructure, allowing such fine tuning of access and roles, is not currently available on the shelf.[3]

The analysis of the TATA-BOX method as a socio-technical device needs to be deepened in order to fully explain the roles, their interactions, and the critical passages in the documents. The objective is not that this modelling automatically be translated into numeric terms, but rather that it help to apply the TATA-BOX method by allowing on the platform certain actions and interactions typical of roles, especially the countless TPD items that actors are constantly releasing and re-injecting into the discussion.

Priority 2: Using Topic Maps to Give Access to Verbatim Records and TPD Items

During the whole process, the TPD community's actors are faced with an intensive flow of changing knowledge and with many difficulties to memorize it, and seek to share numerous documents. The more important ones are the oral and transcribed verbatim recordings of meetings (more than 100 h of audio and video material were recorded in the TATA-BOX workshops). It is also necessary to share, index and retrieve secondary documents such as analyses, annotations by actors, and so on. ICT can be helpful in giving participants and other roles the best semantic affordances to their practices, especially at the critical stages of the TPD process.

We propose to use a method derived from the "Document and Item-based Modelling" method (Cahier and Ma 2010; Cahier et al. 2010) allowing *items* of the discussion to be described easily and organised in a NoSQL Web repository. TPD items are items of interest (actual facts, ideas, actors, fictional facts in prospective scenarios, opportunities, etc.) appearing in the collective inquiry on the complex rural situation, when solutions, governance plans, and so on are being devised. These items emerge from discussions and documents and are continuously (re-) interpreted, (re-)evaluated and (re-)used in participant's discussions and plans. The

---

[3]In terms of basic community services, the corresponding functions are usually the functions of invitation, registration or deletion, directories, sponsorship, diary, etc. We do not advise turning these functionalities into packaged solutions (Content Management System – CMS – or Electronic Document Management Image Management System – IMS). Their role structures, stereotyped for common usage, are not appropriate here. In addition, many teleworking, referral or profile functions (see social networking platforms) are also not necessary since the actors are in direct contact through meetings.

proposed ICT Social Semantic Web services on the platform allow items to be described by topics and by valuated attributes (e.g. for the geographical situation of a TPD item), in order to reflect each one's identity, categories and possible associated Web resources. With this method, actors can safely and confidently characterize, name, memorise and share thousands of TPD items.

To facilitate the retrieval of information, we suggest topic cloud and topic map artefacts. They can be constructed collaboratively by using the entire sequence of meetings' verbatim records (audio and video recordings, re-listening, transcription, reading, re-reading, annotation, etc.). With tools proposed to support the "Document and Item-based Modelling" method, actors can easily refer to the documents on their discussions and the annotations made during and after TPD meetings (on maps, on post-it notes, etc.). They can refer to all TPD items, geo-localised or not, as they identify and discuss them continuously ("on the fly") during the TPD process.

Figure 1 shows how items refer to discussed reality and how they rely on the topic map. The right side of the figure, on the cold slope, corresponds to the documents and to the more reified or consensual items shared by the community. The left side, on the warm slope, corresponds to the more emerging items and their more subjective and interpretable aspects, brought up during the TPD collective inquiry. So the "item" concept creates a bridge between the two sides. At a given point in a discussion, a participant may mention a TPD item he/she observed in the rural situation and that made sense to him/her. Even if it has not been stabilized, this item can be created in the repository (path "3" in Fig. 1). But items can also be detected in a document (path "2" in Fig. 1). All items can then be constructed in more detail (paths "5", "6", "7") by identifying its proper name and adding attributes (e.g. localisation), topics and resources. If other actors recognise an item as a relevant element of the shared situations, they can qualify it with complementary (or concurrent) attributes and topics, thereby allowing the community to maintain the co-building of the item and the topic map linked to it.

By using TPD items as mediation, the indexing of documents can be facilitated by the actors themselves. Participants can name and characterize well-known items because they are familiar in their all-day skills or in their documents (Fig. 1, path 2). This folksonomy facilitates the use of the verbatim records, reduces actor's puzzlement within verbatims and all the TPD documents, despite the fact that they are numerous, changing, added by multiple contributors … and frequently controversial (cf. section "Priority 3: making visible and affordable the diversity of the viewpoints").

This approach can be illustrated and implemented by existing prototype tools, such as Argos and Steatite based on REST Web Services, supplemented by the MM-Record tool (Matta and Ducellier 2013) for indexing recorded audio streams from a tablet. These tools will make it possible to implement the proposed method and build up a document base of TPD items easily accessible on the Web from a topic map.

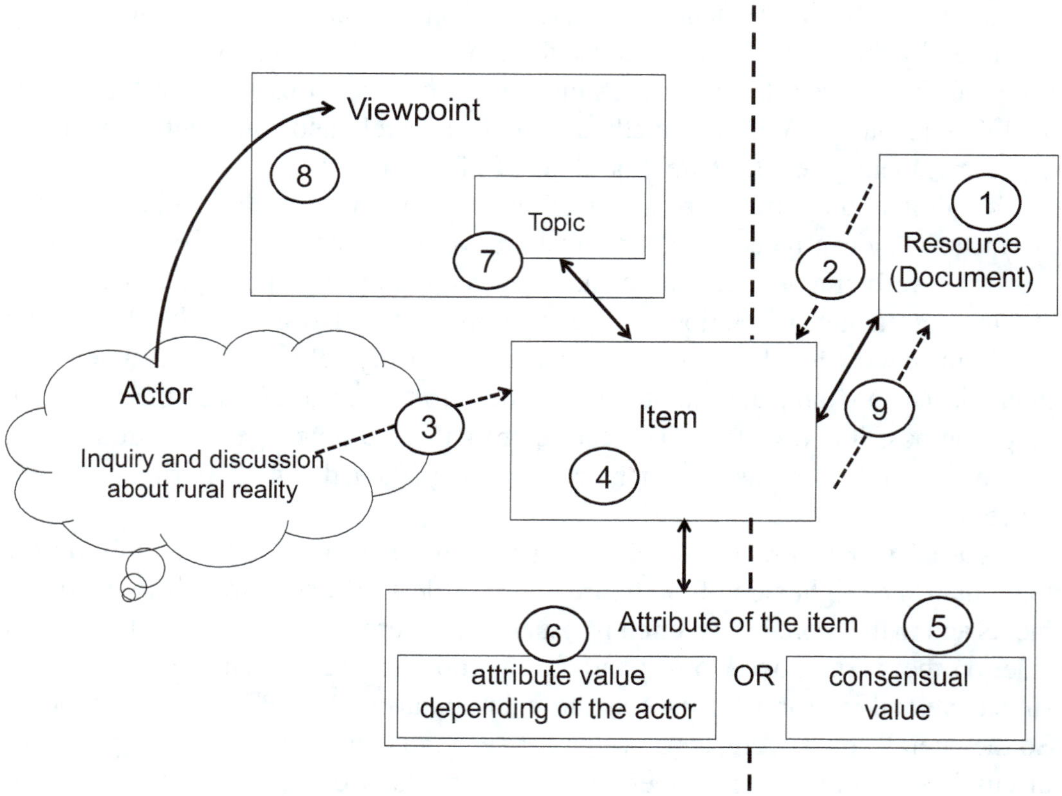

Fig. 1 "Document and Item-based Modelling Method" (using the Hypertopic conceptual frame) applied to TPD collaborative work

Fig. 2 MM-Record tool on tablet (**a**) and its use during a 2017 TATA-BOX design meeting (**b**)

MM-Record (Fig. 2) completes the audio recording of meetings by allowing participants to index the audio content with coloured marks and tags for time, speakers' names, and categories, and to design rationale topics. In the abundant records of the 2017 TATA-BOX meetings (lasting many hours), it helped actors to retrieve oral fragments and tag them. The oral material related by participants was shared and socially annotated. In particular, some tags put onto the recorded audio flow were used to select the more relevant fragments to be transcribed, thus reducing the transcription cost.

Priority 3: Making Visible and Affordable the Diversity of the Viewpoints

One of the key challenges for the platform is to allow the multiple actors to benefit from a multi-viewpoint approach in terms of content categorisation and how to find fragments more easily through a repository that gives meaning to these fragments. This also applies to the primary contents of the design (e.g. traces of the discussions) or derived contents (e.g. annotations, tagging and threads of discussions outside the verbatim records).

Given the specificities of the AET in a given territory, it is important to make visible and integrate the different points of view. To technically implement a multi-viewpoints anchorage structure into the content, we propose that the topic maps artefacts mentioned previously (cf. section "priority 2: using topic maps to give access to verbatim records and TPD items"), using the Hypertopic model (Zhou et al. 2006) as a background. The structure of hypertopic actors and viewpoints can be organised to help actors to make traceable and visible the interpretations from their own perspectives (cf. the upper part of Fig. 1: paths 3-4-8-7 and 1-2-4-7-8 for example). Thus, actors can elicit the TPD items and qualify them, not only at a "reference" level (recording consensus when it exists) *but also* at the heuristic and inter-subjective levels (Bénel et al. 2010), considering that items always stay, to a greater or lesser degree, in design and in debate.

In this way, in order to carry out the necessary tests, we recommend that existing multipoint tools based on the Hypertopic model, such as Argos, Agorae, Porphyry (cf. http://hypertopic.org) that can be used on the Web, be industrialised and extended according to the specific context of TPDs. These tools allow one to build and compare views on the items of the domain. Cassandre and Lasuli, also tools of the Hypertopic suite, can be used by multiple analysts to carry out qualitative analyses of the same corpus, and to compare their categories (Bénel et al. 2011; Lejeune 2011).

Actors may use various views on items in the TPD cooperative work and debate about them. Multi-viewpoint social tagging can be applied with ease by end-users without any knowledge in Information Sciences. It is designed specifically to be used in communities whose members are faced with an intensive flow of changing knowledge and with many difficulties to represent it visually on a topic map, as well as many conflicts between actors, especially conflicts of interpretation.

Note that this approach does not require a set of predefined viewpoints or categories. On the contrary, it offers freedom in the choice of perspectives from which to interpret and organise information. Depending on the way in which a TPD community decides to use these tools, in particular according to the phases of collaborative design, the points of view can be those of the stakeholders (corresponding to their field of competence, their opinion, etc.). Stakeholders may also participate in games of analysis, the dimensions of which are decided by mutual agreement, such as the game of the three major "structural" dimensions advocated by the TATA-BOX method. Thanks to its underlying Hypertopic model, the proposed platform remains agnostic in semantics and does not influence the daily languages of the actors, whether they are experts or not. This flexibility is important for TPD communities, which can thus, for example, balance the expression of minority actors with that of more powerful stakeholders.

ICT to Support the Implementation of a Territorial Agroecological Transition

The previous section discussed the possible utility of ICT during the TPD. The present section explores the roles IC tools can play after the design step, when the transition is practically implemented by the actors of a territory. Reix et al. (2016) classify ICT according their use: communication, functional use, knowledge management, and decision-support. The digital technologies that could economically, socially and environmentally support the development of a territorial AET are classified in this section according to this usage typology, with different practical illustrations for each class.

Communication Systems

Communication technologies facilitate the circulation of data, information and knowledge inside an organization or from an organisation to its targeted public. This includes generic tools such as websites, videoconferencing systems or exchange of digitized data systems (Soulignac 2012).

Communication is essential for all sectors and not only for agriculture. But it is a particularly important stake for agroecology: constructing a territorial transition requires coordination between a wide variety of actors that do not usually communicate but should all be involved for the functional implementation of an agroecological food system. During the TATA-BOX Project, communication was mainly during workshops. In order to maintain and even improve the network dynamic generated during those workshops, digital communication tools should lead up to classic physical meetings during the implementation stage. These tools can notably

Box 1: A Catalogue of Theoretical and Practical Trainings in Agroecology

The French platform Osaé (Osons l'agroécologie, in English "Dare agroecology") was developed to foster the implementation of agroecology. It presents farmers' testimonies and technical syntheses, as well as an agenda listing the future agroecology meetings in France, from colloquia to field visits and technique trainings (Osaé 2017).

Link: www.osez-agroecologie.org

help to communicate synchronously or asynchronously, to increase the visibility of the project, to interact with actors that were not identified a priori (Agroecology in Action 2017), to disseminate results of observatories, or to organize "face-to-face" or digital knowledge exchanges (Box 1).

Functional Tools

Functional technologies include all the applications supporting the processing of recurrent tasks that can easily be digitised. In agriculture, their first applications were accounting tools designed for the economic management of a farm, but they also include commercial support tools (Soulignac 2012).

By enabling almost instantaneous contact between farmers, or between farmers and consumers, functional technologies could support the development of an agroecological food system in at least four ways. First, websites and mobile applications are efficient ways to organise exchanges of agricultural goods such as straw, livestock manure, compost, or even household and communal organic waste (FourrageFWA.be 2017). By exploiting the complementarity of the diverse farming systems present in a territory in terms of inputs and outputs, those exchanges are one of the keys for recycling biomass and minimizing resource losses (Moraine et al. 2016, 2017a, b; Therond et al. 2017, chapter "An Integrated Approach to Livestock Farming Systems' Autonomy to Design and Manage Agroecological Transition at the Farm and Territorial Levels"). Second, the same kind of commercial agricultural websites can facilitate the trade of animal and vegetal (seed) varieties adapted to local/territorial conditions, which is a way to foster genetic diversification and to enhance the value of local resources. Third, organising a sharing system for mechanical material can be facilitated by a digital application (WeFarmUp 2017). This collective practice is a good way to try agroecological alternative practices requiring specific machines at a moderate cost, both economically and environmentally (fewer machines are manufactured). Finally, websites are powerful tools for the construction of re-localised food systems co-managed by both producers and citizen-consumers. The ICT connection between farmers and

Box 2: "Agrilocal", a Platform to Connect Local Farmers and Public Authorities

Developed initially by the General Council and the Chamber of Agriculture of the Drôme and Puy-de-Dôme *départements* (France) in 2012, Agrilocal is a free website that allows local farmers to directly contact public procurement services with a collective catering mission (schools, hospitals, retirement homes, etc.). The platform shows the buyer all the products that are available locally and that correspond to their needs, and each seller has a personal page to present its farm. Agrilocal is now operational in 24 French *départements* (Alim'agri 2015).

Link: www.agrilocal.fr

Box 3: Crowdfunding Applied to Agricultural Projects with "BlueBees"

BlueBees is a participative funding platform dedicated to sustainable agriculture and nutrition projects that are ecological, economically viable and a source of employment and social links (Bluebees 2017). By presenting the projects on the basis of their geographical location, the website helps project leaders to gather a community of contributors that will not only finance their initiative (by lending or giving money) but also probably support its proper functioning once it is launched, for example by buying its products or by participating in consumers' general assemblies.

Link: https://bluebees.fr/

citizens can facilitate both the practical selling process (Box 2) and the financial support for local food production and transformation projects by future consumers (Box 3). This latter website's role is important since a democratic governance of food issues making peasants more autonomous with regard to dominant market forces is a crucial principle in the AET in a territory (Stassart et al. 2012).

Knowledge Management Systems

Knowledge management (KM) technologies include all the tools that serve to create, stock, disseminate and update knowledge (Soulignac 2012).

The development of an AET fostering a "strong" ecological modernisation of agriculture (Horlings and Marsden 2011) requires the implementation of agricultural practices in favour of agro-biodiversity and ecosystem services at different ecological, spatial and temporal scales (Kremen and Miles 2012; Duru et al. 2015b).

Box 4: "Geco", a Collaborative Web Tool for Constructing Knowledge in Agroecology

Geco is a KM web application dedicated to agroecology that was jointly developed by INRA, ACTA and IRSTEA as part of the French Ecophyto plan for pesticide-use reduction in agriculture. The website is divided into two spaces: one is a "knowledge base" enriched collaboratively by contributors from the whole farming community and recognized by their peers. This base presents knowledge in the form of pages classed by concept such as alternative practices, crops, pests, material, pest auxiliaries, etc. The second space is a forum in which anyone can create a discussion topic related to a particular knowledge page or concerning a subject that is not yet treated in the knowledge base. To organize and structure all the information and knowledge available in the base and in the forum, and to enable their effective use during research, the website integrates a semantic model allowing the creation of links between pages (Soulignac et al. 2017). For example, a contributor can define the following relation between concepts by linking pages: the Crop (page) "lentil" – "is attacked by" – the Pest (page) "lentil weevil".

For now, the knowledge base mainly contains pages about innovative agricultural techniques, written by contributors from the inter-technological network "RMT SdCI" (Guichard et al. 2015).

Link: http://geco.ecophytopic.fr/

This implementation can be supported by KM technologies in at least two ways that are described in the next paragraphs.

First, practices combining agricultural production and natural resource management are subject to many uncertainties (Williams 2011; Chapter "A Plurality of Viewpoints Regarding the Uncertainties of the Agroecological Transition"). In order to minimize the risk taken by the farmers adopting those practices, it is crucial to lower their unpredictability. We can distinguish two levels of action for lowering uncertainties: supporting experience-sharing, and creating new knowledge via data analysis.

- **Supporting experience-sharing**: in recent years, many websites assisting agricultural practitioners in accessing and sharing knowledge about agroecology have emerged. Reflecting the socio-economic principles of agroecology defined by the GIRAF (Stassart et al. 2012), these websites value and disseminate a diversity of forms of knowledge (local know-how, empirical knowledge, etc.), and are sometimes designed to foster networking and debate. The range of technologies they use is fairly wide, since some of them simply bring inexperienced and experienced practitioners into contact with one another (Agricool 2017; Agrifind 2017), while others collect and archive stories of successful agroecological experiences (Farmers2Farmers 2017), or collaboratively capitalize on knowledge about sustainable agriculture (DicoAgroecologie 2017), sometimes with the help of semantic technologies (Box 4). It should be noted that besides

Box 5: "AGROSYST", Capitalising on Knowledge About Low Pesticide-Consuming Farms

Developed as a part of the national ECOPHYTO plan, the DEPHY network encompasses more than 2000 farms and experimental sites trying to minimise their dependence on phytosanitary products by modifying their agricultural practices. In order to facilitate the valorisation and transversal analysis of the results obtained with this programme, the information system AGROSYST was developed to collect, store and exploit data from the participating farming systems. The system integrates the following functions: acquisition and hosting of various data (crop rotation, cultivation operations, decision rules, measures and observations, economic margin etc.), calculation of synthesis variables, editing of decision schemes, and interoperability with other information systems (Bournigal 2016).

those websites dedicated to agriculture, generic social networks like Facebook or Twitter are becoming places where farmers gather to informally share their experiences.

- **Creating new knowledge via data analysis**: so much agricultural data already exist that the technical term "big data" is often used to describe them. These data come from drones, sensors, connected objects, satellite images, traceability and crop management softwares (Box 5), or even biodiversity voluntary surveys and other crowdsourcing practices. They are often characterised by spatial and temporal dimensions (#DigitAg 2017). In addition to these structured sources, many agricultural unstructured documents contain precious information that can be extracted and homogenised to produce exploitable data, for example through the use of semantic technologies (Box 6). ICTs are used for the creation, transfer, storage, structuring, sharing and finally exploitation of such data. This process can help to better understand the agro-ecosystem thanks to technologies such as machine learning (less understanding but better prediction), visual analytics, data statistical analysis, data mining or integration of imperfect knowledge (#DigitAg 2017). These advances in the understanding of agricultural phenomena could allow the development of new decision-support tools (Bournigal 2016) that can be more or less near to the principles of agroecology (cf. section "Decision-support systems").

Second, the redesigning of a food system involving agricultural practices that foster agro-biodiversity and ecosystem services needs to be supported by a learning system adapted to the stakeholders. For the past few years, agroecology has been part of the teaching programmes in some agricultural schools (Alim'agri 2016). Master's degrees and certifications in agroecology have been developed throughout Europe (Agroecologie.fr 2017; Certificat-agroecologie 2017; Master-agroecologie 2017). These initiatives are not however sufficient. As vast numbers of people, with

Box 6: Facilitating Access to the Regional Pest Alert Bulletins via Semantic Technologies

During the VESPA project, financed by the national ECOPHYTO plan, the institutes INRA and IRSTEA examined the usefulness of the French crop epidemio-surveillance programme. To do so, thousands of regional pest alert bulletins published during the last 50 years were collected and digitised, and an open data platform was created to access them freely. To facilitate the research through this corpus, three classes of semantic annotations were used to describe the content of each document: spatial annotations (the region concerned), temporal annotations (the publication date), and thematic annotations (the principal crop concerned). These annotations can be enriched by the organisations that published the bulletins and supplemented by links to other resources such as weather reports (Roussey et al. 2016). This perennial access point facilitates the observation of spatio-temporal dynamics for epidemiological modelling, and can lead to the identification of locally efficient crop protection practices if other data such as crop diversity or hedge density are available for the same regions and periods (EcophytoPIC 2017).

Link: www.pestobserver.eu

Box 7: The First Massive Open Online Course (MOOC) on Agroecology

With the technological support of the French Numerical University (FUN), the Montpellier SupAgro engineering school built a MOOC presenting the different approaches of agroecology. More than 12,000 participants from 100 countries enrolled in the first edition of the course. In a participatory training dynamic supported by the social and geographical diversity of the actors, the MOOC proposes to build an agroecology approach at the interface between agronomic, ecological and social sciences. The course content will be available on the platform under a Creative Commons license (FUN MOOC 2017).

Link: https://www.fun-mooc.fr/

a wide range of profiles in terms of nationality, time availability, profession, etc., are interested in such courses, ICTs are already part of the agroecological education plan. Their original forms are interesting for fostering the construction of innovative teaching methods adapted to the holistic and systemic nature of agroecology (Box 7) (Pollen 2017; Supagro 2017; UVAE 2017). Aside from the interest of their original format, virtual teaching is also a good way to facilitate interactions and knowledge-building between the different scientific disciplines involved in research and teaching in agroecology (Stassart et al. 2012).

Decision-Support Systems

Technologies for decision-making are able to process data in order to provide advice. They have several levels of complexity, from simple dashboards presenting data in an organized way, to tools providing clear operation orders according to predetermined objectives and action rules (Soulignac 2012). Decision-support tools are based on two resources: instantaneous data providing information on the present situation, and previous knowledge on which their decision models are built.

Most agricultural decision-support technologies presently concern simple, mono-task operational decisions. They are used in domains such as precision agriculture – the right dose/action, at the right place, at the right time – or pest evolution monitoring. They often reduce the observation and/or action time required from field actors, which is helpful insofar as those activities are generally time-consuming when a new agroecological farming system is implemented. Besides those simple tools, many scientists are already working on decision-support systems that go beyond the mono-task stage, concerning mainly integrated pest management (AGIR 2017), but also other subjects like multi-species meadow design (Box 8). In the future, interdisciplinary research based on the analysis of data available thanks to knowledge management technologies could lead to a better understanding of local agro-ecosystems and of the reaction they can have to global changes in farming systems. This potential acquisition of systemic and contextualised knowledge could open the door to decision-support tools better suited to assist the agroecological redesigning of food systems, for example in the form of a classic (face-to-face) territorial board game in which the players' actions are guided by a software calculating the economic, agronomic, environmental or social impacts of different agricultural and commercial choices (Duru et al. 2015b).

Box 8: "Capflor", an Agroecological Tool for Designing Meadows with Diverse Flora

Capflor is a free decision-support tool that recommends associations of forage species for meadows, based on soil and climatic conditions and on the intended use of the forage (mowing, grazing or mixed). During a research project called Mélibio, the decision model was constructed both in a multi-disciplinary and in a multi-actor way since it synthesised agronomic and ecological criteria and hybridised researchers', advisers' and farmers' knowledge. The INRA team now in charge of the tool has organised a collaborative network with livestock farmers and advisers who give them feedback from the field, thus allowing for a continuous enrichment of the tool. While Capflor is primarily intended for livestock farmers and agricultural advisers, it can also be used as a teaching tool in agricultural schools, to introduce students to multi-species meadows (Capflor 2017).

Link: http://capflor.inra.fr/

ICT and Agroecology: What Are the Challenges?

Research Needs and Development Conditions

Since July 2017, the French research, teaching and industrial organisations working on digital agriculture have gathered around the #DigitAg Convergence Institute to design ICT that will effectively be used by the public aimed during the design (that is not always the case), from the acquisition of data to their exploitation (Alim'Agri 2017). To meet its different challenges, which include the development of IC tools of service to the AET (#DigitAg 2017), the Institute has identified six areas (hereinafter called "axes") in which multidisciplinary research is needed (Table 1). Axes 3–6 correspond to the classic "collecting – organizing – visualizing and mining – modelling" chain of digital data technologies, and mainly concern natural sciences and technology. Axes 1 and 2 highlight the social and economic dimensions and the fact that they must not be forgotten if we want to bring ICT research from the lab to the field.

The first axis examines issues about the social and economic impacts of ICT, like "how do ICT technologies contribute to improving farm-level management and territory governance?" and "How do ICT-enabled new services change the role of agricultural actors, including advisory services?" (#DigitAg 2017). The second axis considers social, legal and management matters such as "How do we build technical and organizational digital innovation that will successfully be adopted by the farmers?" and "How do we address the legal and ethical issues of intellectual property of data and knowledge, and what are their consequences on value share?" (#DigitAg 2017). All these questions are still to be investigated and no complete answers are currently available, but some elements are already interesting to present concerning IC tools adoption and data ethics.

Table 1 The six main axes organizing the #DigitAg scientific communities, their objective and the major scientific disciplines they involve (#DigitAg 2017)

Axes	Objective	Major disciplines
1	Understanding ICTs' influence on rural societies	Economics, management, social science
2	Building ICT-based innovation: Technological, social and legal issues	Law, social science, management
3	Fostering the development of appropriate sensors and data acquisition systems, including crowdsourcing	Physics, optical science, electronics, digital science
4	Making progress in agricultural information system design	Computer science
5	Designing new data-mining methods, appropriate to agricultural data, to extract actionable knowledge	Data science, computer science
6	Exploring new ways for model integration/ qualification	Agronomy, mathematics, computer science, artificial intelligence

The adoption of IC tools by farming communities depends on practical criteria such as profitability, simplicity of use, and efficiency of the designer's communication, and must be analysed with the help of various disciplines like sociology, ergonomics or management (#DigitAg 2017). For example, an ergonomist and an agronomist (Cerf and Meynard 2006) studied the various uses that farmers and advisers made of fertilisation and pest management decision-support tools. Finding that there was a considerable gap between the use that developers had planned and the way the tools were used on the field, they developed the concept of advisory and information systems (AIS), a "temporal and spatial network of humans and material arrangement which allow information to be developed and disseminated in order to make decisions about local monitoring of agroecological processes". They propose to develop a conception methodology taking AIS into account in order to include the users' creativity and needs in the early stages of the design process, which should ease the IC tool adoption by farming communities. The decision-support tool Capflor (Box 8) is a good example of technology co-conceived with its users and continuously adapting to their needs and practices. In the same vein but concerning knowledge management tools, a French knowledge management Club (Club-gc 2017) developed a set of questionnaires allowing communities to better define their needs concerning knowledge. These questionnaires can be adapted to agricultural communities, and help to answer questions like "what technical knowledge is critical or missing concerning this practice?", "how do we diffuse and validate knowledge in our community?" or "which tools would be the most efficient to learn this particular practice?". These examples of design methods integrating users' behaviour and needs are consistent with the agroecological principles of participatory research and final-users inclusion in research, determined by the interdisciplinary group GIRAF (Stassart et al. 2012).

The question of data availability will also become crucial in the next few years, since all the potential of ICT is based on an access to a substantial number of field references. At present, this access does not seem to be guaranteed: although an abundance of data has already been produced, these data belong to a wide range of actors from both sectors – private (farmers, tractor manufacturers, weather station sellers, etc.) and public (Common Agricultural Policy declarations, experimental farms, etc.) –, and collecting them will be a huge challenge (Bournigal 2016). The collection of agricultural data, even for research purposes, furthermore raises questions about intellectual property, privacy, traceability, and freedom, or even about the monetary value of such data (#DigitAg 2017). The future of data is already splitting up into several directions. Some companies try to acquire sufficient amounts of data with their own sensors to develop good decision-support tools. Others are starting to buy agricultural data from diverse origins in order to sell them in a worldwide data market (Dawex 2017). The French government is considering the development of a national public portal to collect and store all the agricultural data. The idea will be that suppliers give their data to the governmental data portal, and in exchange they can visualize these data and the data from other suppliers integrated in the portal as maps, statistics etc... And if the government creates decision tools thank to these data, suppliers will have access to these tools too (Bournigal 2016). The co-existence of these different models would lower the scientific value of each of them: the less data you have, the further from field reality your models are. However, choosing the most relevant model is really complicated, especially because the various data suppliers can have divergent opinions on the purposes those data should serve, their monetary value, and the public they should be accessible to. For the development of an agroecological system that includes ICT as support tools, rather than as a techno-centric model, it will be crucial to construct a transparent governance of data, including field actors as important decision-makers.

In the End, Will ICT Save or Consume Non-renewable Resources?

This chapter has demonstrated that at a territorial scale, many IC tools can support AET as regards their environmental, social, economic and methodological dimensions, during both the design and the implementation phases. However, these demonstrations never considered the steps existing outside the scope of the farmer or territorial community: the construction of the tool, its maintenance and its recycling or treatment as waste. Whatever ICT is considered (sensor, radar, database, website, software, etc.), those steps have considerable non-virtual costs since they all consume energy and non-renewable resources.

From its beginnings, agroecology has been based on the principle that farming systems should use as little fossil-sourced inputs as possible (Altieri 1995). Applying this principle to local farms by using tools that have the opposite impact in other places does not seem very logical.

If ICT designers do care about the energetic cost of their tools, this mainly impacts the steps where an economy of energy is directly profitable (in terms of time and money) for the final user. For example, the COPAIN team of the IRSTEA institute of Clermont-Ferrand is working on minimising the energetic consumption of agricultural sensors by creating programmes optimising activity time for the acquisition and transmission of data (Irstea 2017). The question of the global energetic and environmental impact of a food system using IC tools is not currently covered and should be researched by scientific teams working on life cycle assessment (LCA). The results of LCA studies could help to distinguish between IC tools that have a positive energetic and/or environmental impact globally, from those with a positive impact for the final user, albeit one that is not significant enough to compensate for the negative impacts of the tools' construction, maintenance and treatment as waste. The generalisation of ICT environmental impact assessment via LCA could stimulate developers to favour criteria like circular economy and energetic sobriety in the design of their own future tools. It is moreover worth noting that life cycle scientists do not only work on environmental criteria; matters of economic and social impacts are more and more often addressed via the concepts of life-cycle costing assessment (LCCA) and social life-cycle assessment (SLCA) (ElsaPact 2017). These disciplines would also allow an interesting approach for evaluating the relevance of using various ICTs in an agroecological food system.

Conclusion

ICTs are actually everywhere. As we have seen, these tools can be used in the design of an AET of rural areas: supporting multiple points of view, allowing desynchronised forums or debates, creating topic maps to share knowledge, and so on. However, in this section we have outlined only a few. A lot of work is still necessary to propose functional tools to support these transitions of territorial agroecological system.

Many forums on agroecological systems are available, using Internet either to communicate or to manage knowledge. Sensors are available to provide data on various subsystem elements (plant, soil, pests, etc.) and computer and decision support system tools may help in making better management choices. This digital revolution will require proof of concept, acceptance and training. The role of the social sciences will be fundamental to understand, guide and propose schemes on the use of this new diversity of tools. Therefore, this is not only a digital story but also a human story. The French #DigitAg research project will integrate this social point of view.

References

#DigitAg (2017) #DigitAg: La Recherche. Online. http://www.hdigitag.fr/fr/la-recherche/. Accessed 22 Sept 2017

AGIR (2017) PEST: Analyse et modélisation des effets des pratiques agricoles sur les pressions biotiques et les dommages induits en grande culture. In INRA – AGIR – Equipe VASCO. https://www6.toulouse.inra.fr/agir/Les-equipes/VASCO/Recherche/PEST. Accessed 26 Sept 2017

Agricool (2017) www.agricool.net: on ne fait plus labour mais on sème toujours. http://www.agricool.net/forum. Accessed 19 Sept 2017

Agrifind (2017) Agrifind: "l'expertise terrain partagée". https://connexion.agrifind.fr/infos/about. Accessed 21 Sept 2017

Agroecologie.fr (2017) Agroécologie: Formations diplômantes. http://www.agroecologie.fr/formation.html. Accessed 22 Sept 2017

Agroecology in Action (2017) Agroecology in action: Carte des initiatives. In Agroecology in Action Online. http://www.agroecologyinaction.be/spip.php?rubrique13. Accessed 19 Sept 2017

Alim'agri (2015) Avec "Agrilocal", mettre en relation producteurs locaux et acheteurs publics. In Alim'agri: site du Ministère de l'Agriculture et de l'Alimentation

Alim'agri (2016) "Faites de l'agro-écologie": des lycées agricoles engagés! In Alim'agri : site du Ministère de l'Agriculture et de l'Alimentation. http://agriculture.gouv.fr/faites-de-lagro-ecologie-des-lycees-agricoles-engages. Accessed 22 Sept 2017

Alim'Agri (2017) Agriculture numérique: inauguration de l'Institut Convergences #DigitAg. In Alim'agri : site du Ministère de l'Agriculture et de l'Alimentation. http://agriculture.gouv.fr/agriculture-numerique-inauguration-de-linst. Accessed 27 Sept 2017

Altieri MA (1995) Agroecology: the science of sustainable agriculture. Westview Press, Boulder

Bénel A, Zhou C, Cahier JP (2010) Beyond web 2.0... And beyond the semantic web. In: Randall D, Salembier P (eds) From CSCW to Web 2.0: European developments in collaborative design, computer supported cooperative work. Springer, London, pp 155–171

Bénel A, Cahier JP, Tixier M (2011) LaSuli: un outil pour le travail intellectuel. In: Actes du 14e colloque international sur le document électronique (CIDE). Europia

Bergez JE, Soulignac V, Cahier JP, et al (2016) ICT to help on participatory approaches for the agroecological transition of agriculture. In: 12th European IFSA symposium 2016. Harper Adamn University

Bluebees (2017) BlueBees: financing tomorrow's agriculture. https://bluebees.fr/en/. Accessed 18 Sept 2017

Bournigal JM (2016) AgGate: Portail de données pour l'innovation en agriculture. Report for the French Ministries of National Education, Superior Education and Research, of Agriculture, Agri-food Industry and Forest, and of Industrial and Numerical Economy. http://agriculture.gouv.fr/un-portail-de-donnees-pour-linnovation-en-agriculture-la-synthese-du-rapport. Accessed 22 Sept 2017

Cahier JP, Ma X (2010) Document and item-based modeling: a Hybrid Method for Socio-Semantic Web In: Proceedings of the tenth ACM symposium on document engineering (DocEng) ACM, pp 243–246

Cahier JP, Zaher LH, Isoard G (2010) Document et modèle pour l'action, une méthode pour le Web socio-sémantique. Doc numérique, RSTI, Hermès-Lavoisier 13:75–96

Cahier JP, Benel A, Salembier P (2013) Towards a "non-disposable" software infrastructure for participation. Interact Des Archit J – IxD&A 18:68–83

Cahier JP, Brébion P, Salembier P (2016) Using and supporting explicit viewpoints in territorial participatory design meetings. In: 17th European Conference on Knowledge Management (ECKM 2016), 1 – 2nd September 2016. ACPI, Belfast (UK), pp 136–145

Capflor (2017) Capflor: outil agroécologique d'aide à la conception de prairie à flore variée. http://capflor.inra.fr/. Accessed 27 Sept 2017

Cerf M, Meynard JM (2006) Les outils de pilotage des cultures: diversité de leurs usages et enseignements pour leur conception. Natures Sci Sociétés 14:19–29

Certificat-agroecologie (2017) Certificat en agro-écologie et transition vers des systèmes alimentaires durables: une formation continuée inter-universitaire. http://nmktebld.preview.infomaniak.website/. Accessed 22 Sept 2017

Club-gc (2017) Club Gestion des Connaissances. http://www.club-gc.asso.fr/. Accessed 28 Sept 2017

Dawex (2017) Agriculture. In Dawex: la place de marché pour monétiser et acquérir des données. https://www.dawex.com/fr/industries/agriculture/. Accessed 28 Sept 2017

DicoAgroecologie (2017) Dictionnaire d'agroécologie. http://dicoagroecologie.fr/. Accessed 9 Nov 2017

Duru M, Therond O, Fares M (2015a) Designing agroecological transitions; a review. Agron Sustain Dev 35:1237–1257. https://doi.org/10.1007/s13593-015-0318-x

Duru M, Therond O, Martin G et al (2015b) How to implement biodiversity-based agriculture to enhance ecosystem services: a review. Agron Sustain Dev 35:1259–1281. https://doi.org/10.1007/s13593-015-0306-1

EcophytoPIC (2017) VESPA: Valeur et optimisation des dispositifs d'épidémiosurveillance dans une stratégie durable de protection des cultures. In EcophytoPIC: Le portail de la protection intégrée des cultures. http://www.ecophytopic.fr/tr/innovation-en-marche/ecophyto-recherche/vespa-valeur-et-optimisation-des-dispositifs-d. Accessed 26 Sept 2017

ElsaPact (2017) ELSA PACT: Chaire industrielle en évaluation de la durabilité du cycle de vie. http://www.elsa-pact.fr/language/fr/. Accessed 29 Sept 2017

Farmers2Farmers (2017) Farmers2Farmers, Greenpeace. https://fr.farmers2farmers.org/. Accessed 21 Sept 2017

FourrageFWA.be (2017) FourrageFWA.be: Rechercher votre fourrage en Wallonie. http://fourrage.fwa.be/. Accessed 19 Sept 2017

FUN MOOC (2017) FUN MOOC: Montpellier SupAgro, Agroécologie. https://www.fun-mooc.fr/courses/Agreenium/66001S02/session02/about. Accessed 21 Sept 2017

Guichard L, Ballot R, Halska J et al (2015) AgroPEPS, un outil web collaboratif de gestion des connaissances pour Produire, Echanger, Pratiquer, S'informer sur les systèmes de culture durables. Innov Agron 43:83–94

Herrmann T, Jahnke I, Loser KU (2004) The role concept as a basis for designing community systems. In: Proceedings COOP '04, 6th international conference on the design of cooperative systems, May 11–14, 2004,. Hyères, France, pp 163–178

Horlings LG, Marsden TK (2011) Towards the real green revolution? Exploring the conceptual dimension of a new ecological modernisation of agriculture that could "feed the world.". Glob Environ Chang 21:441–452. https://doi.org/10.1016/j.gloenvcha.2011.01.004

Irstea (2017) Projets de recherche nationaux – Equipe COPAIN. In Irstea: Institut national de recherche en sciences et technologies pour l'environnement et l'agriculture. http://www.irstea.fr/la-recherche/unites-de-recherche/tscf/systemes-information-communicants-agri-environnementaux/projets-nationaux. Accessed 28 Sept 2017

Kremen C, Miles A (2012) Ecosystem services in biologically diversified versus conventional farming systems: benefits, externalities, and trade-offs. Ecol Soc 17:25. https://doi.org/10.5751/ES-05035-170440

Lejeune C (2011) An illustration of the benefits of Cassandre for qualitative analysis. Forum Qual Sozialforsch = Forum Qual Soc Res [FQS] 12:19. https://doi.org/10.17169/12.1.1513

Master-agroecologie (2017) Master interuniversitaire en Agroécologie. http://www.master-agroecologie.eu/. Accessed 22 Sept 2017

Matta N, Ducellier G (2013) Memory meetings : approach to keep track of project knowledge in design. In: 5th international conference on knowledge management and information sharing (IC3K/KMIS). p 12

Merle F, Bénel A, Doyen G, Gaïti D (2012) Decentralized documents authoring system for decentralized teamwork: matching architecture with organizational structure. In: Proceedings of the 17th international conference on supporting group work (ACM GROUP). pp 117–120

Moraine M, Grimaldi J, Murgue C et al (2016) Co-design and assessment of cropping systems for developing crop-livestock integration at the territory level. Agric Syst 147:87–97. https://doi.org/10.1016/j.agsy.2016.06.002

Moraine M, Duru M, Therond O (2017a) A social-ecological framework for analyzing and designing integrated crop–livestock systems from farm to territory levels. Renewable Agric Food Syst 32:43–56. https://doi.org/10.1017/S1742170515000526

Moraine M, Melac P, Ryschawy J et al (2017b) Participatory design and integrated assessment of collective crop-livestock organic systems. Ecol Indic 72:340–351

Osaé (2017) Osaé, osez l'agroécologie: platerforme d'échanges pour la mise en pratique de l'agroécologie. http://www.osez-agroecologie.org/. Accessed 19 Sept 2017

Pollen (2017) Tour du monde Agro-écologie. In Pollen: Le partage des innovations pédagogiques de l'enseignement agricole. http://pollen.chlorofil.fr/documentation/tours-du-monde-agroecologie/. Accessed 21 Sept 2017

Reix R, Fallery B, Kalika M, Rowe F (2016) Systèmes d'information et management. Vuibert, France

Roussey C, Bernard S, Pinet P, et al (2016) Gestion sémantique des bulletins de santé du végétal dans le projet Vespa. Atelier IN-OVIVE @IC 2016, 12 p.,. In: IC2016: 27es Journées francophones d'Ingénierie des Connaissances, Jun 2016. Montpellier, France

Soulignac V (2012) Un système informatique de capitalisation des connaissances et d'innovation pour la conception et le pilotage de systèmes de culture durables. doctoral thesis, Ecole Doctorale Sciences pour l'Ingénieur, Université Blaise Pascal, Clermont-Ferrand

Soulignac V, Pinet F, Lambert E, et al (2017) GECO, the French web-based application for knowledge management in agroecology. Elsevier Comput Electron Agric 29

Stassart PM, Baret P, Grégoire JC, et al (2012) L'agroécologie : trajectoire et potentiel. Pour une transition vers des systèmes alimentaires durables. In: Van Dam D, Nizet J, Stassart PM (eds) Agroécologie: Entre pratiques et sciences sociales, Educagri E. France, pp 25–51

Supagro (2017) PARMI: Promoting AgRoecology deMands Innovation in education. http://www.supagro.fr/ferme_wiki/wikis/Parmi/wakka.php?wiki=PagePrincipale. Accessed 22 Sept 2017

Therond O, Duru M, Roger-Estrade J, Richard G (2017) A new analytical framework of farming system and agriculture model diversities. A review. Agron Sustain Dev 37:21. https://doi.org/10.1007/s13593-017-0429-7

Tosi L, Bénel A (2017) Authenticity in a digital era. In: Proceedings of the 2017 ACM symposium on document engineering – DocEng '17. ACM Press, New York, New York, USA, pp 109–112

UVAE (2017) UVAE: Université Virtuelle d'Agroécologie. http://www6.inra.fr/uvae. Accessed 21 Sept 2017

Warner JF (2006) More sustainable participation? Multi-stakeholder platforms for integrated catchment management. Int J Water Resour Dev 22:15–35. https://doi.org/10.1080/07900620500404992

WeFarmUp (2017) WeFarmUp: Louez un matériel agricole maintenant! https://www.wefarmup.com/fr/. Accessed 19 Sept 2017

Wenger E (1998) Communities of practice: learning, meaning, and identity. Cambridge University Press, Cambridge

Williams BK (2011) Adaptive management of natural resources—framework and issues. J Environ Manag 92:1346–1353. https://doi.org/10.1016/j.jenvman.2010.10.041

Zacklad M (2005) Innovation et création de valeur dans les communautés d'action : les transactions communicationnelles symboliques. In: Entre connaissance et organisation : l'activité collective. La Découverte, Paris, pp 285–305

Zaher L, Bénel A, Cahier JP, et al (2007) Digital identities and management of identifiers for a socio-semantic web. In: Proceedings of the 4th international conference on sciences of electronics, Technologies of Information an Telecommunication, SETIT. pp 1–9

Zhou C, Lejeune C, Bénel A (2006) Towards a standard protocol for community-driven organizations of knowledge. In: Proceedings of the 13th international conference on concurrent engineering (ISPE CE'06). IOS Press, p 12

TATA-BOX: A Model for Participatory Processes?

Sylvie Lardon

Abstract Is TATA-BOX a model worth following, an example or a reference to use in designing other participatory processes or devices? To answer to this question, I went through the book to see if I could find the three main properties of participatory processes: the expression of viewpoints; the justification of reasoning; and the creation of new development models for territories. Three frameworks of analysis collectively guided my viewpoint in this critical analysis: the mechanisms of the construction of a participatory process; the potential of territorial governance; and the connection between collective and territorial actions. The first section discusses representation tools seen as intermediary objects between researchers and actors in co-construction devices. The second section successively considers participation, governance, and collective action as three facets for elucidating the process. The last section answers the crucial question: is TATA-BOX a model worth following? Hybridisation, integration, and inter-territorialisation are conditions for inventing the territories of the future (This is the title of the research project PSDR INVENTER (https://www.psdr4-auvergne.fr/PSDR-4/Les-4-projets/INVENTER), which aims at formalising the evolving dynamics of rural and metropolitan territories, drawing and designing the support for food governance changes.) and for managing territorial transitions.

Foreword

When people hear about the TATA-BOX project, they are immediately impressed. It sounds good. Pleasant and even friendly images come to mind. Why?

When you look at the outputs of the TATA-BOX process – its comic strip (Audouin et al. 2018b), the small, fully-illustrated support guide (Audouin et al. 2018a) –, its magic is still at work. Is this owing to the appealing look of the colourful diagrams presenting the complexity of the territorial transition process?

S. Lardon (✉)
Territoires, Université Clermont Auvergne, AgroParisTech, INRA, Irstea, VetAgro Sup, Aubière, France
e-mail: sylvie.lardon@agroparistech.fr

J.-E. Bergez et al. (eds.), *Agroecological Transitions: From Theory to Practice in Local Participatory Design*, https://doi.org/10.1007/978-3-030-01953-2_13

So, when the TATA-BOX authors ask you to do a critical analysis of their complete work, it's impossible to resist the desire to dive into the manuscript.

The design of the process was collective, along with the wish to review it, but the time allotted made that impossible. I therefore took up my pen of my own accord, although from the angle of a few collective viewpoints stated within the UMR Territoires in Clermont-Ferrand,[1] with the intention of asking a few interdisciplinary questions on the work.

I do not claim to have been exhaustive in my statements, nor even to have fully examined the main question: is TATA-BOX a model worth following? At least not blindly like a cooking recipe (although cooks are also creative), but as an example or a reference to use in designing other participatory processes or devices.

To do so, I went through the book to see if I could find the three main properties that I attribute to participatory processes (Lardon 2013): the expression of viewpoints; the justification of reasoning; and the creation of new development models for territories.

I paid particular attention to the representations present in the book, whether systemic representations (Lemoigne 1990) presenting the complexity of the situation studied; spatial representations (Brunet 1986) anchoring processes in the territory; or simply representation models for an enhanced experience of the world (Legay 1986).

Three analysis frameworks collectively guided my viewpoint in this critical analysis. That of Gouttenoire et al. (2014) analyses the mechanisms of the construction of a participatory process. That of Lardon et al. (2014) raises questions around the potential of territorial governance. That of Amblard et al. (2018) connects collective and territorial action.

My paper consists of three sections. The first section discusses representation tools seen as intermediary objects (in the sense of Vinck 2009) between researchers and actors in co-construction devices. The second section successively addresses participation, governance, and collective action as three facets for elucidating the process. The last section answers the crucial question: is TATA-BOX a model worth following?

From One Representation to the Other

By now, you will have understood that I will not be discussing agroecology, the transition, or rural territories, to paraphrase the subtitle "Agroecological transitions: from theory to practice in local participatory design", but rather their representations in the book to see if the toolbox is effective.

[1] The UMR Territoires is a multidisciplinary body that associates humanities and social science researchers (geography, economics, management science, political science) with biotechnical science researchers (agronomics, systemic zootechnics). It carries out research on livestock farming models, rural and peri-urban territorial dynamics, collective coordination of actors, whether public or private, and support for change in territories.

The first two representations provide the example to be followed. They are taken from Duru et al. (2015).

TATA-BOX is a complex, three-dimensional system. Natural resources, farming systems, and supply chains are the three corners of the triangle. The perspective is based on the formal and informal behaviours of the system of actors.

TATA-BOX is a part of a complex methodological itinerary, in the form of a loop, from analysis of the system to governance mechanisms, through design and prospective analysis.

Diagrams provide a model for clarification to convey equally complex thoughts. We let ourselves be guided along the path and plunge into the viewpoint. This is true for myself as a reader, but is it true for the authors of other chapters? Not all people share the same synthetic and systemic viewpoint.

The Conceptual Framework(s)

"Territorial Agroecological Transition at a concept crossroads" is the first section of the book. It is less ambitious in its representation methods than these first two diagrams, even if its contents are more ambitious.

Various representation methods support the announcement of the conceptual frameworks of the first section:

Special mention goes to the **location maps** of the actors involved in the processes (cf. chapter "An Integrated Approach to Livestock Farming Systems' Autonomy in Designing and Managing Agroecological Transition at the Farm and Territorial Levels") and to positioning the farmers interviewed along **dual gradients** of farming and commercialisation practices (cf. chapter "The Key Role of Actors in the Agroecological Transition of Farmers: A Case-Study in the Tarn-Aveyron Basin").

Tables cross-reference the common worlds from Thévenot et al. (2000)'s economies of worth (cf. chapter "Socio-economic Characterisation of Agriculture Models"). This analysis can focus on levers for action (Thénard et al. 2014) based on livestock farming and feeding practices (cf. chapter "An Integrated Approach to Livestock Farming Systems' Autonomy in Designing and Managing Agroecological Transition at the Farm and Territorial Levels").

Spider graphs reveal the compromises, performances, and complementarities (cf. chapter "An Integrated Approach to Livestock Farming Systems' Autonomy in Designing and Managing Agroecological Transition at the Farm and Territorial Levels").

Typologies of farming methods (Therond et al. 2017) are projected around two pillars – from the global to the local, and from inputs to ecosystem services – and are described in terms of contemporary farming methods in Western countries (cf. chapter "Socio-economic Characterisation of Agriculture Models"). The two pillars of private/public goods and the short-term/long-term reveal the diversity of coordination methods (cf. chapter "Towards an Integrated Framework for the Governance of a Territorialised Agroecological Transition"). The analysis of the role of actors shows that their involvement in the agroecological transition (AET) is inversely proportional to the degree of their influence on farmers (cf. chapter "The Key Role of Actors in the Agroecological Transition of Farmers: A Case-Study in the Tarn-Aveyron Basin").

System diagrams present the conceptual framework for analysing the autonomy of farming systems (Madelrieux et al. 2017) from three angles: the embeddedness of farms, their dependency and their footprints (cf. chapter "An Integrated Approach to Livestock Farming Systems' Autonomy in Designing and Managing Agroecological Transition at the Farm and Territorial Levels"). Others present the MLP approach to transitions (Geels 2004) and an alternative approach consisting in aligning pathways with the socio-technical regime (Elzen et al. 2004) (cf. chapter "Agroecological Transition from Farms to Territorialised Agri-Food Systems: Issues and Drivers"). Others establish the interrelation between the socio-technical system and the socio-ecological system, and their interactions with public policies (cf. chapter "Agroecological Transition from Farms to Territorialised Agri-Food Systems: Issues and Drivers").

Cognitive maps were created alongside farmers (cf. chapter "An Integrated Approach to Livestock Farming Systems' Autonomy in Designing and Managing Agroecological Transition at the Farm and Territorial Levels").

Pathways illustrated by the points of view of the interviewees (according to Coquil et al. 2013) present farmers' decisions to engage in the AET (cf. chapter "The Key Role of Actors in the Agroecological Transition of Farmers: A Case-Study in the Tarn-Aveyron Basin").

Exchange networks between farmers and other actors show evolution in the establishment of coherence between innovation logics (cf. chapter "The Key Role of Actors in the Agroecological Transition of Farmers: A Case-Study in the Tarn-Aveyron Basin").

Lastly, **systems of relations** between the farmers interviewed and the other actors met, depending on whether they are central or outlying actors, explain farmers' degrees of engagement in agroecology (cf. chapter "The Key Role of Actors in the Agroecological Transition of Farmers: A Case-Study in the Tarn-Aveyron Basin").

A box explains the learning dynamic of students involved in analysing strategies and actors' games (cf. chapter "The Key Role of Actors in the Agroecological Transition of Farmers: A Case-Study in the Tarn-Aveyron Basin").

By contrast, the plurality of viewpoints among researchers around the uncertainties of the AET (cf. chapter "A Plurality of Viewpoints Regarding the Uncertainties of the Agroecological Transition") is not represented.

As a result, several types of illustration support the researchers' reasoning. First, there are the classic representations presenting scientific results: tables, typologies based on pillars, radars, etc. Then there are the systemic diagrams that can be expected in a book of this type. Lastly, there is a set of innovative representations (pathways, relationships) that go even further, presenting both the results from the field, including verbatim quotes from actors, and interpretations seen from the angle of theoretical frameworks, so as to understand actors' games and their contribution to change decisions.

The Design Methodology

Unsurprisingly, the second section of the book, "Support methodology for territorial agroecological transition design and feedback from the TATA-BOX project experience", presents a plethora of representations that are each as valuable and explicit as the next.

The participatory methodology (cf. chapter "Participatory Methodology for Designing an Agroecological Transition at Local Level") goes over the principles underpinning the construction of TATA-BOX (3-D diagram and round-trip pathway), before positioning farming types in Midi-Pyrénées (map), locating actors in the dialogue on local AET (diagram), and undertaking to analyse all work sessions between researchers, between stakeholders, and at the interface of the TATA-BOX process from 2014 to 2018. A diagram of the actors' interaction strategies throughout the participatory process, along with photos of mind maps produced during sessions, the construction of playing cards to detail the transition pathway, the global cognitive map, summaries of the productions of the process, action plans, and so on, all constitute varied illustrations of the design and redesign of an interactive and didactic process. It is valid to say that the process generated multiple outputs.

In its own way, the reflexive analysis (cf. chapter "Towards a Reflective Approach to Research Project Management") illustrates the reflexive interventions throughout the process, as well as the participatory exercises highlighting the separate stages of constructing a common world, of project management, feedback and debriefing, and so on. It describes the scales of reflexivity and the levels of resistance to a new proposal. Its results are varied, from the representation of the roles imagined by the participants, to the evolution of participants' postures. It can be said that the actors were really engaged in the process.

The evaluation of the operationality of the process (cf. chapter "Evaluation of the Operationalisation of the TATA-BOX Process") draws from the classic canons of both quantitative analysis (staff tables, spider graphs, bar charts) and qualitative analysis (direct citations). It could be completed by a more detailed description of exchanges in order to qualify the collective dimension of the design work. Unsurprisingly, the actors recognise the proliferation of shared information and the positive consequences for individuals (acknowledgement of their capabilities?) but find the ideas to be unoriginal and their implementation to be difficult, for reasons related to funding and human resources. This is often the case in participatory processes, in which participants tend to highlight the utility of the dialogue as opposed to the originality of actions.

The present study, based on participants' discourses, will be completed by a direct analysis of the exchanges carried out during workshops. This should lead to a more detailed characterisation of exchanges, allowing to better qualify the collective dimension of the design work backed by the TATA-BOX process.

Elements to improve participants' representativeness, such as mobilising relays, organising shorter and more numerous meetings, and insertion within existing processes, should be explored in more detail. Those intended at improving creativity and reducing tension between innovation and realism may include the establishment of common ground upstream, understanding of the logics underpinning the different viewpoints, and potentially the diversification of participants' profiles. The fact remains that "*the exercise of defining the path may have been carried out more to detect the favourable conditions as a whole than to select the means that will effectively have to be activated*". TATA-BOX was clearly a design device that should be applied to use (Béguin and Rabardel 2000).

The idea of coexistence, in the field, of agricultural development models and agroecology is outlined, but not developed.

Viewpoints

Finally, a note on Information and Communication Technologies, which are supposed to help in the application of agroecological principles. They are considered to be "*a range of tools for different agriculture models*" (Therond et al. 2017). The proposed design process platform is open and modulable. It is organised around the concept of items, which can be constructed on the basis of an objective (from a document) or subjective (a viewpoint) piece of data that is characterised (attributes) and elucidated (topic) iteratively. Items are recorded and tagged by participants. This "item-based modelling method" is collaborative, around "topic maps" to make the diversity of viewpoints visible and palpable. It is designed to support the implementation of the transition following the design phase, by increasing the communication of information, using functional tools, managing knowledge, and offering decision aid tools. Yet, is there not a contradiction between the intensive use of new technologies (that consume a great deal of energy) and the paradigm of agroecology, which aims at consuming less fossil fuel? This is a dilemma facing the authors.

This perspective remains open for the most part, as the demonstration of its use has not been carried out around TATA-BOX itself.

Cross-Analysis

What can another group of researchers[2] involved in other processes and other devices have to say about the TATA-BOX process and its analysis by the researchers involved in this participatory process? Let's go over a few collective analysis frameworks for participation, governance, and collective action stemming from the UMR Territoires in order to provide a more interdisciplinary analysis of the book.

[2] The UMR Territoires' AVEC Collective focuses on support in the territorial change process. Participants are involved in supporting actors through participatory devices, action research (and/or training), support for reflexivity… and/or analysing these devices. Theoretical framing, exchanges, and the collective production of knowledge on support are at the heart of the scientific facilitation process.

The Key Points of a Participatory Process

Following a long reflexive undertaking over several years,[3] Gouttenoire et al. (2014) proposed an analysis framework to facilitate exchanges between researchers around participatory research projects. It qualifies participation processes.

The analysis framework is structured into six items. **Researchers' and actors' expectations** compared to what is achieved together are a result of the production of knowledge and actionable knowledge, the exchange of information or discussion, and so on. It is necessary to have clarified this, so as not to confuse the objectives. The course **of the interactions between researchers and actors** is described in successive stages. Interactions include both periods of explicit joint participation (such as workshops or interviews), and the interactions that may take place upstream or downstream (for example, upstream to decide on the actors to invite, or downstream to recover the outputs of participatory processes). For each of these stages, the goals and the concrete facilitation mechanisms are specified. The **characteristics of groups of participating actors** demonstrate their diversity or homogeneity, whether intended or not. A crucial point is that of the **actors' engagement**: what are they engaging in by participating in the process, and to what extent are they engaged? In parallel but not identically, the **role of actors** and the **mechanisms of researcher intervention** are specified during each stage of interaction. The role of actors can be to provide information, to give their opinion, or to co-construct or even drive forward the action; that of researchers can be to facilitate participation, formalise productions, or even co-construct. This distinction between researchers and actors in turn links back to the expectations of each of them, and can be subject to different mechanisms, depending on the stage.

What About the TATA-BOX Participatory Process?

If we apply this framework to TATA-BOX, we can say that, on the one hand, part of the framework is still relevant as a framework for analysing TATA-BOX; and on the other hand, that TATA-BOX is effective from the perspective of this framework and enriches it. Below we justify these different points.

In TATA-BOX, researchers' expectations are clear: their intention is to support the TAET. The conceptual frameworks to achieve this are varied, and go beyond the MLP approach to transitions (Geels 2004). An *ad hoc* framework is built for a complex system considered dynamically in its full complexity (Therond et al. 2017), at the articulation between socio-technical systems and socio-ecological systems, and aimed at achieving the autonomy of farming systems (Madelrieux et al. 2017).

The actors' expectations are vaguer and are revealed by their assessments of the operationality of the process: they find the ideas to be unoriginal and difficult to implement. Lastly, we may wonder how the researchers manage to enrol them. This raises the question of the actors' engagement. Are they there to please the researchers, or are they seeking solutions for their future? Are they aware that they are experimenting with new paths, or do they just feel good about contributing to a

[3] Originally a researcher workshop on participation, organised in 2017 by the INRA-SAD.

trendy subject? The ties existing between researchers and actors upstream of the process are also a representation to discuss further.

TATA-BOX gives a remarkable representation of the course of the interactions between researchers and actors, whether through the conceptual diagram of the methodological itinerary, from the analysis of the system to governance mechanisms, including prospective design and analysis, or the variety of illustrations of the design and redesign of an interactive and didactic process. The small guide to support the collective design of an AETS (Audouin et al. 2018a) is the most palpable intermediary object, and can in itself be a model to follow for the quality of representations and the quantity of information transmitted.

The features of the actors participating (or not) are equipped by original representations, whether through the positioning of the actors interviewed, the analysis of their central or peripheral role in the system of interactions, their cognitive maps, or their illustrated pathways. These tools can be generalised to other situations in other processes. They can be a source of inspiration.

Yet the actors play a somewhat silent role. If they are key actors in the system, are they explicitly called upon as such, and do they consciously contribute to this intermediary role?

Conversely, do all researchers have the same activity in the process, and do they intervene homogenously? This is apparent in the articles that they produce: one theorises the process, another evaluates it; one produces useful knowledge, another uses it; one is reflexive, another is visionary. Some are both simultaneously. While the collective learning process is examined critically with respect to the actors, it would also be useful to question it with respect to the researchers, or even the students involved in the action-research-training devices (Lardon 2009), supposing the experiment renews not only knowledge but also capabilities (Lardon et al. 2015).

More recently, the researcher workshop entitled "Science and participatory research: practices and epistemology" (*Les sciences et recherches participatives: pratiques et épistémologie*),[4] organised by FormaSciences, bolstered the relevance of questions around the engagement of actors and researchers. This included questions that explain the reasons for the failure of such processes (Gonzalo-Turpin et al. 2009), as well as those on the place of researchers in processes, as experts contributing scientific knowledge or as facilitators of interactions between actors, through their active role in the formalisation and transmission of knowledge, or by establishing distance to guarantee the neutrality of the device. While participatory research processes are widespread at the INRA, they do not have the same assumptions or the same mechanisms. Diversity is a must.

[4] This research workshop took place in Pont à Mousson on 9–13 October 2017. Bringing together around 30 researchers, it aimed at understanding the diversity of types of participatory projects at the INRA; sharing the experiences, tools, and methods used as a part of participatory processes; and situating participatory projects within an epistemological and ethical framework.

Governance of the Transition Process

Governance was addressed in two different ways by a group of researchers involved in the DATAR's research in 2008–2010, evaluating "rural excellence hubs" (PER, *pôles d'excellence rurale*)[5] (Lardon et al. 2016). The first consisted in analysing the socio-spatial configurations making up territories engaged in PER projects. The second came into play through the actors-stakeholders involved in creating and facilitating PER to analyse development pathways.

PER had to constitute spaces for discussing the project and defining the boundaries of cooperation. By putting emphasis on public-private partnerships and support for the initiatives of small project drivers, the device was primarily intended to encourage joint territorial governance practices. In the end, 670 PER were certified between 2006 and 2009.

The analysis framework for socio-spatial configurations reveals the development potential of PER projects. It specifies the dynamics of social relations necessary to construct collective action (Angeon et al. 2006), and sheds light on their territorial anchoring and the spatial models resulting from territorial dynamics (Lardon and Piveteau 2005). Socio-spatial configurations are characterised by the spatial distribution – whether concentrated or diffused – of actors and actions within the project territory, and the ties – whether established or not – with external actors. These are cross-compared with the way that the different types of actors (institutional, consular, association, corporate, or research actors) get involved in building the PER project. The models created in this way demonstrate the contribution of actions to the territorial project and reveal the synergies at play between actors and their room to manoeuvre in order to execute their project.

The dynamics and changes marking the development pathways of PER (Milian and Bacconnier-Baylet 2014) can be interpreted through the continuities, shifts, and even discontinuities that take place in the (re)definition of the resources on which a territorial system bases its development (activation and valorisation of these resources), as well as its organisational and structural reorganisation. The idea is to retrace the evolution of the local system's capacity for initiative, and to reconstruct the way in which it became involved in "innovative" processes, thus contributing to the construction or consolidation of territorial governance. This leads to a study of the facilitation and decision-making system at work around the territory's development pathway, and the innovations and compositions that it promoted or catalysed.

Cross-analysing these two approaches highlights possible levers for action for territorial development. Two key elements must be noted with respect to territorial governance. The first is the need to facilitate places and moments of dialogue between the different actors involved, with a view to co-constructing development processes. The second is to ensure support for intermediary actors that were not necessarily foreseen at the outset, but which emerged during the course of the process.

---

[5] http://www.datar.gouv.fr/programme-de-recherche-evaluative-sur-les-poles-dexcellence-rurale-2009

Is Reflection on Territorial Governance Projects Applicable to TATA-BOX? TATA-BOX uses pathways to represent actors' logics and to line these up with transition phases. It analyses social configurations by representing technical, economic, and material exchange networks between actors. However, spatial configurations are not taken into consideration. At the most, the spatial dimension serves to qualify the biotechnical and decision-making dimensions through its anchoring, dependence, and imprint (cf. chapter "An Integrated Approach to Livestock Farming Systems' Autonomy in Designing and Managing Agroecological Transition at the Farm and Territorial Levels"). The map representation is first used to locate the groups involved in the process in terms of types of production system (cf. chapter "An Integrated Approach to Livestock Farming Systems' Autonomy in Designing and Managing Agroecological Transition at the Farm and Territorial Levels"), and then later used to locate the fields of study in Midi-Pyrénées in terms of types of production system (cf. chapter "Participatory Methodology for Designing an Agroecological Transition at Local Level"). Moreover, while the shared vision of local agriculture by 2025 for the territory of Midi Quercy is a detailed representation of 65 targets, at times localised, symbolised within the three concentric circles of the farm, the territory, and the country, none of these analyse either the specific spatial configurations or interrelations between spatial scales. Would this not therefore constitute a loss of development potential for territories, through the deprivation of tools for analysing socio-spatial configurations (Lardon 2015)?

TATA-BOX stresses the diversity of the actors already involved and to involve in order to make it possible to steer action. From an analytical point of view, this consists in identifying the boundaries of the system and the stakes of actors-stakeholders. From an operational point of view, governance relates to the mechanisms that these actors use to make the system evolve towards the desired state. These boundaries are within a space of overlap between the socio-technical and the socio-ecological system. Issues surround the public and private coordination of goods. In a guide on implementing governance to support the sustainable development of territories, Hélène Rey-Valette et al. (2011) proposes a tool to strengthen territorial governance engineering. Its key features are: sharing knowledge around the topic of territorial governance and complementary forms of partnership; the need to make scales of observation evolve along the way; and partnerships, due to both the multi-level nature of the processes and the complexity of contexts. These key features convey and compare the diversity of possible forms for implementing this territorial governance. The operationalisation of territorial governments is based on five properties: participation; the organisation of steering; interdisciplinarity; assessment; and continuous improvement. Participation, the organisation of steering, and assessment position the framework of devices and actions carried out. Interdisciplinarity and continuous improvement are more general goals that raise questions in order to integrate these principles into procedures. They result in adaptive processes involving the consideration of the long term and requiring collective learning processes in view of sustainable development (Valette et al. 2008). These proposals go further than what is done in the TATA-BOX process, even if the same goals have been defined.

Collective Action for Territorial Development

Amblard et al. (2018) provide a summary of the contributions of a territorial approach to collective action, in terms of the production of scientific knowledge and knowledge for action. It is based on the authors' analysis of work carried out by researcher members of UMR Territoires, and is compared to a review of national and international literature. The authors' analysis identifies three lines of questioning that structure the contributions of research on collective action in relation to the territory:

1. To what extent **do territories determine the development of collective actions?** What is the respective contribution of territorial and supra-territorial factors in the development of collective action? The strategies of the actors involved as well as the procedures and governance modes in place are often central in the analysis of the role of collective action. Seen on the territorial scale, these analyses shine light on the territorial factors determining the development of collective action. Research also highlights the role of factors related to socio-economic and political contexts, beyond the borders of the territory, as well as the interactions between the characteristics of the territory and contextual factors.
2. **Collective actions take on different forms in territories**. How are these forms related to territorial particularities? How is it possible to build common ground between the actors involved? What is the contribution of the different actors to the collective dynamic? The apparent diversity of forms adopted by collective action in territories raises the question of the logics underpinning these organisational configurations in relation to the issues of each territory. Defining the boundaries of the groups involved leads to questions on the power relations or construction of common worlds between participants in the action on the territorial scale. The analysis of collective action in territories reveals the specific role played by certain actors in an intermediary situation within the groups in question.
3. Lastly and reciprocally, **how does collective action contribute to territorial development**, generate new resources, and make new territories emerge? The question of the impact of collective actions on the territory constitutes a major component of research on the subject. Some publications analyse the effects of these actions on the economic development of territories; others emphasise their indirect effects on the transformations of territories, through the creation or activation of new resources. The entanglement of collective actions, which can either compete with or complement one another, influences their capacity to transform territories.

Two conceptual frameworks were put to the test (Amblard et al. 2018). That of socio-ecological systems (Ostrom 2009) aims at broadening the institutional approach to collective action so that it takes into consideration the characteristics of the ecological system considered, along with interactions with social systems. It simultaneously allows us to understand the role of territorial factors, modes of governance, and contextual factors. The processual approach (Mendez et al. 2010) introduces the possibility of identifying the territorial resources activated over time, and thus aims at understanding the combination of the dynamics at the bases of processes. But what tools support collective action in territories and what is their heuristic scope?

We have answered this question, such as through the example of the "territory game", which is an expression game that promotes the construction of a shared viewpoint of the territory (Lardon 2013). It constitutes an instance of interventional research, in the sense of Hatchuel (2001), providing its capacity for establishing distance and formalising results. The production of knowledge for action is carried out as a part of an iterative process, in which the knowledge of actors and of researchers is shared (Béguin 2007) and appropriated. A collaborative design process takes place (Brassac 2004), in which each person finds his or her place and makes use of the collective production. In this sense, the game establishes dialogue between territorial actors (elected officials, inhabitants, professionals, etc.) and researchers around different intermediary objects (Vinck 2009), such as the maps produced during the process of knowledge production and the valorisation of experiences. These outputs are analysed from the angle of the concerns that they address (Lardon et al. 2016).

What Is the Case for TATA-BOX?
As indicated in chapter "A Plurality of Viewpoints Regarding the Uncertainties of the Agroecological Transition" on the uncertainty of the AET, "*research on the analysis and support of the AET has mainly sought to describe change processes. However, it is necessary to consider the obstacles and levers involved in these changes on different organisational levels (production systems, supply chains, the territory, etc.), as well as the trajectories and pathways of the transition, and in doing so, to consider methodologies for supporting actors in this transition.*"

Chapter "Participatory Methodology for Designing an Agroecological Transition at Local Level" presents the participatory methodology by highlighting transition trajectories and pathways. This could be the case in chapter "Information and Communication Technology (ICT) and the Agroecological Transition", which establishes a platform based on information and communication technologies, if it were to go beyond the stated challenges.

When considering this scientific publication, a slight sense of dissatisfaction remains in terms of the contribution to participatory research for action, even though quite the opposite is true when we look at the short companion guide (Audouin et al. 2018a).

TATA-BOX, a Tool for Action and a Participatory Research Model

Based on my interdisciplinary study of the TATA-BOX project, I have three comments and three proposals.

First, the project's scientific outputs are remarkable, with more specifically the formalisation of conceptual frameworks upstream and along the way, the reflexive feedback and assessment downstream, the diversity of disciplinary viewpoints provided, and the involvement of researchers in the 40 or so workshops and sessions over the course of the 4 years of the process. The AET is modelled; the practices of farmers and other actors are analysed; the governance and management of uncertainty are discussed. However, even though a diversity of viewpoints is expressed, the logic of reasoning is justified, and the invention of new reference frameworks is presented, this is in relation to the scientific component. Ultimately, there is no space where actors are able to appear as knowledge producers, whether actionable knowledge or not. Actors are similarly categorised by the researchers into their livestock farming typologies (such as in chapter "An Integrated Approach to Livestock Farming Systems' Autonomy in Designing and Managing Agroecological Transition at the Farm and Territorial Levels") or based on their exchange networks (in chapter "The Key Role of Actors in the Agroecological Transition of Farmers: A Case-Study in the Tarn-Aveyron Basin"). Alternatively, their behaviours or exchanges are evaluated by the researchers (such as in chapter "Towards a Reflective Approach to Research Project Management").

One suggestion would be increasingly to hybridise expert knowledge and situated knowledge, because, as Luc Gwiazdzinski (2016) put it: "*The territory is at the heart of these recompositions and hybridisations that engage the sensible and the ephemeral. New figures emerge, new scenes, and new ways of cooperating appear on different scales and according to plural mechanisms. To address issues, crossed actions take place; hybridisations become possible*".

The operational production of the process is promising: a diversity of tools was created, whether to represent action logics, trajectories, networks of actors, strategic pillars, transition pathways, levers for action, or other. These tools present the complexity of the systems studied without reducing it, not to facilitate appropriation but to multiply forms, to better transmit them. There is however no analysis of what these tools produce beyond the words of actors around individual contributions. What of the knowledge produced on dynamics, the collective dimension, and the development of territories?

One proposal to increase actors' ability to master the processes involving them (Deffontaines et al. 2001) would be to deepen the analysis of these intermediary objects that constitute boundaries (Vinck 2009) between researchers and actors. They are integrative objects (in the sense of Schmid and Hatchuel 2014) that articulate portions of knowledge and take on meaning in the future. The territory is the place of the interconnection between public policies and local initiatives, between the past and the future, between activities and uses. All of this must be taken into consideration in a systemic and operational approach.

Reflexivity is at the heart of the device; it is manifest in the feedback on the device as well as in the anticipation of future tools, made possible by information and communication technologies, and in the valorisation of past knowledge, which becomes a resource in the present. While participatory processes are adaptive by nature, in order to address the needs expressed in the field, they are also performative, because an objective presides over the performance, the process over the results, and the meaning over achievements.

One proposal to have a more detached view of experiments of this type would be to ensure their traceability, to retain records of them, and to send the knowledge acquired and capabilities developed to be mirrored. The territories will not do so alone; they will do so in relation to neighbouring territories, integrating the local and the global. While Martin Vanier (2008) announced this over a decade ago, when it comes to multi-scale territorial recompositions, the urgency of inter-territoriality is currently taking on its full meaning. The territorial project remains applicable, but ties with neighbouring territories, involvement in broader scales, and acknowledgement of territorial differences are all interactions that should be taken into account, in order to open up a territory to the world instead of solidifying it within its boundaries.

Hybridisation, integration, and inter-territorialisation are conditions for inventing the territories of the future[6] and for taking charge of the territorial transitions underway (Lardon 2017), whether agroecological or not. This is a paradigm shift!

Let's draw inspiration from Albert Jacquard's (1991) *Voici le temps du monde fini*.

References

Amblard L, Berthomé GEK, Houdart M, Lardon S (2018) L'action collective dans les territoires. Questions structurantes et fronts de recherche. Géographie, économie, société 20:227–246. https://doi.org/10.3166/ges.20.2017.0032

Angeon V, Caron P, Lardon S (2006) Des liens sociaux à la construction d'un développement territorial durable: quel rôle de la proximité dans ce processus? Développement durable Territ. https://doi.org/10.4000/developpementdurable.2851

Audouin E, Bergez JE, Choisis JP et al (2018a) Petit guide de l'accompagnement à la conception collective d' une transition agroécologique à l' échelle du territoire. Toulouse

Audouin E, Bergez JE, Therond O (2018b) TATA-BOX, Neuf métaphores des concepts clefs des démarches participatives pour la transition Agroécologique

Béguin P (2007) Innovation et cadre sociocognitif des interactions concepteurs-opérateurs: une approche développementale. Trav Hum 70:369. https://doi.org/10.3917/th.704.0369

Béguin P, Rabardel P (2000) Designing for instrument mediated activity. Scand J Inf Syst 12:173–190

Brassac C (2004) Action située et distribuée et analyse du discours: quelques interrogations. Cah Linguist française, Univ Genève 26:251–268

Brunet R (1986) La carte-modèle et les chorèmes. Mappemonde 4:2–6

Coquil X, Lusson JM, Beguin P, Dedieu B (2013) Itinéraires vers des systèmes autonomes et économes en intrants: motivations, transition, apprentissages. In: 20eme Rencontres Recherches Ruminants. France, Paris, pp 1–4

[6]This is the title of the research project PSDR INVENTER (https://www.psdr4-auvergne.fr/PSDR-4/Les-4-projets/INVENTER), which aims at formalising the dynamics of the evolution of rural and metropolitan territories, drawing support from food governance, and designing the support for change.

Deffontaines JP, Marcelpoil E, Moquay P (2001) Le développement territorial: une diversité d'interprétations. In: Lardon S, Maurel P, Piveteau V (eds) Représentations spatiales et développement territorial. Bilan d'expériences et perspectives méthodologiques. Hermès, Paris, pp 39–56

Duru M, Therond O, Fares M (2015) Designing agroecological transitions; a review. Agron Sustain Dev 35:1237–1257. https://doi.org/10.1007/s13593-015-0318-x

Elzen B, Geels FW, Green K (2004) System innovation and the transition to sustainability: theory, evidence and policy. Edward Elgar Publishing, Cheltenham

Geels FW (2004) From sectoral systems of innovation to socio-technical systems. Res Policy 33:897–920. https://doi.org/10.1016/j.respol.2004.01.015

Gonzalo-Turpin H, Couix N, Hazard L (2009) Rethinking partenerships with the aim of producing knowledge with pratical relevance: a case study in the field of ecological restoration. Ecol Soc 13:53

Gouttenoire L, Taverne M, Cournut S et al (2014) Faciliter les échanges entre chercheurs sur les projets de recherche participative: proposition d'une grille d'analyse. Cah Agric 23:205–212. https://doi.org/10.1684/agr.2014.0703

Gwiazdzinski L (2016) L'hybridation des mondes. Territoires et organisations à l'épreuve de l'hybridation. L'innovation autrement, Elya Editions, France

Hatchuel A (2001) Quel horizon pour les sciences de gestion? Vers une théorie de l'action collective. In: David A, Hatchuel A, Laufer R (eds) Les nouvelles fondations des sciences de gestion. Vuibert, Paris, pp 21–45

Jacquart A (1991) Voici le temps du monde fini. Seuil, Paris

Lardon S (2009.) Former des ingénieurs-projets en développement territorial) Un itinéraire méthodologique pour faciliter la participation des acteurs. In: Béguin P, Cerf M (eds) Dynamiques des savoirs, dynamiques des changements. Octarés, Toulouse, pp 209–227

Lardon S (2013) Developing a territorial project. The « territory game », a coordination tool for local stakeholders. FaçSADe, Res Results 38:4

Lardon S (2015) L'agriculture comme potentiel de développement des territoires périurbains. Analyse par les configurations socio-spatiales. Articulo. https://doi.org/10.4000/articulo.2673

Lardon S (2017) L'aménagement du territoire au prisme des transitions territoriales: un triple processus à l'œuvre. Pouvoirs Locaux 110:81–86

Lardon S, Piveteau V (2005) Méthodologie de diagnostic pour le projet de territoire: une approche par les modèles spatiaux. Géocarrefour 80:75–90. https://doi.org/10.4000/geocarrefour.980

Lardon S, Milian J, Loudiyi S et al (2014) Du potentiel à l'action: la gouvernance territoriale des pôles d'excellence rurale. Norois:69–81. https://doi.org/10.4000/norois.5380

Lardon S, Albaladejo C, Allain S et al (2015) Dispositifs de Recherche-Formation-Action pour et sur le développement agricole et territorial. In: Torre A, Vollet D (eds) Partenariats pour le développement territorial. QUAE, Versailles, pp 47–57

Lardon S, Marracini E, Filippini R et al (2016) Prospective participative pour la zone urbaine de Pise (Italie). L'eau et l'alimentation comme enjeux de développement territorial. Cah Geogr Que 20:265–286

Legay J (1986) Méthodes et modèles dans l'étude des systèmes complexes. Cah la Rech 11:1–6

Lemoigne JL (1990) La modélisation des systèmes complexes. Dunod, Paris

Madelrieux S, Buclet N, Lescoat P, Moraine M (2017) Écologie et économie des interactions entre filières agricoles et territoire : quels concepts et cadre d'analyse?

Mendez A, Bidart C, Brochier D et al (2010) Processus: concepts et méthode pour l'analyse temporelle en sciences sociales. Academia-Bruylant, Paris

Milian J, Bacconnier-Baylet S (2014) Requalifier les territoires de l'action locale. Territ en Mouv:54–67. https://doi.org/10.4000/tem.2399

Ostrom E (2009) A general framework for analyzing sustainability of social-ecological systems. Science 325:419–422. https://doi.org/10.1126/science.1172133

Rey-Valette H, Pinto M, Maurel P et al (2011) Guide pour la mise en œuvre de la gouvernance en appui au développement durable des territoires

Schmid AF, Hatchuel A (2014) On generic epistemology. Angelaki 19:131–144. https://doi.org/10.1080/0969725X.2014.950868

Thénard V, Jost J, Choisis JP, Magne MA (2014) Applying agroecological principles to redesign and to assess dairy sheep farming systems. In: Options Méditerranéennes. Série A. ICAMAS/CIHEAM, Paris, pp 785–789

Therond O, Duru M, Roger-Estrade J, Richard G (2017) A new analytical framework of farming system and agriculture model diversities. A review. Agron Sustain Dev 37:21. https://doi.org/10.1007/s13593-017-0429-7

Thévenot L, Moody M, Lafaye C (2000) Forms of valuing nature: arguments and modes of justification in French and American environmental disputes. In: Lamont M, Thévenot L (eds) Rethinking comparative cultural sociology: repertoires of evaluation in France and the United States. Cambridge University Press, Cambridge, pp 229–272

Valette HR, Lardon S, Chia E (2008) Editorial: governance- institutional and learning plans facilitating the appropriation of sustainable development. Int J Sustain Dev 11:101. https://doi.org/10.1504/IJSD.2008.026506

Vanier M (2008) Le pouvoir des territoires. Essai sur l'interterritorialité, Paris

Vinck D (2009) De l'objet intermédiaire à l'objet-frontière, Vers la prise en compte du travail d'équipement. Rev d'anthropologie des connaissances 3:51–72

Review and Critique of the TATA-BOX Model

Charles A. Francis and Geir Lieblein

Abstract The TATA-BOX model provides a practical and operational set of methods for helping multiple stakeholders design an agroecological transition at the local level. It provides a framework for researchers to study current methods and design future systems based on ecological principles. In contrast to extreme technology-driven agriculture and globalisation of food systems, this strategy returns control to the local level. Location-specific planning provides an imaginative model for sustainable development, building potential to overcome crippling bureaucracy of governments and tyranny of narrowly vested interests that result from a "productionist model" that has prevailed over the past century. The new model based on "post-normal science" is defined in terms of participatory, integrated assessment, involving all stakeholders in local development, working in transdisciplinary teams. The process addresses "wicked", complex current and future challenges today, problems involving multiple people and incommensurate goals. This requires careful study, reasoned discussion, and thoughtful compromise to reach common ground. Principles include holistic thinking, whole systems focus in the local context, involving multiple stakeholders, local autonomy, linking production with consumption in local food systems, and broad-based governance at the community, landscape, and *terroir* levels. This can assure participation in decision making, and action, leading to adoption of transformative systems.

C. A. Francis (✉)
Department of Agronomy and Horticulture, University of Nebraska-Lincoln, Lincoln, NE, USA

Department of Plant Science (IPV), Norwegian University of Life Sciences, Aas, Norway
e-mail: cfrancis2@unl.edu

G. Lieblein
Department of Plant Science (IPV), Norwegian University of Life Sciences, Aas, Norway
e-mail: Geir.Lieblein@nmbu.no

J.-E. Bergez et al. (eds.), *Agroecological Transitions: From Theory to Practice in Local Participatory Design*, https://doi.org/10.1007/978-3-030-01953-2_14

Foreword

The TATA- BOX model includes "*an operational set of articulated methods for supporting stakeholders to design an agroecological transition at local level*", and provides guidelines for researchers to design future systems based on ecological principles following strategies currently used in education and research. These currently fall under the umbrella term **agroecology**, defined variously as "*the ecology of food systems*" (Francis et al. 2003) or "*a science, a set of practices, or a movement*" (Wezel et al. 2009). As an overview, whatever the definition, the concepts and applications of ecological principles in design of agricultural production and food systems have come to predominate among today's progressive researchers and educators. The book's authors present this model as an alternative that provides stark contrast to the current dominant globalisation of agriculture and food, and articulate a strategy that returns control to the local level. What is presented in this book is an imaginative and practical alternative model for sustainable development, one that can overcome the bureaucracy of governments and tyranny of narrow vested interests that have resulted from the "productionist model" that has grown to prevail over the past century. It is a pleasure to evaluate the details of this new model and reinforce the authors' emphasis on education at all levels in moving its implementation.

The authors describe their approach as rooted in "post-normal science" which is defined in terms of its paradigm in "participatory integrated assessment" and which is involved in solving the "wicked challenges" (Ostrom 2009) of tomorrow that can only be addressed by multi- or transdisciplinary teams. These challenges are often considered to be incommensurable, in that no solution may be available that will meet the needs of all involved stakeholders, and thus it is essential to compromise and spend major effort in finding common ground. Such solutions to problems are at the other end of the spectrum from the "technical packages" of seed, fertiliser, and pesticide that were the basis of the successful Green Revolution which increased production in the most favourable areas for agriculture. Because of narrow focus on short-term production and economic gains, scientists and development experts created a number of secondary problems due to inattention to distribution of benefits, and how this use of technology in agriculture may have contributed to greater economic and social disparities in many places where it was applied with apparent and immediate success. The philosophy behind the TATA-BOX model includes anticipating these challenges in order to contribute to sustainable development in farming and food systems.

A logical method of approaching a critique and evaluation is to move through the document one section and chapter at a time, and then to provide an overview at the end. The TATA-BOX strategy is a valuable contribution to the future of sustainable development, although it is important to not create yet another strict model that would introduce a more broad-based yet still top-down method with strong government or regional control. It is important to remember that the phenomena in farming and food systems are always richer than the lens we put up to observe them. The farm is always more than the model. A valuable feature of the model presented is the emphasis on shared and broad governance. It is important to avoid "monoculture thinking", and to avoid another one-size-fits-all methodology that is the opposite of what is learned from ecology and the special potentials and characteristics of every agroecological niche, including its biophysical and socioeconomic uniqueness compared to others. At the same time, it is recognised that there are processes and methods that can be generalised, based on ecological principles, as long as these are not squeezed into some artificial framework that becomes a comfort zone that can cause those in agricultural development to think the job is done now that a new and special model has been developed. Here is an evaluation of each of the sections.

Introduction

The authors begin with an overview of major challenges that have resulted from a productionist philosophy and focus on improving agricultural practices and the efficiency of input use in the process. This is a laudable strategy that they call a "weak ecological modernisation", one that appears frequently in the literature as "eco-intensification" and is often used as a term to describe improvements in fine-tuning today's production systems. It should be noted that E.U. countries in general appear to have progressed more rapidly in the implementation of this strategy than other countries such as the U.S. where pressure by large-scale farm owners supported by narrow-thinking federal support programs continue to encourage maximum production, in spite of diminishing returns to inputs and a glut in the market for major commodity crops. In contrast, the authors here propose a "strong ecological modernisation" that includes a broader and more comprehensive set of goals such as improving the understanding of system properties and interactions. These go beyond the biophysical and short-term economic evaluation of system outputs to include their environmental and social impacts. They emphasise social values, improvement through learning and experimentation, encouraging participation by all actors involved in food systems, and recognising the importance of governance at all levels of scale. The authors discuss the importance of diversified and integrated systems that build on synergies among the components, and discuss the Socio Ecological System (SES) and Socio Technical System (STS) frameworks that deal with governance of resources and the dynamics of innovation in what are essentially "*human activity systems*" (Checkland and Scholes 1999) (term not included in the chapter). There is strong emphasis placed on uniqueness of place, recognition of boundaries around specific areas of inference for application of new technologies, and the "adaptive governance" needed to adjust decisions to an increasingly volatile natural and economic climate. There is much value in this approach.

What are perhaps ignored in this overview are other important dimensions of the food system that must be considered. There is an abundance of food produced on a global scale, but 30–45% of this is not consumed by people due to losses in the field (more often in Third World) or wasted in the system of marketing, transport, or consumption (more often in First World). Another inefficiency not considered in the introduction is the massive production of animal protein in confined operations that use inordinate amounts of grain, water, and other resources, especially those producing meat from ruminant livestock that did not evolve to consume that type of diet. Another misallocation of cereal grains is production of biofuels, a luxury that may raise prices a small amount to help farmers but a strategy far less desirable than conservation of fuels and development of renewable resources to replace fossil fuels. Yet another example of poor allocation is the documented obesity epidemic worldwide, with more people overweight than undernourished at this point in time. The importance of spatial availability of food, and the lack of access by those in poverty conditions must be considered in the assessment matrix of any food system, when this is viewed at the global or any level.

Those who developed the TATA-BOX have done this in the context of Europe, and that is commendable. Yet there are broader issues that are obvious in the developing South that are also important in the developed world that need to be considered in any future strategy that will lead to people being able to produce or access food with a degree of sovereignty or control of their own diets. These are wide nutritional and food issues, and it would be valuable to put the new techniques into proper perspective at local as well as regional and global levels.

TATA-BOX at a Glance

The context of the project is briefly reviewed, before the authors move into a literature review about the history of design techniques based on conceptual frameworks of food system development. The idea of dividing "*materials resources*" into those from natural resources, farming systems, and supply chains may be confusing to some readers, since the second and third of these in most languages are considered human constructs that are quite different from the biophysical input resources that go into production agriculture. It may be a translation question, but perhaps a better term than "*material resources*" here could be "natural and human systems resources" to be more inclusive and to help understanding by a broader audience.

Aside from this semantic challenge, the thinking process described and model illustrated both show the connectivities of social and biological components and importance of diversity in each provides a strong conceptual foundation for understanding complexity. The idea of "backcasting" from a future desirable situation to today's reality, and the multiple paths that could lead to success mirror the approach we have taken in agroecology education and research in Norway and Nebraska (Lieblein et al. 2001, 2004; Francis et al. 2016). Depicting this in a colourful visual is an effective device for communication and building understanding.

Details on the scope and organisation of the project are illustrative to the reader, and help to build confidence in the results, but are not particularly germane to the flow of the discussion. It is recognised that these are essential components of a project report and should be maintained in the text if that is one of the purposes of the book.

Socio-economic Characterisation of Agriculture Models

This useful chapter explains a new typology that includes six models based on "*social principles and moral values*", in contrast to most categorisations in the literature that only include two types that greatly oversimplify the discussion. The authors begin with descriptions of a number of contemporary models that illustrate this contrast, citing references to "*shallow and deep sustainability*", "*weak versus strong multifunctionality*", "*weak versus strong ecological modernisation of agriculture*", and "*life sciences versus an agroecology vision*". They rightly recognise the simplicity of creating such dualism in characterising systems, and how such constructs for comparison tend to set up a "straw-man" system against which a better one may be compared, in the opinions of each author. Although they see use in such models, the authors feel that these often ignore some of the most important elements or characteristics that include the impact of social values on decision making, and that moral values and social principles should have a higher profile. To meet this challenge, they propose that multiple analysis frameworks should include agronomic and physical characteristics to be combined with socio-economic criteria, and thus a need for more models of greater complexity and improved inclusivity of measures that really matter.

The first section deals with sustainable agriculture in terms of socio-economic factors and how social values should be considered as a broader approach to assessing the worth of a system and how these influence decisions. The authors build on a construct from political philosophy that categorises cities based on wealth, efficiency, equity, honesty, grace, and fame, and add their own category of cities "*based on the principle of good intentions directed at the environment*" (Chapter 3, p. 4). People who live in these "cities" or communities of common interest use such values to "organise collective action" and support institutions to design activities consistent with values. They recognise that goals are often incommensurate, and that compromise is essential for harmonious action. Criteria for evaluating these contrasting "cities" or world views are summarised in an illustrative table, and the process then related to agriculture and the choice of practices or strategies. As these relate to the current interest in multifunctional landscapes and agriculture as a key component, the categories provide a framework for evaluation and how this will impact the continued potential success of chosen systems.

A second section describes models that interface production and food systems, and introduces the essential element of local social dynamics in the complex process of choosing internal versus external inputs, deciding on importance of ecoservices, and putting emphasis on local versus global orientation of the food system – at any level of scale. They set up a somewhat tenuous comparison of sources of inputs versus ecoservices as indicators of the degree of sustainability. The authors rightly describe systems based on internal resources and clever design to make use of natural processes as more complex than the "domination and control" model of using pesticides and fertilisers to run a successful production system. And they bring this idea full circle when suggesting that managing systems using natural diversity and considering ecoservices as an input rather than a consequence of systems performance. It's a complicated reasoning that is difficult to grasp, at least in a practical agronomic sense. The authors move on to describe the contrast between (1) local sources to purchase inputs and global markets for product sale and (2) the integrated systems now in vogue of agroforestry, multi-cropping and mixed farming systems incorporating animals, and conservation agriculture. They rightly maintain that global and local markets are complementary, and few would argue that in most places these will continue to exist and interact, especially in the developed world.

The last section characterises models based on socio-economic criteria and contrasts six alternatives to the "historical-conventional system" or (1) "productionist model" currently prevalent in the North and in some favoured areas with good resources and infrastructure in the developing world. This is compared to (2) a technology-intensive model, (3) a techno-domestic model, (4) a circular model, (5) a diversified-globalised model, (6) a diversified local model, and (7) a diversified integrated landscape model. All include different degrees of technology use, integration with "global" input sources and markets, and dependence on natural systems and local ecoservices for driving the system. It jumps out to us that model (7) is the preferred ideal from the agroecologists point of view, and that this is the "gold standard" to strive for in achieving long-term, lasting, and equitable outcomes that will benefit all stakeholders in the agriculture and food system.

The chapter concludes with describing the value of these models and how understanding them could inform decisions on public policy. Without making a value judgement on which systems are necessarily most desirable for the future, it is clear that the choice of any one system or combination of elements from several systems will depend on the world views of the stakeholders in a given location. The authors point out the importance of making explicit the goals of people in a particular situation, recognising that the socio-economic reality of each place may be different, and that development of effective policy will be most successful if values are known and made explicit. The ideas of multi-criteria assessment, incommensurable goals of the players in each place, and difficulties in clearly envisioning the constraints that will face people in designing future agriculture and food systems are brought into focus in the chapter.

One is faced with the ultimate question of, "So what"? Is there a framework or series of potential frameworks here that can provide some clarity to analysing systems, making the interactions of their components more understandable, and ultimately leading to more appropriate actions? Are there vested economic and political interests that are so powerful in many locations that make this type of analysis moot in the face of larger pressures that complicate decisions and present naïve strategies that are unlikely to be useful. As with any analysis, there are often more questions raised than answered, and it is up to the reader to add meaning and potential applications of the ideas and categories in the chapter to the complexity of the "world out there" where decisions are made and people need to have food.

An Integrated Approach to Livestock Farming Systems' Autonomy in Designing and Managing Agroecological Transition at the Farm and Territorial Levels

Challenges in agriculture that have arisen as a result of the wide application of industrial agricultural technology are clearly described in the introduction to this chapter on integrated farming practices, using agroecology as a guide at different levels of hierarchical scale.

An important contribution of the chapter is the focus on hierarchy of scale, with emphasis moving from farm level to landscape or "territory", and the discussion of collective action in planning and decision making is valuable. There is careful attention to the issues of resilience and autonomy, and an appropriate perspective of examining organisation at the "territory" level in contrast to most analyses that are done at the farm or regional levels. Several key concepts from agroecology include **closing cycles**, using **biotechnical autonomy** as a guide to management, and **coordinating organisational strategies** at each appropriate scale. The terms of **levers** or positive forces and **locks** as those that may impede progress provide a useful classification similar to what others call a "force field analysis" in development. A key to success is seen in the integrated efforts of players in the area, and how these are much more useful in causing lasting and sustainable change than when farmers act alone.

To illustrate the principles a case study of production of sheep milk is presented in detail. The participatory approach to identifying key forces is shown through figures and tables developed by the farmers in the area, and they begin with the typical cataloguing of biophysical components and activities and their principal interactions. Perhaps the most compelling information relates to the organisational activities and how these are essential to success. This is an important contribution because of its focus at the "territory level" and the emphasis on people working together for the good of the community. The chapter is long and repetitive in English, but the ideas are sound and will prove valuable to interested readers.

Agroecological Transition from Farms to Territorialised Agri-food Systems: Issues and Drivers

Any transition from the current dominant industrial system to one that is guided more by agroecological principles and practices must consider the reality that any transformation is often more impacted by markets, including influence of support programs, and regulations at different levels in the spatial hierarchy. Transformations at levels above the farm are more difficult because of the number of additional stakeholders and influences at the "territory" or larger geographic area. For this reason, the authors emphasise the need for transitions and incremental changes rather than a "new revolution". This point would be debated by some who feel frustrated by the inertia of the current agri-business and market-driven paradigm in food systems, and the need for a more global "transformational change". In our view there is a need for a revolution, but that this should be a "revolution in the mind", to be followed by a "transition in the field".

The authors categorise potential changes into "weak" and "strong" transitions, with the latter characterised by more radical modifications, substantial increases in system biodiversity, and the need to focus on what will lead to long-term sustainability in the whole food system. They rightly point out that this must engage farmers and broader-based groups at territory level to make meaningful change. What sets the chapter apart is emphasis on need for change in basic values, as these inform not only attitudes but also individual practices or crops or animal species. Many of our programs are imbedded in a current culture that is "locked in" to establish systems that in fact hinder creativity needed to make change. There is such an investment in the currently modal industrial paradigm, and so many vested commercial interests that depend on the *status quo* for short-term financial gain, that change is extremely difficult. We find in teaching, especially among undergraduates in the agricultural sciences, that there is a "monoculture mentality" that pervades the thinking of many. And with our farmer clients some of the same attitudes prevail. To be sure, there is value in conservatism and resistance to change, especially as the climate and economic future remain so uncertain and profits are narrow in agriculture. But one of the clear options presented in the chapter is to think and make decisions at spatial scales larger than the farm, and to cooperate at the territory or landscape levels. This will benefit everyone, rather than just a few, and help to sustain our rural families as well as their communities.

A Plurality of Viewpoints Regarding the Uncertainties of the Agroecological Transition

One of the many challenging problems of the present "command and control" industrial model of agricultural management that makes it untenable for the long-term future is an inability to deal with uncertainty and ambiguity, as well as not providing flexibility to handle situations of imperfect and incomplete information. In addition to this thesis presented in the chapter, we would add the complexity of incommensurate goals among the multiplicity of players in today's agriculture and food system environment. The authors appropriately point to the importance of adaptive management to deal with what they term a "*diversity of uncertainties*". They further describe decisions with incomplete information as "wagers on the future", a compelling introduction to what is developed as a treatise on how agroecology and systems thinking can solve these seemingly intractable problems.

The chapter explores the inadequacy of well-meaning but simple production-consumption approaches, here placed in the category of "*weak ecological models*" as practiced by small and select groups of practitioners that depend on accumulation of information on today's reality to support action in development. The authors propose that "*strong ecological models*" are essential, involving a wide group of actors, and incorporating ways to recombine knowledge that are relevant across the temporal and spatial continua. For example, they point out that merely increasing technological efficiency is useful but not sufficient to explore the consequences of uncertainty, and that experiments that combine biophysical elements with social behaviour and decision making are needed to develop new and effective models for the future.

The authors describe the well-known food chains from production to consumption, and explain the "contractual chains" that guide their organisation. We suggest that the concepts of "food chains" or even "value chains" could logically be replaced by "food webs" and "value webs" in our quest to apply ecological principles to design and management of food systems. Further, we suggest that the authors consider the current thinking about vertical integration in supply chains as a neoclassical construct similar to much of our linear thinking in describing systems. It could be more transformative to again insert the term "webs" and to conceptualise lateral, vertical, and multiple dimensions of integration in the interactions of inputs, outputs and socioeconomic consequences of food systems, such as equitable distribution of benefits, as a broader way to think beyond the typical biophysical components and short-term outputs of a system.

When considering "sensemaking in management" the authors describe the importance of dealing with and reducing ambiguity, yet realise how this in itself may not reduce misunderstandings of system function. Reduction of ambiguity through increased transdisciplinary activity might help stakeholders to move towards an understanding of the situation as a whole. The authors point out the importance of considering who are the participants, what are their goals, what is the relevant place or territory, how does one establish a relevant time frame, and what are the most important criteria for evaluating a system. This leads to discussion of the value of modelling, using solid data on systems, probabilities of conditions such as weather and economics, and other parameters that drive the systems. They also describe the difficulty of incorporating ecoservices into the models, often not clearly defined and even more often not monetised, and the challenges of deciding what outcomes to optimise. The chapter does a good service to the reader by raising many questions. This is often more difficult that providing clear answers, but that is the nature of dealing with ambiguity, resulting from disciplinary specialisation.

Towards an Integrated Framework for the Governance of a Territorialised Agroecological Transition

The introduction makes two key points, that shared and participatory governance is a key factor in preventing excesses of influence by one leader or a small group of influential interests, and that critical issues include the recognition of system boundaries, generally spatial/geographic, and identification of the players or stakeholders who are included. The authors recognise a major challenge to be the willingness of leaders and all players to dedicate enough time and energy, as well as the resources needed, to design and implement an effective form of governance that will help everyone in the territory realise a relevant and transformative transition. They note the importance of recognising importance of both the" Socio Ecological System (SES)" and the"Socio Technical System (STS)" and the vital need to seek convergence of the two in promoting a successful Territorialised Agroecological Transition (TAET).

The chapter continues with discussion of the complexity of designing governance to encourage TAET when there are multiple goals, and with the difficulty of measuring most ecoservices. There are options that include both incentives (payments) or sanctions (penalties) for adopting new practices, and education that leads to a positive approach is certainly favourable as a strategy more palatable to stakeholders. The mix of market incentives with certain regulations provides wide opportunity for creative alternatives in policy, and the authors suggest that a "*reflexive approach to governance*" that includes negotiation and involvement of all stakeholders will assure a smoother route to adoption. There is recognition of the importance of regular communication and the need to suppress power relationships, and we think that these are especially critical when vested political and financial interests come into the equation.

Lastly and perhaps most important is attention in the chapter to a higher order issue, and that is interest in both the results of creative governance and the process of getting there. In any educational endeavour, the value of process cannot be underestimated as this is what will prepare people to deal with uncertainty in the future, whether this is in deciding on farming practices, methods of local marketing, or strategies for the formidable and transformative activities in implementing TAET. The authors conclude that "*environmental, economic, and social processes do not stop at the boundaries of a given territory*" and that the recognition of flows of materials, connections to markets, and many interdependencies require a scaling up of thinking and establishing connections. Perhaps the term "right scaling" would be preferable to "up or down" scaling, as this will be unique to a given territory and its circumstances. As more is learned about how to quantify and reward ecoservices, and environmental impacts are better understood, it will become increasingly apparent that the marketplace is not the only or perhaps even the primary factor driving the system. The process of governance will take this into account in the future, and the results will be territory-specific.

The Key Role of Actors in the Agroecological Transition of Farmers: A Case-Study in the Tarn-Aveyron Basin

There is an immediate dichotomy established between those who support and those who oppose AET, a useful but simplified description of reality. We often find an array of interests that fall on a spectrum between the extremes. And a critical observation at the outset is that a mere fraction of research funding has been spent on agroecology approaches. Thus any comparison of the systems must take into account the paucity of data that is essential to improve practices using ecological principles. The authors rightly recognise that increasingly our systems are hybrids of the extremes, and that thoughtful farmers are using ideas from both. The two extremes can be summarised as the authors said, that one group considers agroecological methods as moving back to the past, while proponents describe this as moving into the future. We agree with the latter, of course recognising the value of knowing about past experiences, especially from the pre-industrial era.

In contrast to other chapters based primarily on the literature, this one is partly based on empirical data derived from interviews with farmers, with results used to explore the importance of past experience as well as how they see the current networks helping to move toward TAET. One in-depth interview provided insight on a conventional farmer who cited his large investment in the technology paradigm, and who saw no rational reason to change. Another provides a profile and history of an active group member who is making the transition, and has experienced ups and downs during the process. He is working toward autonomy and takes a long-term view, while considering himself a "technical farmer" who has embraced the philosophy of agroecology including strong respect for the environment. In the second year of the study the focus was on food systems, especially ways to add value to products before sale. Here is where contacts and support in the groups were seen as vital to getting information and finding new and practical ideas. In summary the authors describe the importance of major conventional farmers and marketing organisations in hindering progress toward goals of introducing agroecology, yet the number of farmers moving in this direction demonstrates its viability economically and its consistency in agroecology principles. This conclusion is similar to what we have observed in Norway and elsewhere, that the move toward transformational change is a powerful one and is stimulated by a number of educational programs in universities and practical training activities in agriculture.

Participatory Methodology for Designing an Agroecological Transition at Local Level

The next major section of the book deals with recommended methods and tools for the TAET, the importance of reflection and evaluation during the process, and the details of the TATA-BOX approach. The ideas here are based on the previous chapters, and are meant to provide a "road map" to implementation. This chapter describes an essential and complex "*co-evolution of technical, social, economic, and institutional dimensions*" that is highly dependent on close interactions between "*stakeholders in the farming system, supply chain, and natural resources management*". The consistent emphasis through the book is on local control and commercialisation, and the dependence to the degree possible on natural and renewable resources. As in other chapters, there is focus on importance of integration of efforts through interdisciplinarity and systems thinking promoted by having multiple stakeholders and their ideas represented in the process.

Discussion of implementation starts with the goals of the TATA-BOX tools designed to implement the switch from "*AET design theory to operational and effective practices*". As described in the introduction, the strategy is organised around three domains: the farming system, the socio-technical system involving supply chains, and the socio-ecological system involving territorial resource management. Illustrative figures build understanding of the relationships among these elements. The design process is familiar to us because it is similar to that used in our Agroecology MSc course in student project teams: observe and evaluate current resources, develop visions of a desired future, designing a more desirable future as a result of transformative changes, exploring alternative pathways to reach that future, and implementing a governance strategy to guide the process, as illustrated in Fig. 2.

What contributes to making this perhaps the most valuable chapter in the book is the highly accessible step-by-step process in the development of the TATA- BOX method itself. Accompanied by multiple illustrations, the text transforms a case study in transformative development of a new model into practical language and a logical pathway to apply the method in other situations and territories. It is useful to theorise about change and learn from the literature about experiences of others, but far more valuable to walk through the steps that were used to develop the method described in this book. A number of practical tools are described that were essential parts of the participatory workshops, specific ideas about how to organise focus groups, foster as much participation and ownership as possible, and then move into the implementation of the whole process.

Operational outcomes of the model development workshops included the shared diagnosis of the current situation, the agreed-on goals of the overall project, and specific steps in the action plan as well as a framework for project governance. There is careful thought given to the *a priori* anticipated outcomes and impacts, providing another opportunity to fine-tune the process to help reach common goals. Four workshops were used to reach the eventual, shared-agenda action plan, with the last workshop organised partly by the stakeholders. Although the process appears long and complex, we agree that this is perhaps the most important part of the entire exercise. The chapter on tools and implementation concludes with asking key questions about the process and how it could be improved. Lastly, the workshops resulted in "*ready to use outputs for local stakeholders*", most importantly a shared diagnosis for 2015 and a vision for 2025 of the new territorial plan and a projected action plan to get there. The process represents one described by futurist Joel Barker two decades ago: "*Vision without action is merely a dream. Action without vision just passes the time. Vision with action can change the world*".

Towards a Reflective Approach to Research Project Management

This chapter steps back to take another look at the process used to arrive at the TATA- BOX. It describes the design of AETs to promote transformational changes in agriculture and food systems, with the full participation of local stakeholders. On another level the process includes designing a participatory methodology to define those transitions and a plan to achieve them. Lastly the process includes design of a governance strategy to oversee and catalyse transitions. This chapter outlines how project advisors worked with leaders to reflect on their success and how to continue improvement.

The authors begin with review of the literature on management of activities, and of the design process itself, citing the value of "*Activity-Centred Ergonomics*" as a proven method of engaging players in reflective thought about their own organisational actions. They describe a series of meetings, each with its own goals, that leads to an iterative strategy of evaluation. First is a diagnosis of project management, an assessment of how things are proceeding. This includes identifying key issues, then evaluating how the reflective process helped to resolve them, and an evaluation of how the process worked. A number of quotes from participants enrich the discussion with personal opinions about their experience in the workshops. Most were supportive, while others made specific suggestions on how to improve the process. Although most of them reported success in many ways, one conclusion was that the reflective process was not robust enough to solicit enough people for feedback, and a concern that not all voices were heard.

Evaluation of the Operationalisation of the TATA-BOX Process

An important feature of this book on the TATA- BOX tool and the process through which it was derived is the focus on development of the tool as a case study itself, a process that "*can be defined through two main features: it is participatory and it is intended to support design*". Also important is the willingness of the editors and authors to evaluate the implementation of the process through this unique tool. They have chosen "*design process evaluation frameworks*" from the literature, using metrics that were developed for other applications but here applied to changes in farming and food systems based on agroecology principles.

Semi-structured in-person interviews were conducted with participants in the process to explore expectations regarding workshops and then impressions of how they went, the workshop results, how the process related to other parts of tool development, and the interpretation and eventual influence on territorial development.

General reaction to the workshops was positive, with some critique about over-representation by INRA specialists and need for more representatives from the case territory. Specific tools used in conducting the workshop were viewed as valuable, and participants felt that their voices were heard. There were positive comments about the facilitators, participants felt that the results were objectively summarised and reported, and only a minority of those interviewed felt that the process had not been adequately completed. About half reported that they had learned valuable information and process, and had used this in their own work. Quotes from participants add rich detail to the summarised data from the interviews.

Workshop participants generally felt that there was good communication, and that a wide range of views were expressed and reflected in the report. There was less agreement on the value of the outputs, and how these would result in concrete actions to achieve the stated goals of the project. It was difficult to tell from the summary how enthusiastic the workshop participants were about the entire process, and they seemed to take a "wait and see" attitude about the long-term results of the TATA- BOX tool and methods. From this we conclude that the exhaustive process of multiple workshops, careful design of theoretical constructs for classifying people, places and their characteristics, human intentions, and actions may have been less successful that what was hoped by the organisers. There is no consistent evidence that a state of "shared governance" had been achieved in these workshops that generally were organised and administered in a "top-down" manner, one that contradicts one of the basic tenets of the plans for the entire project. Of course it is difficult to assess success of the venture until the tools are applied in multiple places, and there is a more comprehensive evaluation of outcomes.

Information and Communication Technology (ICT) and the Agroecological Transition

Rapidly expanding capabilities and accessibility of information and communication technologies (ICT) provide a number of tools for the design and implementation of a truly transformative change in agriculture and food systems. Among the capabilities are computer support for multi-criteria guidance in design, constructing and implementing research frameworks, and communication potentials for improved networking with stakeholders. Authors raise the intriguing question, is ICT a useful tool or an invitation to slavery once a person or system is wedded to this technology? We ask further whether this is a viable tool that can be used when appropriate and available, or is it a set of clever technologies that were expensive to develop and with companies searching for applications to pay for the investment? Most of us in science and development agree that there are countless advantages to the computational and communication capabilities of ICT, and few would question our dependence on the equipment and software that currently pervade our workplaces and our lives. We often find it difficult to imagine how we accomplished anything without these tools at our fingertips.

The chapter introduces the term "connected agriculture" to encompass the myriad devices and apps now at our fingertips that allow us to monitor soil moisture and nutrients as well as keep current on local and distant markets and prices, as well as computer updates from purveyors that "most assuredly" would enhance our efficiency and output if we would just "buy the new app". It appears to us that these new tools do indeed facilitate our conceptualising new systems and assessing their potential outcomes in ways that were previously unknown and impossible. Thus the tools could comprise a new model for agriculture based on instant communication and access to the widest possible range of information, and the challenge soon becomes how to interpret this avalanche of data and to glean those few nuggets of wisdom useful to us in designing biodiverse production systems that result in healthy food for the community. ICT provide tools that help connect these nuggets not only for integration of local resources in agroecologically sound farming systems, but to scale these up to watershed and landscape level, collectively called sustainable territorial systems.

ICT is such a large and pervasive presence in today's global economy and community, and is important to each unique location and the design of systems for each place. Among the key issues that stand out in the chapter are the need for an open infrastructure and equal access by all to the power of the internet, related to this the importance of reducing costs of access, and exploiting potentials of adapting power of ICT to local situations, cultures, and constraints, all critical to specific needs in each territory. There is discussion of how open access to capacities of ICT can facilitate the building of consensus in communities of shared interest, and also to facilitate the shared governance described in other chapters. At the same time, ICT capabilities provide potential for tapping into, recording, and integrating diversity of opinions and goals, so that communities and territories can design strategies to meet multiple needs in an equitable way. Among these tools are communications strategies, functional tools that digitise information to improve access, and both data management and decision support systems. Lastly the chapter explores costs and benefits of ICT, and speculates on future applications that will result from this ever-expanding electronic network.

This brief summary barely skims the depth of ideas presented in the book, and especially in the complex and growing world of ICT. As educators, researchers, and development specialists we are hard pressed to keep up with new technologies; often our students are our teachers, as they are "natives in the complex information environment, while many in our generation are 'immigrants in this new land". While we may lag in understanding the bells and whistles of current ICT, those with experience have a special obligation to help younger colleagues seek meaning and wisdom from the oceans of data now available. We expand on this is the overall summary section.

TATA-BOX: Summary and Critique of Tools for Transformational Agroecology in Development

After reading and reviewing the chapters, we have learned much more about the importance of organising participatory strategies to organise, administer, implement, and evaluate a transformative process toward agroecological farming and food systems. The book has valuable information and is clearly written. As this is being published in English, we assume that it is directed toward an international audience interested in agroecology, and providing guidelines to others for improving sustainability and autonomy of food systems at the territory level. While the text is very well written, there are places where it is repetitive, for example when introducing each chapter with much of the same information about the negative impacts of industrial-model farming and food systems. We recommend that readers should skim parts of some chapters to glean what is important, a strategy that will help avoid "reader fatigue". Some of the categorisation and analysis is described in more detail than needed to apply the model.

The small shortcomings in writing do not diminish the valuable ideas about agroecological focus for development. The principles of holistic systems thinking, local and broad involvement of multiple stakeholders, autonomy in production linked to consumption in local food systems, and broad-based governance at territory level bring together biophysical and socioeconomic dimensions of production and consumption of food. These all reflect the current writings about agroecology as a viable organising framework for agriculture, and the importance of involving all the players in planning future directions. When all stakeholders assume ownership of the process, it is much more likely that transformative alternatives will be developed and adopted. These need to consider the goals of all participants, the available local resource base, and the production, economic, environmental, and social implications and outcomes of any new strategy. We commend the authors of this book, and look forward to following the results as the applications of the TATA-BOX model are implemented.

References

Checkland P, Scholes J (1999) Soft systems methodology in action. Wiley, New York

Francis C, Lieblein G, Gliessman S et al (2003) Agroecology: the ecology of food systems. J Sustain Agric 22:99–118. https://doi.org/10.1300/J064v22n03_10

Francis C, Østergaard E, Nicolaysen A et al (2016.) Learning agroecology through Involvement and reflection) Agroecology: a transdisciplinary, participatory and action-oriented approach. In: Mendez V, Bacon CM, Cohen R, Gliessman SR (eds) Advances in agroecology series. CRC Press, Boca Raton, pp 73–99

Lieblein G, Francis CA, Torjusen H (2001) Future interconnections among ecological farmers, processors, marketers, and consumers in Hedmark County, Norway: creating shared vision. Hum Ecol Rev 8:60–71

Lieblein G, Østergaard E, Francis C (2004) Becoming an agroecologist through action education. Int J Agric Sustain 2:147–153. https://doi.org/10.1080/14735903.2004.9684574

Ostrom E (2009) A general framework for analyzing sustainability of social-ecological systems. Science 325:419–422. https://doi.org/10.1126/science.1172133

Wezel A, Bellon S, Doré T et al (2009) Agroecology as a science, a movement and a practice. A review. Agron Sustain Dev 29:503–515. https://doi.org/10.1051/agro/2009004

Opening the TATA-BOX to Raise New Questions on Agroecological Transition

Jean-Marc Touzard, Jean-Marc Barbier, and Laure Hossard

Abstract This chapter is written by researchers at UMR Innovation (Montpellier) who propose an external critical analysis of the TATA-BOX project. Firstly, we highlight the main contributions of both the project and the book, which cover different stages of a participatory research project, by crossing several disciplinary viewpoints. We note multiple outputs that can strengthen the capacities of farmers and researchers. We then develop three questions that echo our own works on innovations and agroecological transition: (i) How to associate agroecological issues with the diversity of practices and projects in a given area which is not necessary in a transition towards "strong ecological modernization"?; (ii) What are the conditions for disseminating/outscaling the TATA-BOX approach? In particular, can the project be replicated without the support of researchers? (iii) How to best integrate the political dimensions of ecological transition at different scales? We conclude that TATA-BOX could become a "political object" that promotes agroecological transition to local authorities, national policy makers and media, and international networks.

Introduction

The urgency of ecological challenges not only calls for changing the ways we produce, exchange and consume our food, but also the ways we are doing research and supporting the processes of change in agriculture. The TATA-BOX project is in line with these perspectives by proposing, experimenting and evaluating a participatory approach to support local actors towards agroecological transition (AET). This project has mobilised a multidisciplinary research team, including agronomists, economists, sociologists, geographers, etc. In this book the team reports its common

Thanks to Roy Hammond for his help in revising English of this paper.

J.-M. Touzard (✉) · J.-M. Barbier · L. Hossard
Innovation, INRA, CIRAD, Montpellier SupAgro, Univ Montpellier, Montpellier, France
e-mail: jean-marc.touzard@inra.fr; jean-marc.barbier@inra.fr; laure.hossard@inra.fr

J.-E. Bergez et al. (eds.), *Agroecological Transitions: From Theory to Practice in Local Participatory Design*, https://doi.org/10.1007/978-3-030-01953-2_15

analytical frameworks, its results and its evaluations, going as far as integrating external critical analysis. This is the aim of our paper, written by researchers at UMR Innovation (Montpellier) who study innovations in agricultural and agro-food systems (Faure et al. 2018) and have discovered the TATA-BOX project by reading the other chapters of this book. Firstly, we will highlight the contributions of the book and then develop three questions that echo our own work on innovations and AET: How to associate agroecological issues with the diversity of practices and projects in a given area? What are the conditions for disseminating or outscaling the TATA-BOX approach? How to better integrate the political dimensions of agroecological transition?

Strengthening the Capacities of Farmers and Researchers

First of all, reading this book generates much interest. The contributions cover different stages of a participatory research project, by crossing several disciplinary viewpoints. Few studies in this area have taken the time to report and analyse their approach with such precision and reflexivity: an updated survey of the issues and principles of AET; a conceptual framework to analyse this transition by combining three systems, usually mobilised separately (farming systems, Socio Ecological Systems, Socio Technical Systems); the proposal of a five-step approach to support the stakeholders of a territory; the operationalization of this approach in two local areas with feedback on the participatory process; a focus on several tools used during the project (surveys, network analysis, graphical tools, etc.); and an evaluation of both local participatory processes and research management.

These contributions refer to recent theoretical and methodological knowledge that fuel the increasing number of participatory research projects dedicated to transitions and innovations in agriculture, particularly in the SAD department of INRA (Barbier and Elzen 2012; Meynard et al. 2012; Prost et al. 2017): the reaffirmation of a systemic framework; the taking into account of the knowledge, the judgments and the interactions of the stakeholders; the articulation of several scales and fields of action; the attention given to pathways more than to the states of the systems; the application of design theory to agriculture; the perspective of "adaptive management", etc. The book is thus a resource to better understand the current research on AETs.

Admittedly, each contribution is not necessarily original and the project remains focused on the co-design phase of an action plan, without going so far as to analyse the actions implemented afterwards. But the combination of these contributions, their confrontations and their reflective dimension offer a precious source of knowledge for all the categories of actors potentially involved in AET: researchers and research managers, farmers, policy makers, local development organizations, etc. The TATA-BOX project thus appears as a real laboratory for the development of participatory research responding to agroecological issues at a local scale. The uncompromising evaluation of the experience (Chap. 11) shows that the project has clearly strengthened the capacity of the stakeholders participating to the local participatory processes. By providing tools, references and ideas, the book may also strengthen the capacity of its readers, and thus contribute to scaling up local initiatives of AET (De Tourdonnet and Brives 2018). But more than that, the interest of the book is also to raise questions that are sometimes evacuated by other researches on AET.

Which Agroecological Transition?

The first question is about the way in which the vision of a sustainable agriculture in a territory can be designed as well as the AET that can lead to it. The TATA-BOX project assumes the objective of favoring a "strong ecological modernization" based on "radical redesign of agricultural systems". This option gives a normative dimension to the participatory process promoted by the researchers, even if it relies on maieutic principles. The objective is indeed to help stakeholders co-design – or "discover"- through their interactions with researchers (i) a shared diagnosis of the current situation, (ii) a common vision of a territorial agroecological system, and (iii) a pathway for the transition. The implementation of such an approach then calls for several remarks:

The approach is applied in two local areas where agriculture still plays an important role and is already well engaged in AET. Indeed some actors and organizations carry individual and collective initiatives that refer to strong agroecology, for instance the development of organic agriculture. They also seem to have a consistent role in the two territories and in the workshops of the project. But what about territories where agriculture is less present or dominated by conventional agriculture or weak agroecology? Should we then support activist groups, though they be marginalized or in conflict in the territory? Should we instead propose work oriented more towards a weak form of agroecology? Should the approach be integrated in a wider territorial approach that takes more into account the evolution of non-agricultural dynamics, whether economic or ecological? In any case, the TATA-BOX approach deserves to be tested in more contrasted territories in terms of their commitment to agroecological transition.

The proposed approach has the merit of taking into account the plurality of views on agroecology (cf. Chap. 3), the uncertainties (cf. Chap. 6) and the diversity of current practices and projects in local agriculture and food systems. Studies in the Tarn French *département* (FADN NUTS III, http://ec.europa.eu/agriculture/rica/) show that many farmers hybridise strong and weak agroecology, or integrate strong agroecology into their practices, networks and conventional agriculture organizations (cf. Chaps. 7 and 8). But the challenge of TATA-BOX is ultimately to obtain convergence on a desirable and shared future, and then to propose an action plan. This future certainly includes the coexistence of a diversity of projects and practices, more or less close to strong agroecology, but it remains linked to the search for a compromise that can limit the range of proposals. This methodological choice is shared with many participatory research projects and facilitates reflections on "the" transition. But we can also follow a prospective approach that holds several scenarios and several possible paths (Delmotte et al. 2017). It can maintain comparability of different options, not necessarily dominated by strong agroecology.

In fact, there is a possible contradiction between the need for a strong Ecological Modernization of Agriculture and the ambition to develop participatory approaches. The TATA-BOX approach begins with a diagnosis of the territory and its agriculture, but participation could well result in more than just the capture of knowledge and data from the stakeholders. What happens if they want to deal with "their" problems and propose solutions that are not referring to "strong ecology" or even "weak ecology"? What happens if there is no possible convergence towards a "Territorial Agroecological System"? We agree that AET cannot be promoted without participation, even with more punitive and coercive public policies, but participation does not necessarily create an AET.

Can TATA-BOX Be Used Without Researchers?

Promoting the development of "strong agroecology" through a participatory approach raises a second question: what is the reproducibility of the approach proposed by TATA-BOX and what are the conditions for its adoption by other groups of local stakeholders?

The characteristics of local agriculture, its territory and its more or less "receptive" and innovating actors have already been mentioned. But we must emphasise another condition for the success of TATA-BOX: the very strong involvement of research. More than 40 researchers have been mobilised around the project and their role has been multiple, as Chaps. 10 and 11 show. Researchers are indeed at the initiative of the process, have set up groups and workshops, have strongly contributed to the diagnosis, have supported stakeholders towards the search of consensus, have ensured the follow-up, the evaluation and capitalizing on the project. The researchers who implemented the TATA-BOX project were themselves accompanied by other researchers who supported them in their reflexivity and analysed the impacts on stakeholders. These circumstances are exceptional, and even a luxury for research. But the implementation of such a device may be expensive, time-consuming, and thus difficult to reproduce. How to ensure that the TATA-BOX approach can be "decoupled" from the networks of actors who created and applied it, and in the first place from the researchers? This question can be addressed by analysing three major contributions of researchers to the project.

An initial contribution of researchers has been the production of references, methods and knowledge on local agriculture, ecosystems and food-chains. This cognitive function contributed to the initial diagnosis and the analysis of the diversity of strategies. It also played a role to legitimise the implementation of the approach and to provide references used by stakeholders during the workshops including during the discussions on action plans. This function could be extended in the longer term through the organization of an "action research project" (Faure et al. 2014). What is the possibility of launching the TATA-BOX approach in other territories without this contribution of scientific knowledge? The introduction of "self-diagnosis" methods and the use of collaboration through student training are often proposed solutions.

A second contribution of researchers is the design and implementation of a combination of tools, methods and actions constituting the global TATA-BOX approach. In fact, these tools and methods already exist separately. The stakeholders themselves considered them "unoriginal" at the end of the three workshops (Chap. 11). But at the same time, they recognized as "very stimulating" the TATA-BOX experience as a whole, because it stimulated collective learning processes. Researchers then played a role of "animation for local development". This function must be implemented autonomously by other actors to ensure the dissemination of the approach. A guide has been produced for this purpose (Audouin et al. 2018). But what is the capacity or willingness of these animators and their organizations to mobilise the TATA-BOX process and, if necessary, readjust the combination of tools according to new local conditions? This question applies to the duplication of design approaches in other territories, but also for the implementation of action plans. Because once conceived, it is not the researchers who will support all the local changes.

The third main contribution from the researchers refers to the evaluation of the approach. TATA-BOX has been monitored and evaluated by researchers, with in particular a survey of a sample of participants. This external evaluation cannot be done each time and could be largely integrated into the participatory process. Indeed the empowerment of the approach calls for proposing a participatory evaluation where the stakeholders could define the objectives, criteria and evaluation methods for the future agroecological systems, the actions they propose and their collective process of design. Here, we can draw attention to the risk that the assessment could be limited to the impacts on local agriculture and ecosystems, whereas all the dimensions of sustainability must be considered, as well as the possible impacts outside the territory (Andrieu et al. 2018).

The diffusion of the TATA-BOX approach without the participation of researchers refers to the principle of autonomy, put forward by agroecology (cf. Chap. 4). AET should not hide a new form of "dependence" on research! The importance of links with research and universities has already been shown in several emblematic examples of agroecology, including to ensure their economic viability, for example by hosting trainees and projects or by providing training (Morel 2016). As suggested by the authors of this book, we must consider the TATA-BOX approach neither as a reproducible "turnkey" solution, nor as a pure product of research not reproducible as it stands, but as a living lab, which allows methods and tools to be tested, provides references and shows clearly that the essential thing is to create spaces of debate and exchange in the territories where agroecology is ongoing.

To Highlight the Political Dimensions of Ecological Transition

Finally, the TATA-BOX project and the book that reports this experience lead us to question and better understand the political dimensions of AET.

First of all it is clear that TATA-BOX results from a political project. By announcing a "preferential option" for strong agroecology and bottom-up/participatory development, the researchers' perspective is to go beyond the mere understanding of the world, to engage in a transformation of the world. In France, in scientific practices, including in social sciences, this political dimension is often overshadowed, for fear that political commitment could call into question the scientific nature of an approach. But as soon as the political purpose is clarified and contextualized, and that the social categories that can benefit from the project are debated, there is room for "research capable of engaging and disengaging" (Callon 1999). In this sense, the research presented in this book goes beyond the vision of an agroecology "detached from political action" that INRA puts forward (i.e. defining it as the study of relations between ecosystems and agriculture). At the same time, this research does not enclose itself in a vision that is undoubtedly too "committed" that Altieri (1995) advocates, even if the defense of family farming makes sense through the TATA-BOX project. At the same time it does not just focus on the defense of organic farming (Lamine 2015). The political project of TATA-BOX tends towards "territorial agroecology", which opens spaces of dialogue and negotiation and leads to the elaboration of an action plan. But to fully assume the political dimensions of this "territorial agroecology", several points undoubtedly deserve to be developed.

Firstly, the inscription of the TATA-BOX process and workshops in the local political system can be discussed. The project has implemented ad hoc groups, bringing together a diversity of stakeholders, but not necessarily having a strong political legitimacy. Another option would have been to work more directly with an existing political body, such as a steering group of mayors or representatives dealing with the agricultural and ecological issues within their different municipalities. This is the choice that has been made by the UMR Innovation in support of the agroecological and food policy of the Montpellier metropolis (Soulard et al. 2017). The political impact is stronger, but probably with less debate; less agronomic knowledge and a lower degree of detail on agricultural specificities. This could compromise the final implementation. It's about finding a tradeoff between political impact and political debate.

Moreover, as noted by the authors of the book, the process of political design calls for continuing the experiment, following the evolution of local political action.

In this respect, one can think that the "weak" ties of TATA-BOX with local political authorities can limit the commitment of concrete actions. One could imagine that the project could play a more diffuse role within the local community, promoting the emergence of solutions, and strengthening local capacity to maintain pressure on policy makers. In any case, there is room for complementary work in political science aimed at analysing and supporting the future evolution of political actions and tools for the AET of these territories. Highlighting the notions of "local resource governance" and "socio-technical systems" provides a favorable framework for inviting political science in the wake of this project.

Lastly, the TATA-BOX approach could be considered as a political object that can be mobilised by researchers and stakeholders for action at different scales. The experience can thus lead to the writing of "policy briefs" for national or European policy makers. The presentation of the lessons of the approach in conferences or professional media is to be continued, in agricultural as well as scientific circles. In fact scientists have also to integrate these approaches to participate in AET. The challenge is then to amplify debates, feedback and questioning in the political field. This is also what can be done at the level of other countries through the promotion of this book which is a reference for all those who want to build an "international AET".

References

Altieri MA (1995) Agroecology: the science of sustainable agriculture. Westview Press, Boulder

Andrieu N, Barbier JM, Delmotte S et al (2018) Co-conception of technical and organisational changes into agricultural systems. In: Faure G, Chiffoleau Y, Goulet F et al (eds) Innovation and development in agriculture and food systems, quae. Versailles, France, pp 151–161

Audouin E, Bergez JE, Choisis J-P, et al (2018) Petit guide de l'accompagnement à la conception collective d' une transition agroécologique à l' échelle du territoire. Toulouse

Barbier M, Elzen B (eds) (2012) System innovations, knowledge regimes, and design practices towards transitions for sustainable agriculture, INRA. Paris, France

Callon M (1999) Ni intellectuel engagé, ni intellectuel dégagé: la double stratégie de l'attachement et du détachement. Sociol Trav 41:65–78. https://doi.org/10.1016/S0038-0296(99)80005-3

De Tourdonnet S, Brives H (2018) Agro-ecological innovation: how to mobilize ecological processes into agricultural systems. In: Faure G, Chiffoleau Y, Goulet F et al (eds) Innovation and development in agriculture and food systems, quae. Versailles, France, pp 71–80

Delmotte S, Couderc V, Mouret JC et al (2017) From stakeholders narratives to modelling plausible future agricultural systems. Integrated assessment of scenarios for Camargue, southern France. Eur J Agron 82:292–307. https://doi.org/10.1016/j.eja.2016.09.009

Faure G, Gasselin P, Triomphe B, et al (eds) (2014) Innovating with rural stakeholders in the developing world: action research in partnership, Quae, CTA. LM Publishers, Versailles, France; Gembloux, Belgium

Faure G, Chiffoleau Y, Goulet F et al (2018) Innovation and development in agriculture and food systems, Quae. Versailles, France

Lamine C (2015) Sustainability and resilience in agrifood systems: reconnecting agriculture, food and the environment. Sociol Ruralis 55:41–61. https://doi.org/10.1111/soru.12061

Meynard JM, Dedieu B, Bos AP (2012) Re-design and co-design of farming systems: an overview of methods and practices. In: Darnhofer I, Gibbon D, Dedieu B (eds) Farming systems research into the 21st century: the new dynamic. Springer, Dordrecht, pp 407–431

Morel K (2016) Viabilité des microfermes maraîchères biologiques. Une étude inductive combinant méthodes qualitatives et modélisation. Université Paris-Saclay

Prost L, Berthet ETA, Cerf M et al (2017) Innovative design for agriculture in the move towards sustainability: scientific challenges. Res Eng Des 28:119–129. https://doi.org/10.1007/s00163-016-0233-4

Soulard CT, Perrin C, Valette E (eds) (2017) Toward sustainable relations between agriculture and the city. Springer International Publishing AG, Cham

Index

J.-E. Bergez et al. (eds.), *Agroecological Transitions: From Theory to Practice in Local Participatory Design*, https://doi.org/10.1007/978-3-030-01953-2

U

V

www.ingramcontent.com/pod-product-compliance
Lightning Source LLC
LaVergne TN
LVHW060056220926
848484LV00008BA/1960
* 9 7 8 1 0 1 3 2 7 5 8 3 8 *